A Handbook to
THE POSNER CENTER FOR
SPECIAL COLLECTIONS,
Carnegie Mellon University

Rare Books & Ancestral Machines

Sam Lemley

Carnegie Mellon University
Press

Tartan Editions
An imprint of Carnegie Mellon University Press
Pittsburgh, PA

1,000 copies of this book were published to mark
the reopening of the Posner Center for Special Collections,
Carnegie Mellon University.

ISBN-13: 978-0-88748-723-1

Library of Congress Control Number: 2025939052
LC record available at: https://lccn.loc.gov/2025939052

Printed and bound in Lithuania

Images on the folding cover panels are from Jacob Bernoulli,
Ars conjectandi (1713). Posner Memorial Collection,
Carnegie Mellon University Libraries, Special Collections
(Call number RC183 .A2 J4).

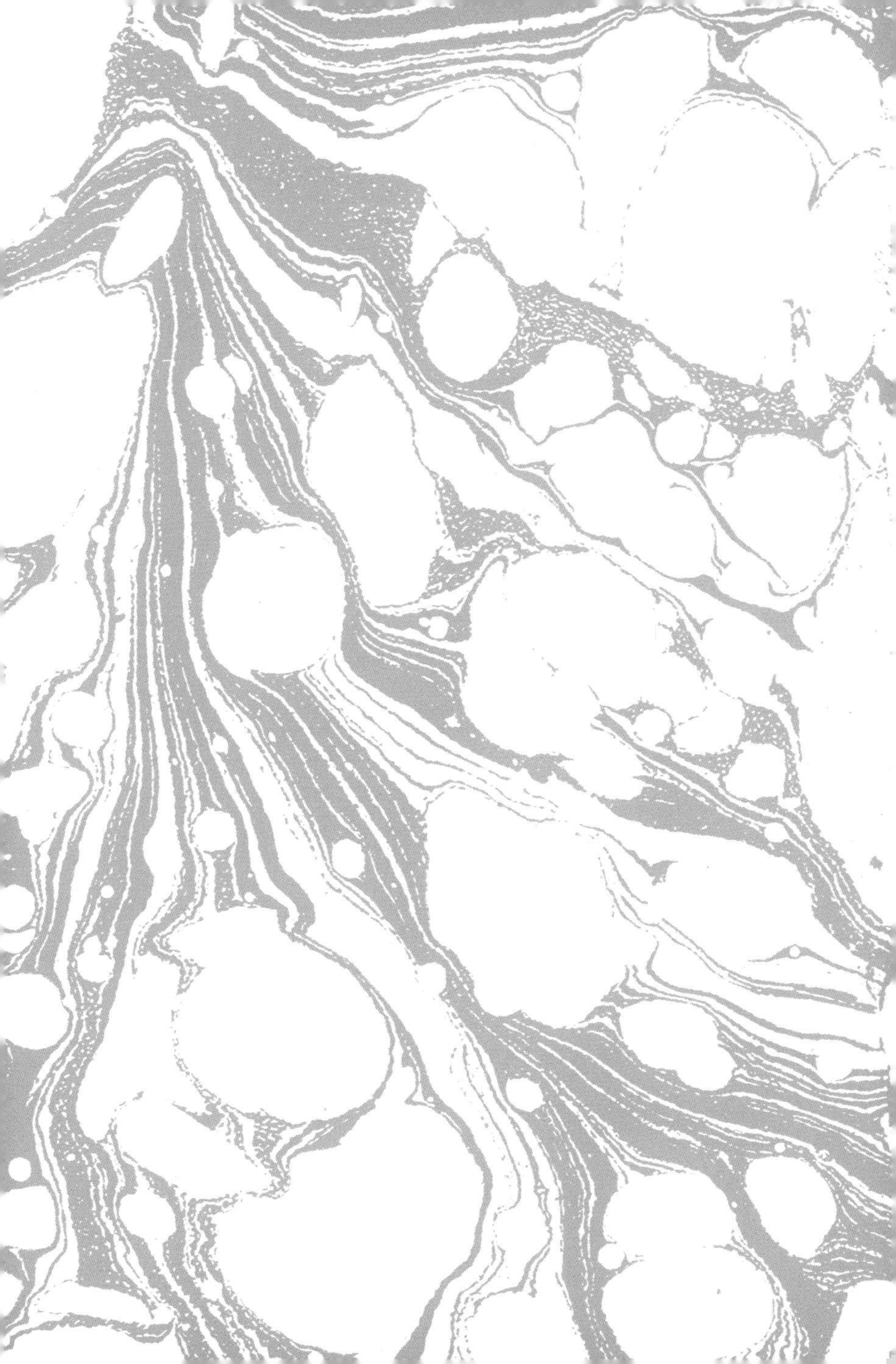

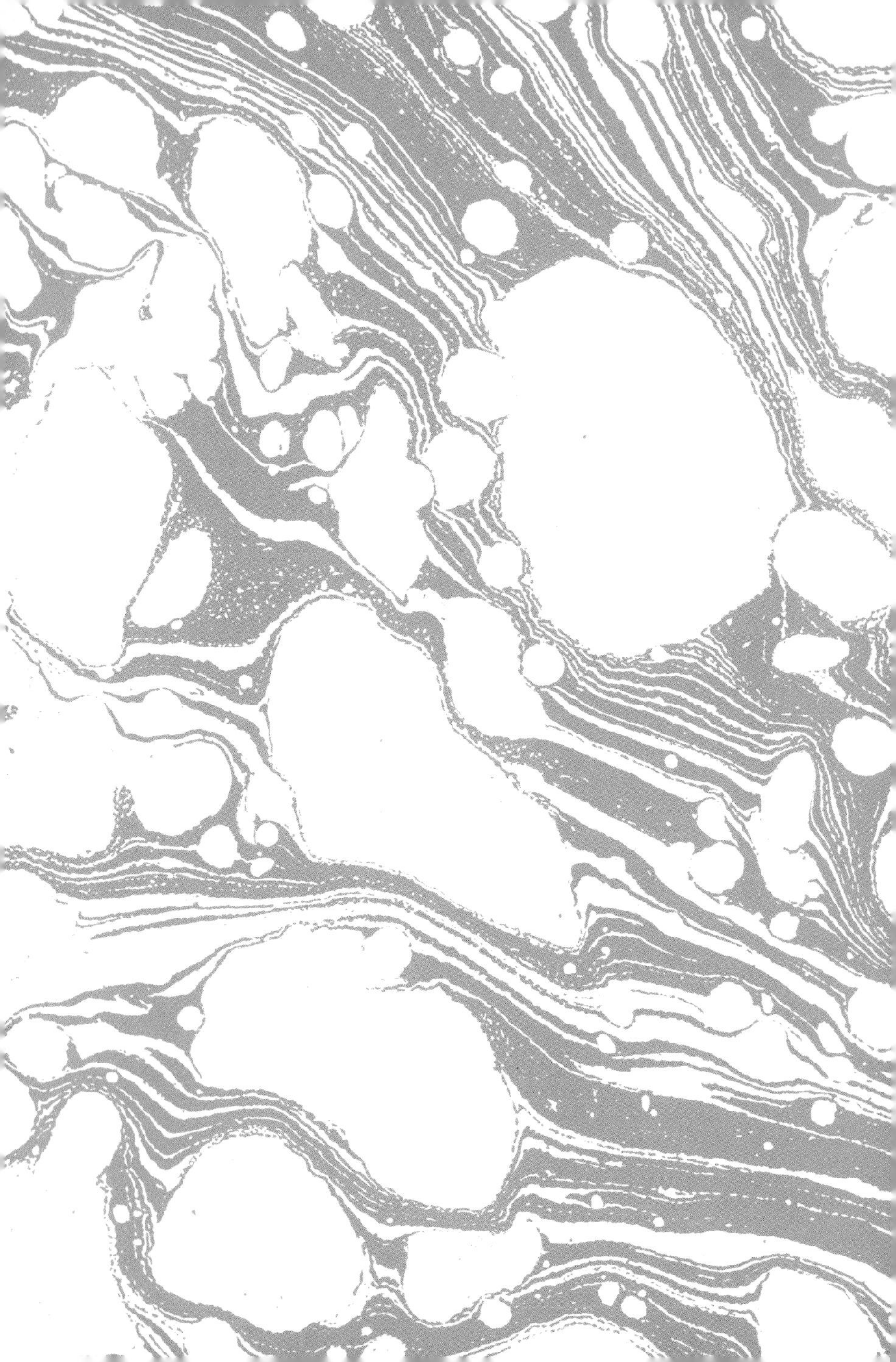

17 Foreword
Keith Webster, Helen and Henry Posner, Jr. Dean of the University Libraries

21 Introduction

35 PART I The Posner Memorial Collection

77 ESSAY Stab-Stitch Holes, Deckles, & Broken Types: Some Notes on the First Official Printing of the US Bill of Rights (1792)

89 ESSAY Jenner's Inquiry (1798) & the Invention of Vaccines

107 PART II Ciphers & Encryption

133 ESSAY On Cipher Discs & Early Encryption

155 PART III Robotics & Automata

173 ESSAY On Renaissance Robots, "Engines of Flesh," & Artificial Life

193 PART IV Artificial Intelligence & Writing Machines

223 ESSAY Artificial Intelligence, Lockean Minds, & Cartesian Machines: The Enlightenment Roots of Machine Learning

243 PART V Mechanical Computing

293 INTERLUDE Ancestral Machines: The Traub-McCorduck Collection of the History of Computing

309 PART VI Scientific Inquiry & Theory Building

349 PART VII Technology & Media

379 ESSAY Old Books, New Technologies: Computational Research in Special Collections

393 PART VIII Old Books, New Technologies

411 ESSAY On The Hinman Collator, Machine Vision & the Mechanical Origins of the Digital Humanities

429 PART IX Miscellanea

535 INTERLUDE The Hunt Botanical Library & Hunt Institute for Botanical Documentation

555 ESSAY On The Four Shakespeare Folios, 1623–2023

575 Bibliography

583 Index of Books & Objects

588 Colophon

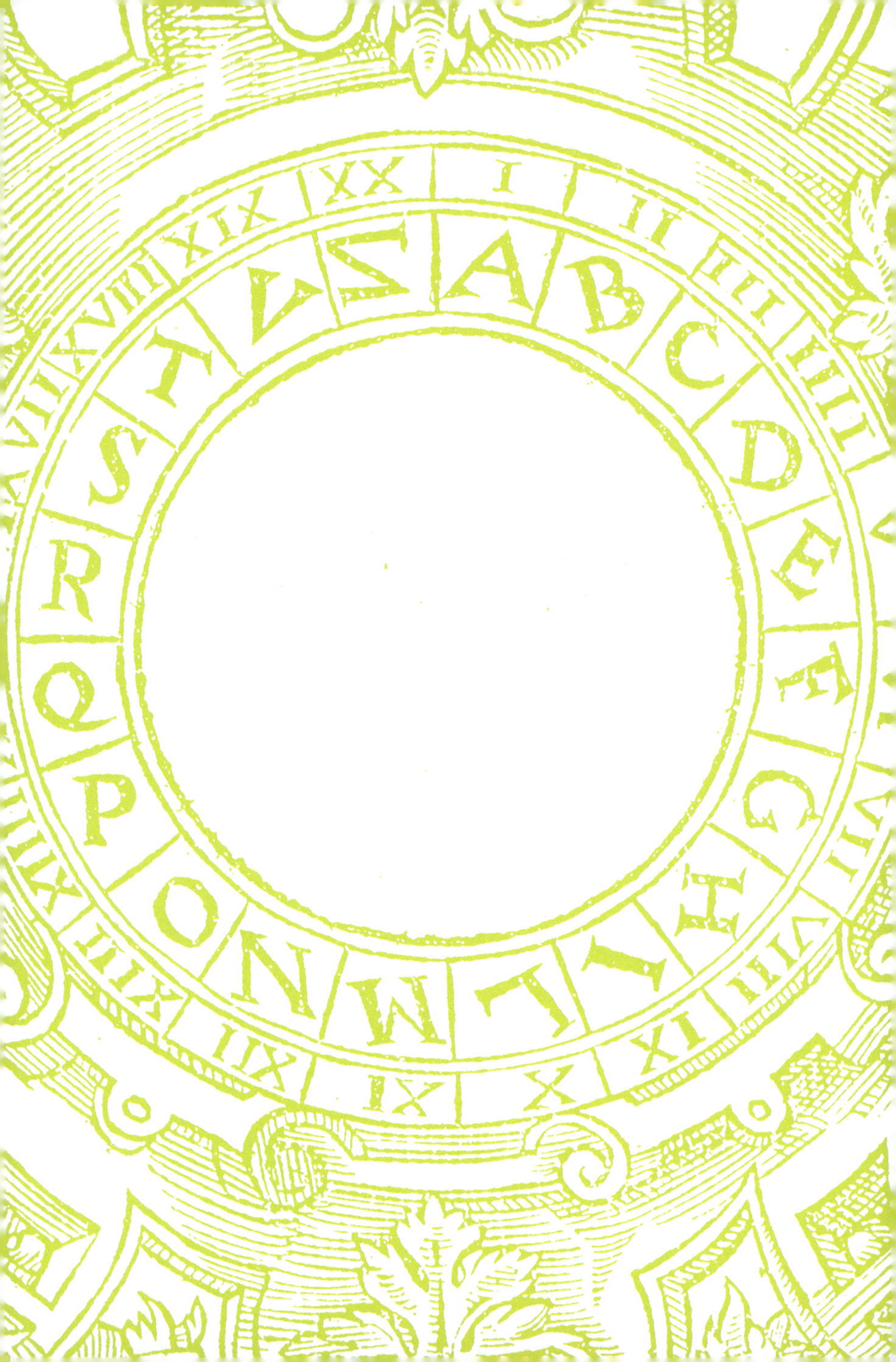
A B C D E F G H I L M N O P Q R S T V Z
I II III IIII V VI VII VIII IX X XI XII XIII XIIII XV XVI XVII XVIII XIX XX

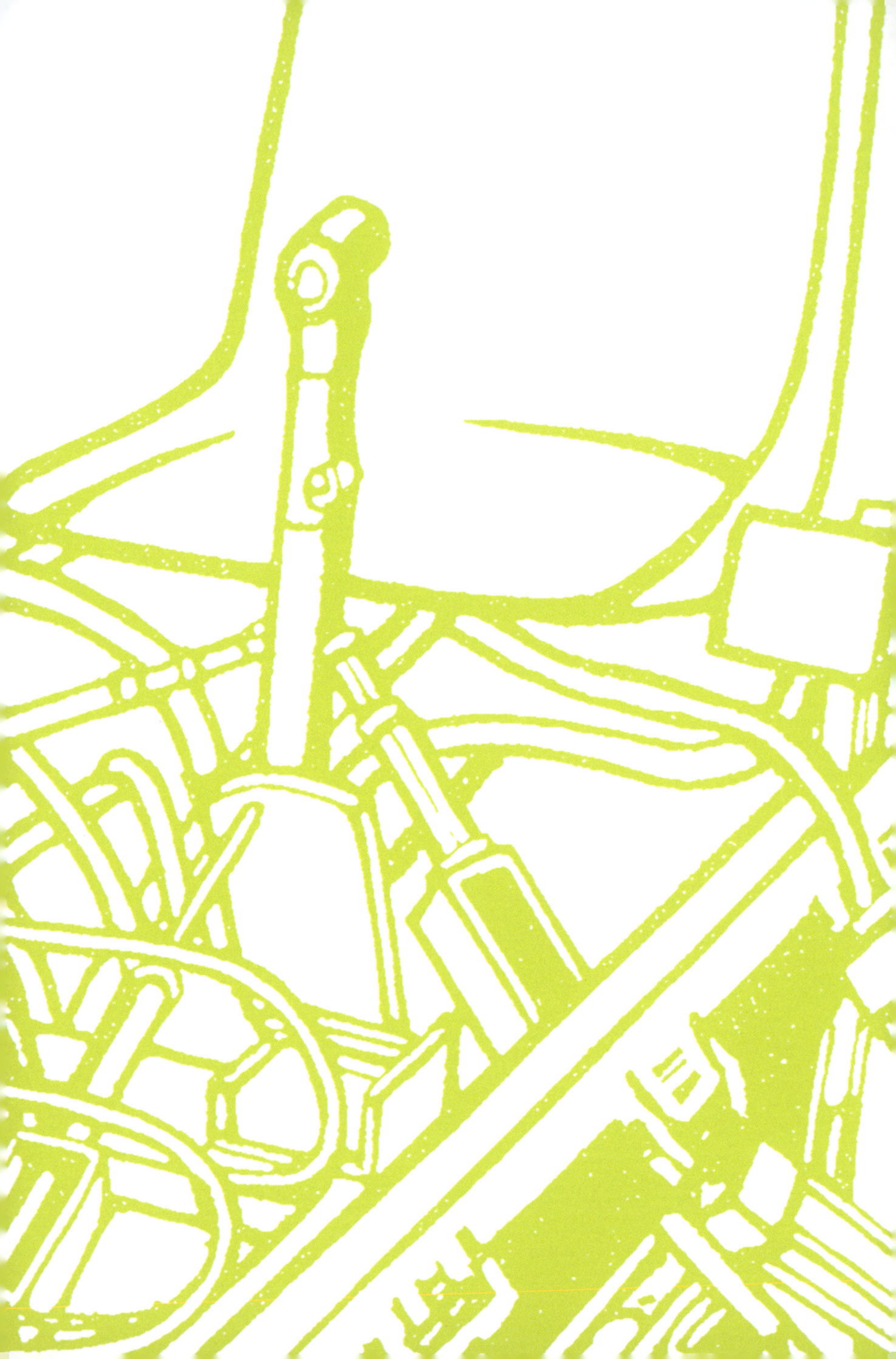

A

B

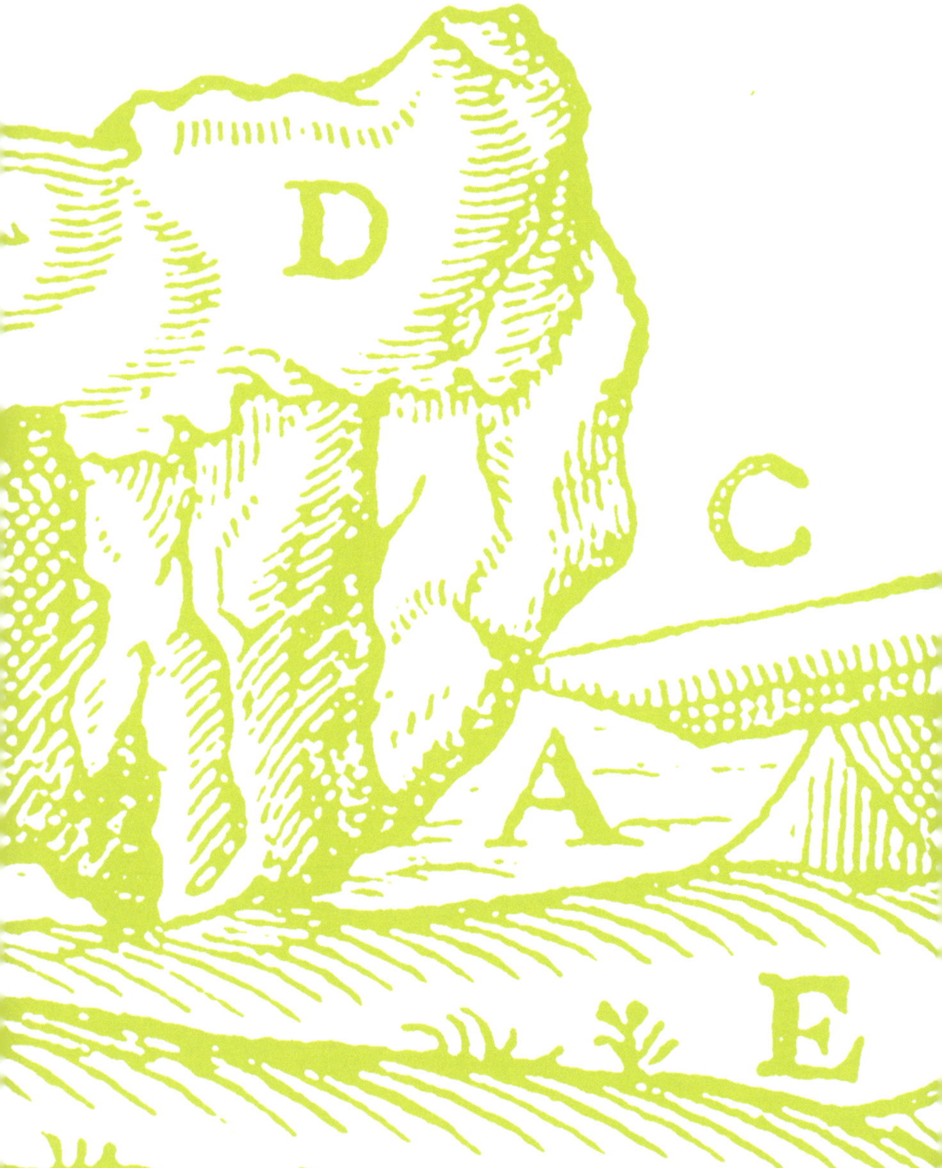
D
C
A
E

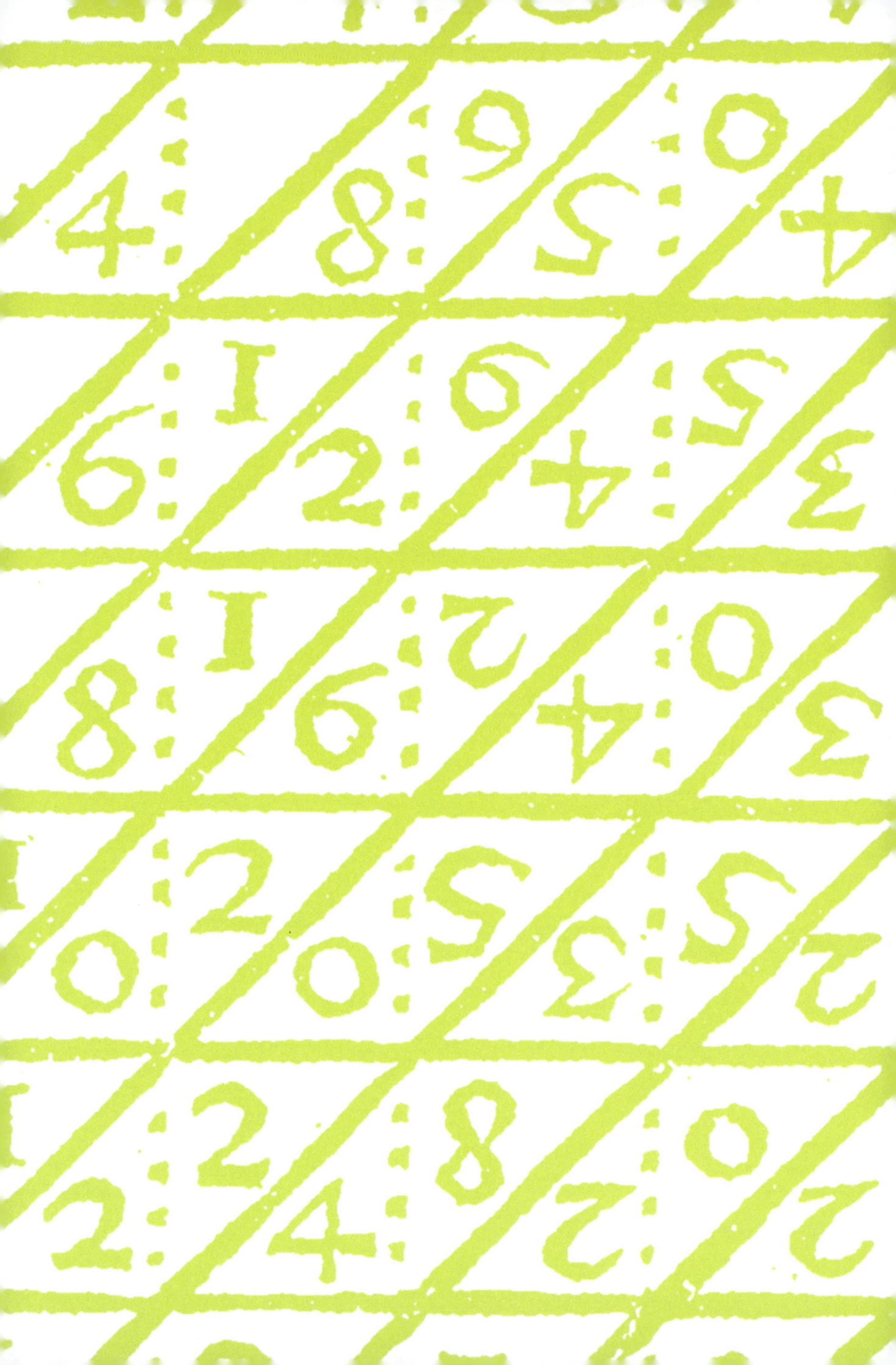

FOREWORD

There are spaces on every university campus that quietly gather meaning over time — places and objects where past and future seem to converge: a reading room, a manuscript, a machine. At Carnegie Mellon University, the newly transformed Posner Center for Special Collections becomes such a place. The Center is a home for rare books and historical technologies, and its opening represents a new era — one marked by openness, experimentation, and an opportunity for engagement with the most urgent questions of our time.

This book, *Rare Books & Ancestral Machines*, is a portable tribute to the collections, people, and ideas that animate the Posner Center. It invites readers into a curated world of objects and essays, offering a constellation of encounters across centuries and disciplines — from Renaissance cipher wheels to World War II encryption devices, from Enlightenment science to artificial intelligence, from the First Folio of Shakespeare to the pioneering works of Marie Curie and Alan Turing. Each artifact is both a physical object and a conceptual bridge, linking the tactile past to our digital present and our computational futures.

Special collections have always had a dual identity. They are sites of preservation — safeguarding the unique, the rare, and the irreplaceable — and they are laboratories of interpretation, inviting students, faculty, and the public to ask new questions of old things. At CMU, this dual identity finds powerful expression in the mission of the Posner Center. Supported by a transformative gift from the Posner Foundation of Pittsburgh, the Center emerges not only as a space for storing and displaying rare materials but also as a dynamic locus for cross-disciplinary scholarship, hands-on learning, and community discovery.

The spirit of innovation that defines Carnegie Mellon University is deeply embedded in this collection. Whether it is the early history of computing, the philosophical roots of machine intelligence, or the typographic artistry of early modern Europe, the holdings of Special Collections remind us that ideas do not arrive out of nowhere. They are the products of minds in motion — experimenting, failing, building, preserving. They are, in a very real sense, ancestral machines: tools built by previous generations to explore what it means to know, to create, and to endure.

This handbook captures the flavor of the Posner Center — not as a complete catalog but as an invitation to further inquiry. It reflects the interconnected work of librarians, scholars, students, and collectors. It honors the legacy of visionary donors and stewards who saw, in these books and artifacts, more than antiquarian interest. They saw teaching tools, cultural catalysts, and windows into the future.

As we celebrate the opening of the renovated Posner Center, this volume serves not only as a guide to its treasures but as a reminder of the values that sustain great libraries: curiosity, care, and the belief that knowledge — in all its forms — matters. I invite you to turn its pages with wonder and with the awareness that you are entering a living archive, still in motion.

Keith Webster
Helen and Henry Posner, Jr. Dean of the University Libraries
Carnegie Mellon University

INTRODUCTION

Rare Books & Ancestral Machines offers a portable and richly illustrated survey of the Posner Center for Special Collections at Carnegie Mellon University. The book is divided into nine parts, each comprising a curated list of rare books and early technologies that reflect key themes in the collection. Printed alongside the book's descriptive labels and illustrations are short interpretive essays that, besides providing more nuanced and wide-ranging investigations into the objects and stories in Special Collections, also contain cross-references and allusions to topics treated elsewhere in the book. Taken together, then, the book's parts, essays, and item labels present a constellated and abridged view — a partial listing in one hundred objects — of the Posner Center's range and scope.

In compiling this book, I have tried to convey the richness of Special Collections without resorting to an exhaustive catalog or list. It is a portable companion, a handbook in size, tone, and intention: it parcels out the collection's many themes and distills its approximately eight thousand books and objects into a volume whose shape fits neatly in the hand. There are inevitably many things left out. This book is therefore best viewed as a cross-section of a richer and more varied whole.

⋆ ⋆ ⋆

Carnegie Mellon University (CMU) Libraries' Special Collections mainly curates objects and books from the history of four related fields: computing, robotics, cryptography, and artificial intelligence. These disciplines present startling histories full of forking paths, abandoned prototypes, and forgotten technologies. Each part of this handbook is addressed to one (or more) of these fields. The first part focuses on the *Posner Memorial Collection*, a unit of Special Collections that consists of rare books from the history of science, including first editions by Galileo, Ibn al-Haytham, Ptolemy, Isaac Newton, Marie Curie, Charles Darwin, Johannes Kepler, and Albert Einstein, among others. Assembled over many decades by Pittsburgh collector and entrepreneur Henry Posner Senior (1889–1976), the Posner Collection forms the core of Special Collections' holdings in the history of science and technology; it also lends the Posner Center its name. Alongside descriptions and illustrations of six items from the Posner Collection are two essays: the first, "Jenner's *Inquiry* (1798) & the Invention of Vaccines," considers the eighteenth-century discovery of vaccinology and the pastoral metaphors invoked by the word *vaccine*. The second essay, "Stab-Stitch Holes, Deckles, & Broken Types: Some Notes on the First Official Printing of the US Bill of Rights (1792)," uncovers new bibliographical evidence in the Posner copy of the first ratified printing of the US Bill of Rights — one of only five copies in existence — that modifies the received account of its printing and distribution.

Ciphers & Encryption (part 2) examines the material and printed history of cryptology — the science and mathematics of building and breaking cipher systems. Described here are four books and objects that illustrate this history, from the first Western treatise on ciphers, printed in Venice in 1568, to a World War II-era Enigma machine marked and patinated by time and use. An essay, "On Cipher Discs," studies Renaissance cipher wheels as paper-and-string precursors to pseudorandom number generators and electronic encryption.

Parts 3 through 5 (*Robotics & Automata, Artificial Intelligence & Writing Machines,* and *Mechanical Computing*) examine objects and books in Special Collections from the overlapping histories of computer science, machine learning, and robotics. Together, these three parts demonstrate that the mathematics of cogwheel calculators evolved to become modern computer science and show that premodern self-moving machines gave way to innovations in modern robotics. The story of this centuries-long progression is riddled with fascinating experiments and unfamiliar niches of practice and theory. One essay, "Artificial Intelligence, Lockean Minds & Cartesian Machines: The Enlightenment Roots of Machine Learning," examines the emergence of artificial intelligence and computational heuristics; another, "On Renaissance Robots, 'Engines of Flesh,' & Artificial Life," revels in the mechanical complexity of early automata and *karakuri* — clockwork technologies and mechanical figures designed and built in Edo Japan. A final interlude in the form of a list describes the Traub-McCorduck Collection of the History of Computing — an important group of mechanical calculators, cryptographic devices, and rare books that came to Special Collections in 2018 (see "A Brief History of Special Collections," below).

Scientific Inquiry & Theory Building (part 6) steps back to examine the intellectual strategies that evolved to become the scientific method. Listing eight books that vary in genre, period, and language, this part runs from the experimental philosophy of seventeenth-century French philosopher and mathematician René Descartes (1596–1650) to the discovery of radioactivity and the emergence of natural selection as a fundamental principle of biology and genetics. The books included in this part remind us that scientific inquiry is iterative: built upon observation and centuries of theory refinement.

Old Books, New Technologies (part 7) offers a shift to method. Special Collections is a space for experimentation and knowledge-making as much as it is a space for the preservation and display of cultural artifacts.

Item labels provide examples of objects in Special Collections that have prompted new discoveries by researchers and students at CMU, while an essay, "Old Books, New Technologies: Computational Research in Special Collections," describes ongoing work by the Print & Probability project, an interdisciplinary team of researchers at CMU and UC San Diego that develops machine learning tools to detect subtle evidence in centuries-old books. This essay argues that bibliographical analysis — a type of scholarly work that examines books and printed artifacts as archaeologists examine the exhumed remains of long-lost cultures — is a powerful tool, though one that's primed for technological disruption and reinvention.

Technology & Media (part 8) opens with the deceptively simple observation that the book is itself a kind of technology. This observation gives way to six examples of written (or inscribed) artifacts and books, all illustrating innovations designed to convey information and promote the exchange of ideas. From a four-thousand year-old cuneiform clay tablet to early experiments in digital type design, part 8 considers how the written word and the printed page have propelled science and culture across cultures and centuries.

The handbook's final part, *Miscellanea*, considers Special Collections in microcosm. It illustrates artifacts and books that lack a single unifying theme while finding parallels and affinities among them. Objects as disparate in tone and style as Lewis Carroll's *Alice* novels and George Boole's treatise on symbolic logic are placed in dialogue, bridging children's literature and formal mathematics. A final essay, "On the Four Shakespeare Folios, 1623–2023," revisits a 2023 exhibition held to mark the four hundredth anniversary of the printing of Shakespeare's First Folio at CMU and Pittsburgh's Frick Museum. As one of Special Collections' chief rarities, the First Folio is emblematic of the collection's quality and depth.

As will be obvious by now, this handbook adopts a number of formal strategies more often associated with museums and literary criticism than research libraries: expository labels, essays, and illustrative figures. But the handbook's peculiarity in arrangement and tone is purposeful, reflecting the Posner Center's status as a kind of museum and laboratory for the humanities. Far from a mere storehouse of curiosities, the Posner Center plays a vital and active role in CMU's mission of teaching, community outreach, and research.

A BRIEF HISTORY OF SPECIAL COLLECTIONS

Special Collections originated with the 1964 donation of Rachel McMasters Miller Hunt's (1882–1963) non-botanical books. Hunt's collection of rare books on plants, gardens, and botany formed the Hunt Institute for Botanical Documentation at CMU. Though specializing in historical botany, Hunt's activity as a collector was wide-ranging: among her books held in Special Collections are late medieval breviaries and books of hours; a large collection of fine bindings; a copy of Francesco Colonna's *Hypnerotomachia Poliphili*, printed in Venice at the press of Aldus Manutius in 1499 and called "the most beautiful book of the fifteenth century;" a leaf from the Gutenberg Bible; and a copy of Hartmann Schedel's *Liber Chronicarum*, an obsessive and intricately illustrated history of the known world up to 1492.

The collection has since grown from Hunt's founding gift through the addition of important bequests. These include the collection of Charles J. Rosenbloom

(1898–1973), who by an unusual arrangement, gifted his books to Yale University — his alma mater — except for titles that Yale's libraries already held. These would-be duplicates instead formed a separate bequest to CMU. From this gift, Special Collections obtained a copy of the First Folio of Shakespeare's plays (1623): Yale owns two, and so the Rosenbloom First Folio stayed in Pittsburgh, along with companion copies of the Second (1632), Third (1663), and Fourth Folios (1685). The Shakespeare Folios are but one example. The Rosenbloom bequest also included a number of other literary firsts, including a fine copy of the 1818 first edition of Mary Shelley's *Frankenstein*; a first edition of Samuel Johnson's *Dictionary* that survives in its original paper-board binding; several rare editions of music by Bach, Beethoven, Schubert, and Rossini; and early printings of English and American literature, including first editions by Melville, Coleridge, Charlotte Brontë, Whitman, and Twain. The bulk of Rosenbloom's collection of art — including etchings and engravings by Rembrandt and Dürer — meanwhile went to Pittsburgh's Carnegie Museum of Art.

In 1978, the Posner family deposited a rich collection of books illustrating the history of science, technology, and political thought with CMU Libraries. The Posner Memorial Collection, named in honor of collector and entrepreneur Henry Posner Senior, includes a number of high points of Western science and now forms the core of Special Collections' holdings in the field. More recently, in 2018 the Libraries received the Traub-McCorduck Collection of the History of Computing, an important grouping of early calculating machines, cryptographic devices, and rare books on computation, number theory, and mathematics. Assembled by Joseph F. Traub (1932–2015; head of CMU School of Computer Science, 1971–1979), the collection was donated to CMU Libraries by Pamela McCorduck

(1940–2021), an authority on the early history of AI and Traub's wife.

Alongside these major donations are smaller but no less significant acquisitions and gifts, including a nearly complete run of books issued by the Kelmscott Press from the collection of Rachel McMasters Miller Hunt; a trove of primary source materials — sketches, paintings, letters, and notebooks — by the Victorian artist Kate Greenaway (1846–1901) from the collection of bibliophile, advertising executive, and journalist Frances Hooper (1892–1986); the personal libraries of artificial intelligence pioneers Allen Newell (1927–1992) and Herbert A. Simon (1916–2001); an important collection of artwork by Harold Cohen (1928–2016), whose AARON was the first AI software used to create art; and the reference collection of type designer and internationally renowned calligrapher, Arnold Bank (1908–1983). Each of these gifts and acquisitions has contributed to Special Collections' growth over its first seven decades.

Now amounting to some eight thousand books and objects, Special Collections is entering a period of transformation made possible by a major gift from the Posner Foundation of Pittsburgh. In 2025 the Foundation gave $8 million to fund the renovation of the Posner Center and to support its day-to-day operation. Special Collections' core programs — collection storage and preservation, student engagement and instruction, research, and exhibitions — are now brought together in a purpose-built, vibrant, and accessible space. While Pittsburgh's cultural history is full of similarly transformative acts of generosity, the Posner Foundation gift demonstrates that this philanthropic tradition remains very much alive. I am deeply grateful to the Foundation and the Posner family for their long-standing support of Special Collections and CMU Libraries.

I close this introduction by naming and thanking the two Special Collections librarians who preceded me at CMU Libraries: Anne Skoog and Mary Catherine Johnsen. All libraries reflect the librarians who built and preserved them. I am fortunate to have inherited the legacy of their commitment and skill. It is to them that I dedicate this book.

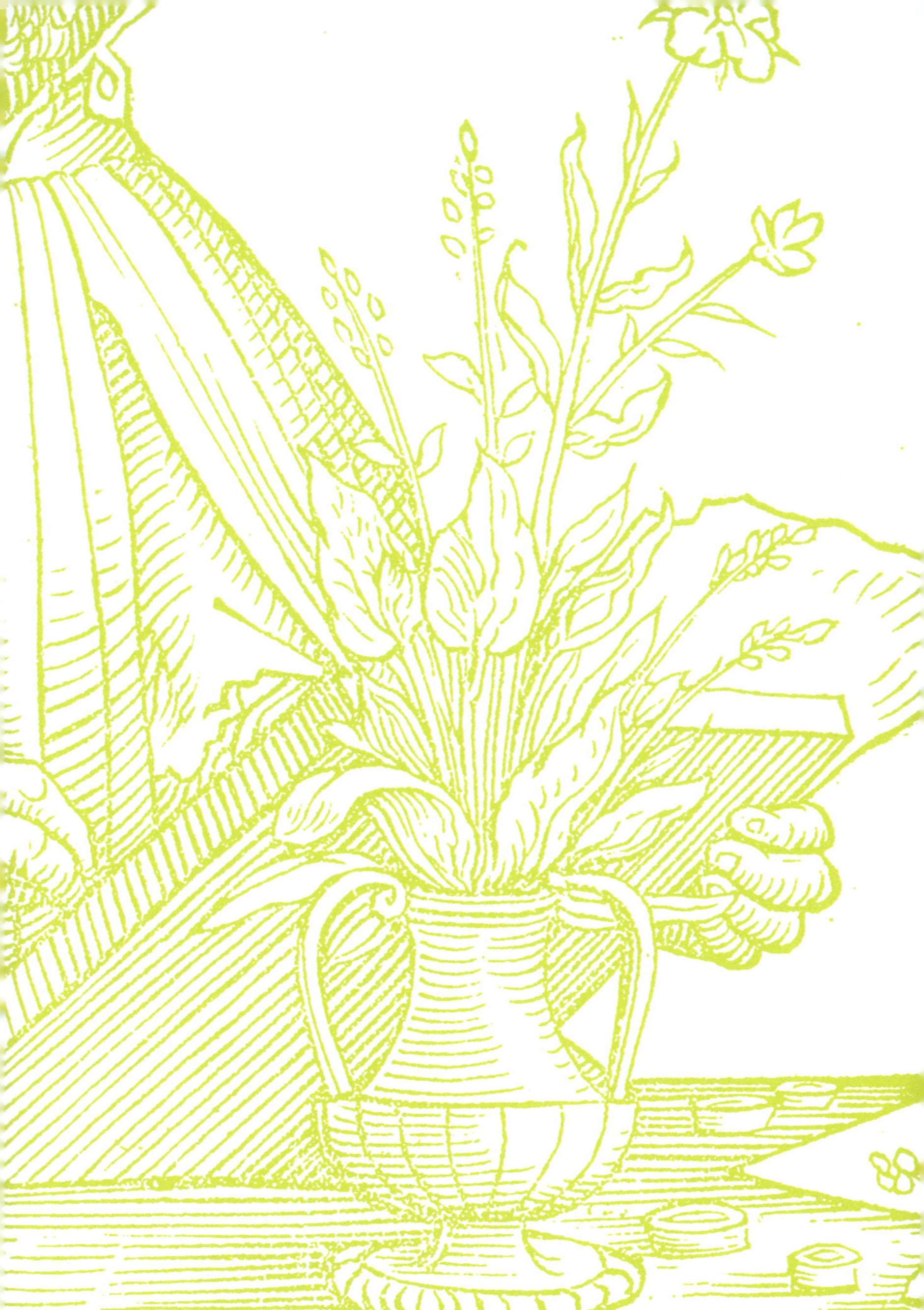

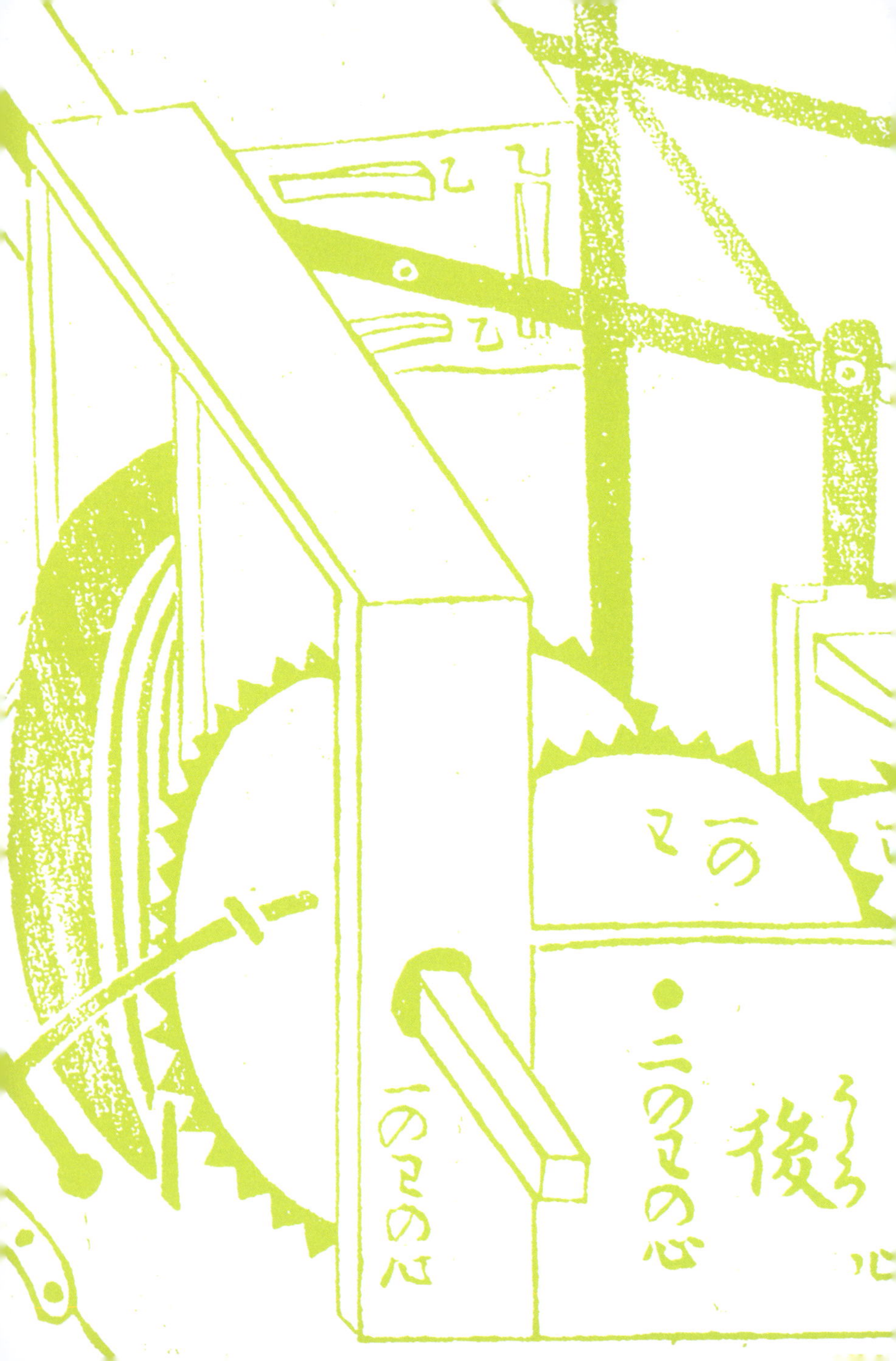

乙
乙
後

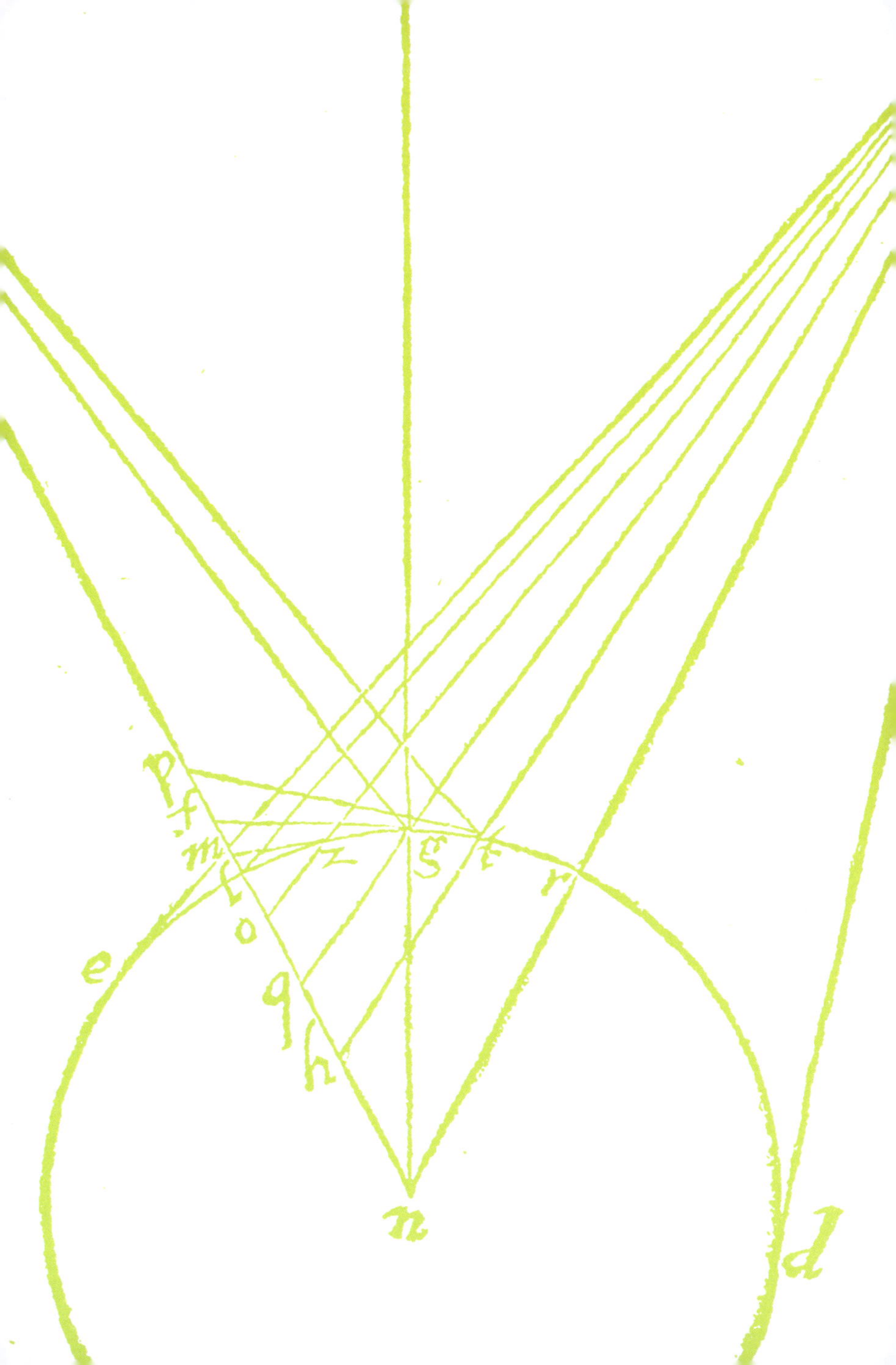
p
f
m
l
z
g
t
r
o
e
q
h
n
d

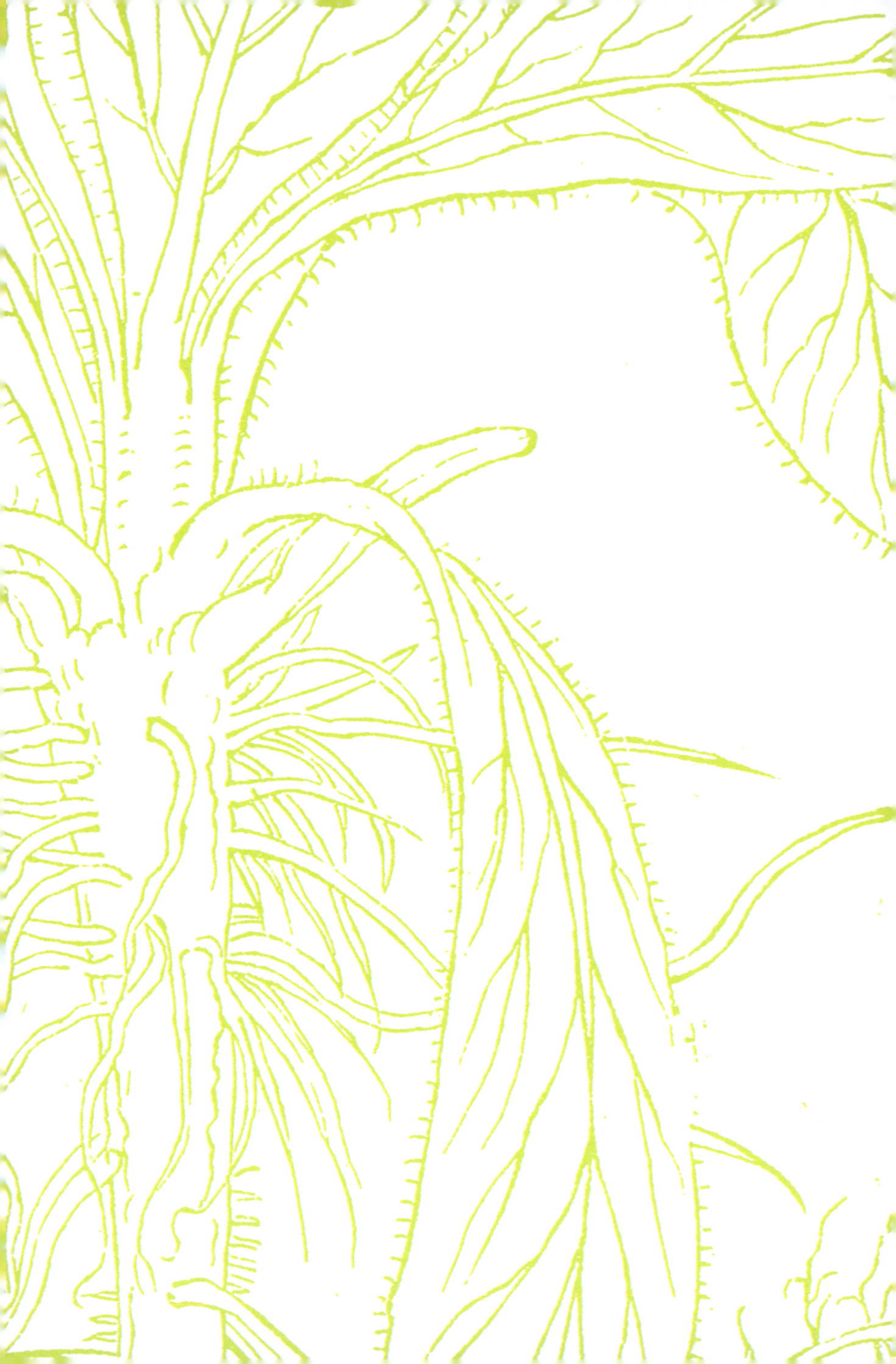

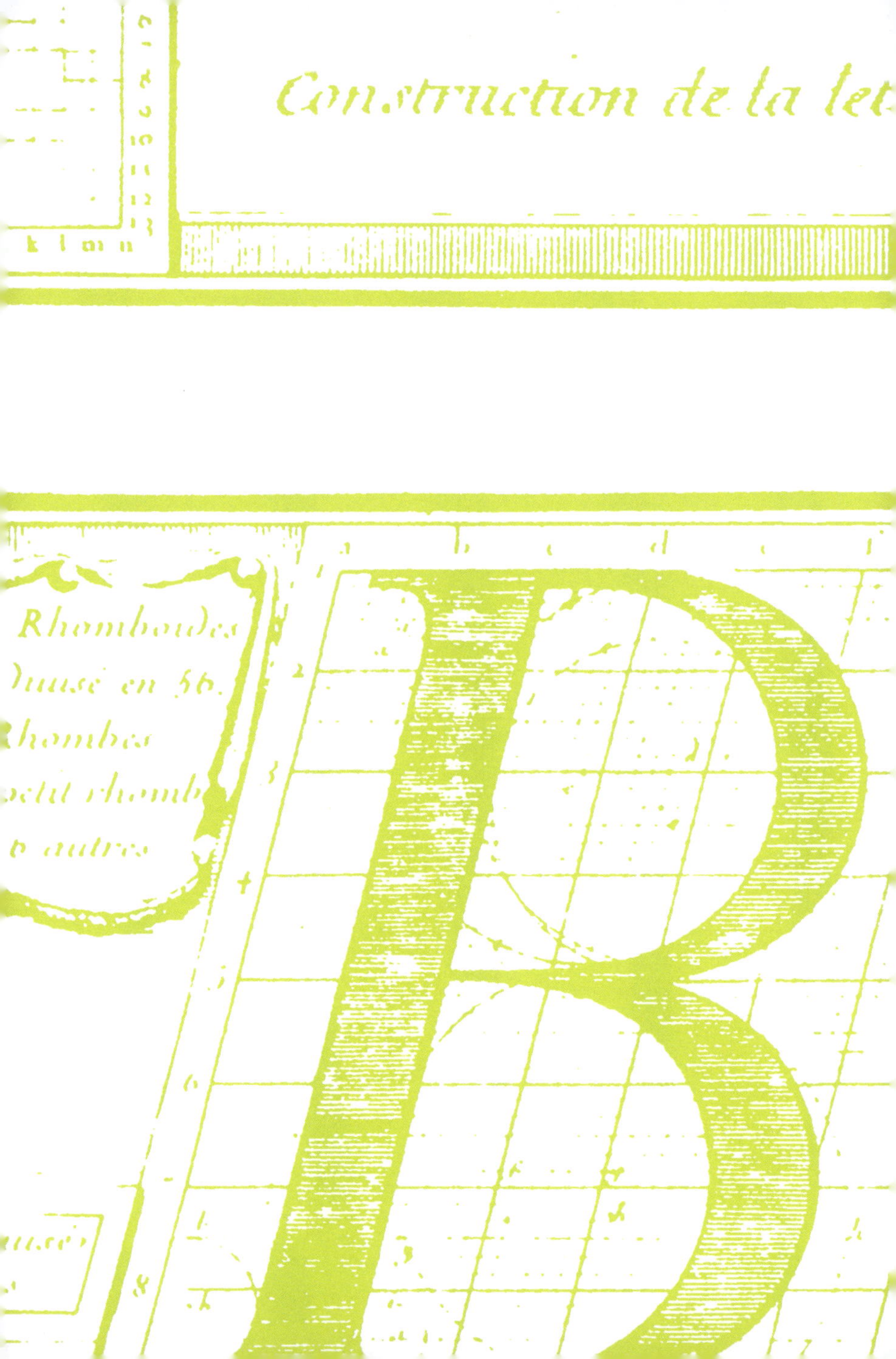
Construction de la let
Rhomboides

PART I

The Posner Memorial Collection

Named for Pittsburgh collector and entrepreneur Henry Posner Senior (1889–1976), the Posner Memorial Collection holds some of the most important works in the history of science: among its 1,000-odd volumes are first editions by Isaac Newton, Galileo, Marie Curie, Charles Darwin, Albert Einstein, and others. It also holds one of only five surviving copies of the first official printing of the US Bill of Rights.

In 1978 Posner's family deposited the collection with CMU Libraries, where it now forms the core of Special Collections' holdings in early science and technology; it also gives the Posner Center its name. This part offers a representative sample of titles from the Posner Collection.

Henry Posner Senior: Entrepreneur & Collector

Over five decades, Henry Posner Senior formed one of the leading private collections of rare books from the history of science and political thought, including a rare suviving copy of the first official printing of the US Bill of Rights.

Born in Warsaw, Poland, Posner emigrated to the United States in 1905. He enrolled at Carnegie Tech in 1912, leaving to start a business in Pittsburgh's East Liberty neighborhood. In 1927, that business — the Alpha Claude Neon Corporation — became the first Pittsburgh patent holder for neon lighting. By 1932, the business was renamed the Pittsburgh Outdoor Advertising Company, which Posner led until his retirement in 1964.

After depositing Posner's collection with CMU Libraries in 1978, Henry and Helen Posner Jr. gifted the Posner Center in memory of their parents in 2004. After twenty years and a major renovation, the Posner Center is CMU's destination for discovery, research, and hands-on engagement with the past.

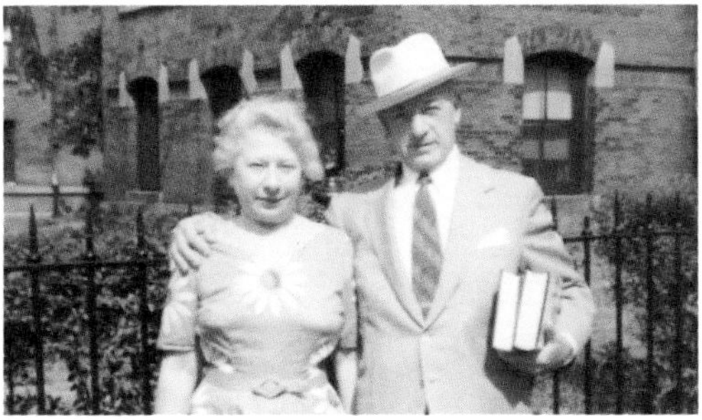

↑ Henry Posner Senior and Ida Posner in 1955

"I generally know what I am after and will not buy anything except what I want, and in first class condition at that."

— Henry Posner Sr.

COPERNICUS, Nicolaus (1473–1543).
De revolutionibus orbium coelestium.
Nuremberg, 1543.

The first work to demonstrate heliocentric (sun-centered) planetary motion.

Before Copernicus, astronomers relied on an Earth-centered model of celestial motion to predict the position of the planets. Copernicus's *On the Revolutions of the Heavenly Spheres* was the first to propose a sun-centered, or "heliocentric" alternative. The Posner copy of the first edition shows that Copernicus's idea was viewed as radical from the beginning: an early reader censored a section on heliocentrism, adding "hypothesis" above the offending passage. This annotation reflects an early and cautious reception of Copernicus's ideas, which would remain contentious for more than a century. Galileo, among the most famous of Copernicus's defenders, was condemned by the Roman Inquisition in 1633 for espousing the heliocentric model.

net, in quo terram cum orbe lunari tanquam epicyclo contineri diximus. Quinto loco Venus nono mense reducitur. Sextum deniq; locum Mercurius tenet, octuaginta dierum spacio circū currens. In medio uero omnium residet Sol. Quis enim in hoc

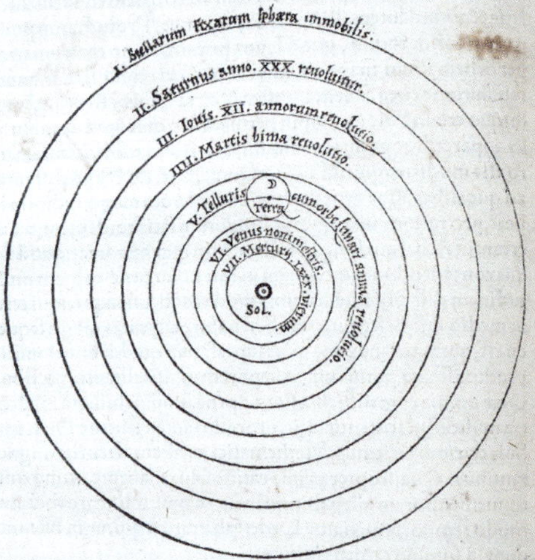

pulcherimo templo lampadem hanc in alio uel meliori loco poneret, quàm unde totum simul posit illuminare? Siquidem non inepte quidam lucernam mundi, alij mentem, alij rectorem uocant. Trimegistus uisibilem Deum, Sophoclis Electra intuentē omnia. Ita profecto tanquam in solio re gali Sol residens circum agentem gubernat Astrorum familiam. Tellus quoq; minime fraudatur lunari ministerio, sed ut Aristoteles de animalibus ait, maximā Luna cū terra cognationē habet. Concipit interea à Sole terra, & impregnatur annuo partu. Inuenimus igitur sub hac

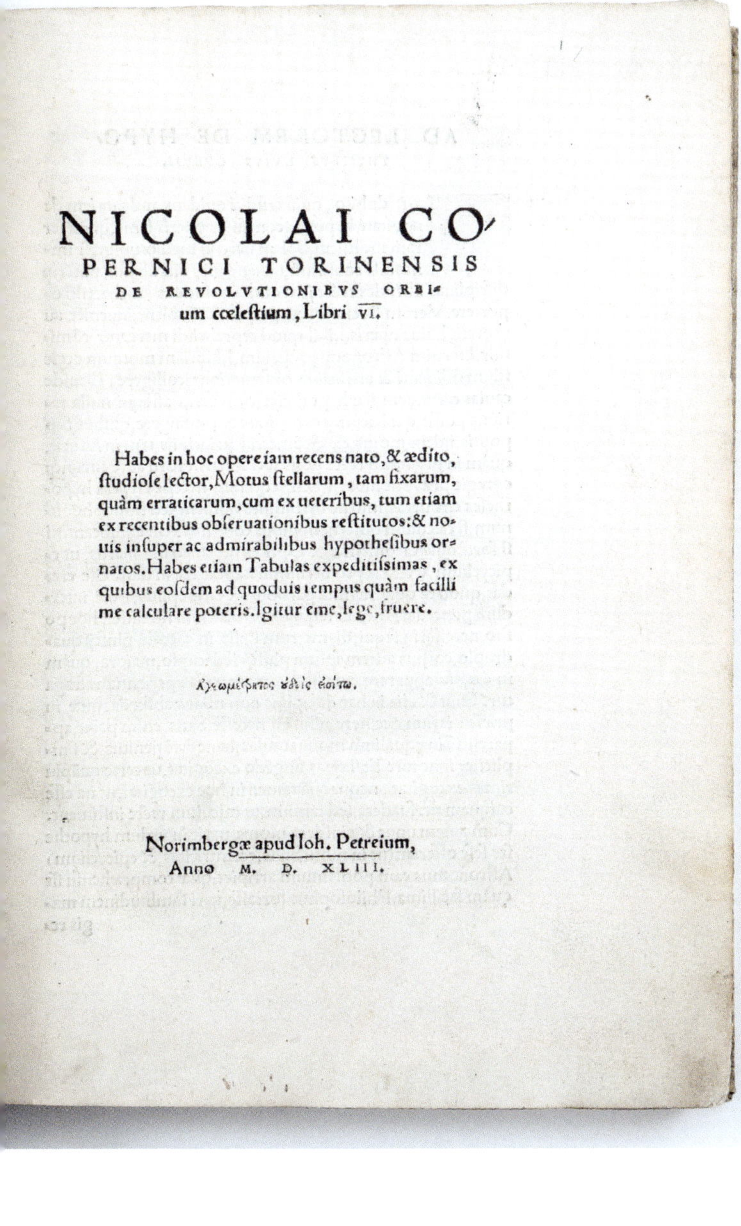

NICOLAI CO-
PERNICI TORINENSIS
DE REVOLVTIONIBVS ORBI-
um cœlestium, Libri VI.

Habes in hoc opere iam recens nato, & ædito, studiose lector, Motus stellarum, tam fixarum, quàm erraticarum, cum ex ueteribus, tum etiam ex recentibus obseruationibus restitutos: & nouis insuper ac admirabilibus hypothesibus ornatos. Habes etiam Tabulas expeditissimas, ex quibus eosdem ad quoduis tempus quàm facillime calculare poteris. Igitur eme, lege, fruere.

Ἀγεωμέτρητος οὐδεὶς εἰσίτω.

Norimbergæ apud Ioh. Petreium,
Anno M. D. XLIII.

terſit, ſcintillantia illorum lumina demõſtrant. Quo indicio maxime diſcernuntur à planetis, quodq; inter mota & non mota, maximam oportebat eſſe differentiam. Tanta nimirum eſt diuina hæc Opt. Max. fabrica.

Hypothesis triplicis motus Terræ eiusque demonstratio

De triplici motu telluris demonſtratio. Cap. XI.

CVm igitur mobilitati terrenę tot tantaq; errantium ſyderum conſentiant teſtimonia, iam ipſum motum in ſumma exponemus, quatenus apparentia per ipſum tanquã hypoteſim demonſtrentur, quẽ triplicẽ omnino oportet admittere. Primum quem diximus νυχθημέρινον à Græcis uocari, diei noctiſq; circuitum proprium, circa axem telluris, ab occaſu in ortum uergentem, prout in diuerſum mundus ferri putatur, æquinoctialem circulum deſcribendo, quem nonnulli æquidialem dicunt, imitantes ſignificationem Græco

c ij rum,

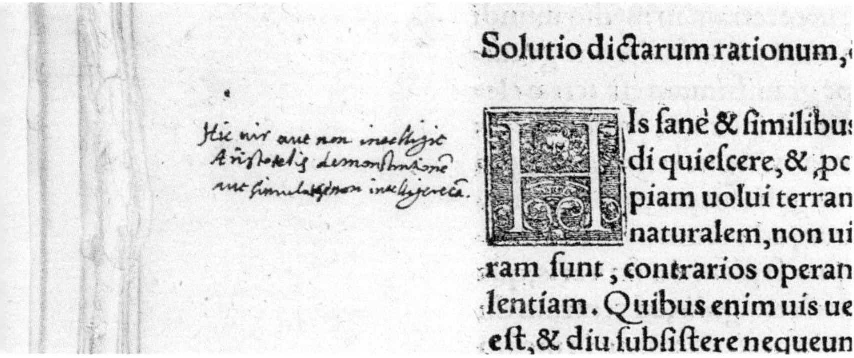
Solutio dictarum rationum,

His sane & similibus
di quiescere, & po
piam uolui terran
naturalem, non ui
ram sunt, contrarios operan
lentiam. Quibus enim uis ue
est, & diu subsistere nequeun

PTOLEMY, Claudius (circa 100–170 CE). *[Almagest] Magnae constructionis, id est perfectae coelestium motuum pertractationis.* Basel, 1538.

The standard model of planetary motion for a thousand years.

Ptolemy's *Almagest* provided the standard model of planetary motion before Copernicus (see previous entry). Written in Greek, it introduced mathematical methods for modeling and predicting the positions of the planets and stars. This 1538 edition, printed in Basel, includes woodcut diagrams illustrating Ptolemy's geocentric system. It also includes Ptolemy's writing on eclipses, a catalogue of stars, and descriptions of scientific instruments.

The provenance of Special Collections' copy of the *Almagest* is well traced and documented: it was owned by the sixteenth-century Flemish scholar and bibliophile Marcus Laurinus (1530–1581) and was later held in the vast personal library of Jean-Baptiste Colbert (1619–1683), French statesman and Minister of Finance under Louis XIV. Henry Posner purchased the copy from Jacob Zeitlin of Zeitlin and Ver Brugge in Los Angeles in 1972. This was Posner's last major acquisition before his death in 1976.

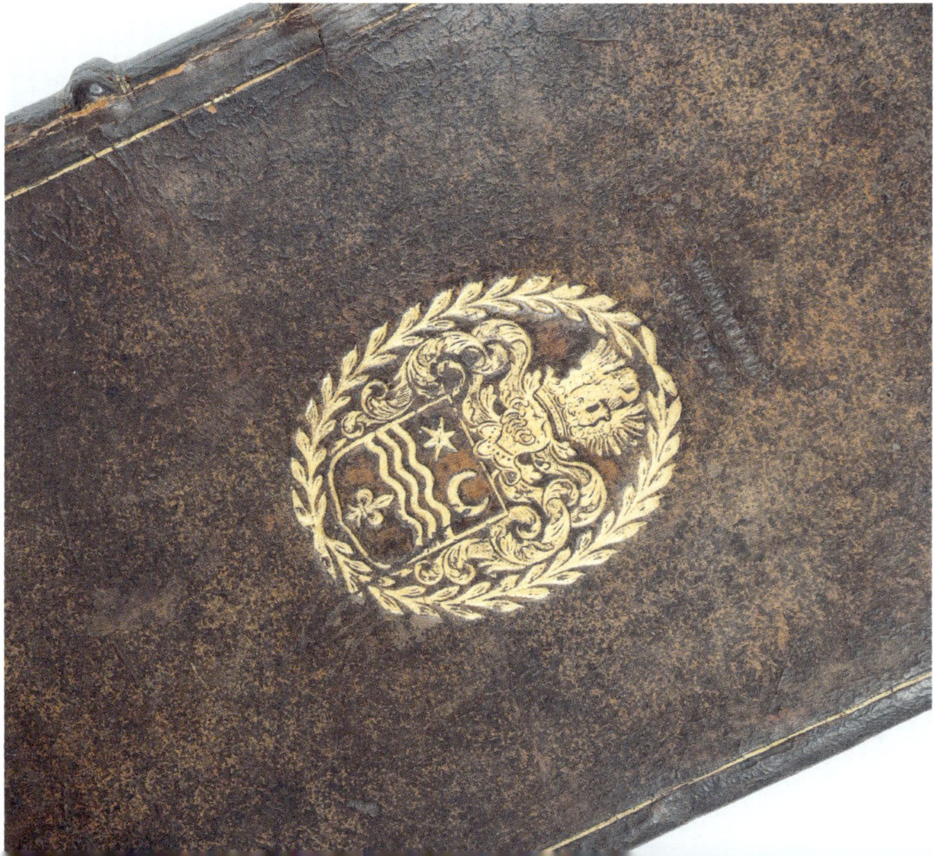

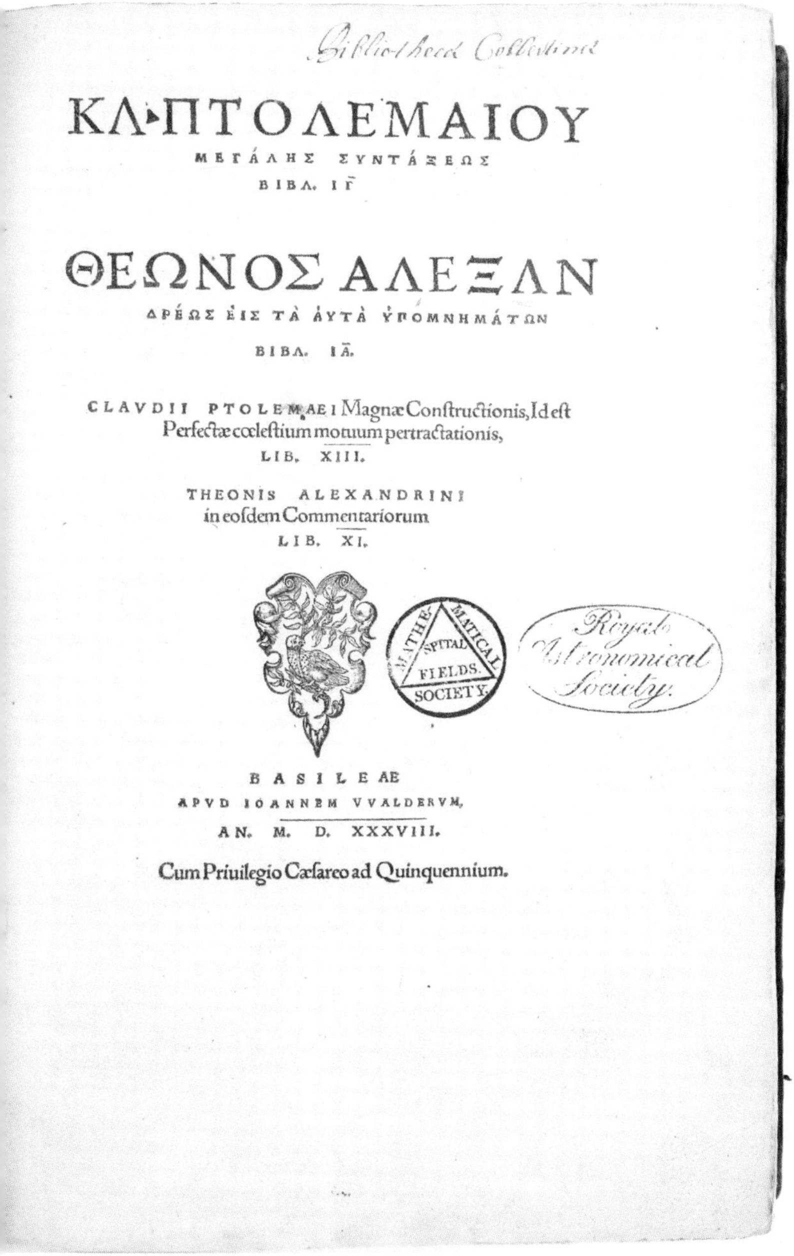

ΚΛ· ΠΤΟΛΕΜΑΙΟΥ

ΜΕΓΆΛΗΣ ΣΥΝΤΆΞΕΩΣ

ΒΙΒΛ. ΙΓ

ΘΕΩΝΟΣ ΑΛΕΞΑΝ

ΔΡΈΩΣ ἘΙΣ ΤᾺ ἈΥΤᾺ ὙΠΟΜΝΗΜΆΤΩΝ

ΒΙΒΛ. ΙΑ.

CLAVDII PTOLEMAEI Magnæ Constructionis, Id est Perfectæ cœlestium motuum pertractationis,

LIB. XIII.

THEONIS ALEXANDRINI in eosdem Commentariorum

LIB. XI.

BASILEAE

APVD IOANNEM VVALDERVM,

AN. M. D. XXXVIII.

Cum Priuilegio Cæsareo ad Quinquennium.

πρὸς τὴν ὑπὸ τὴν διπλῆν τῆς βγ, δοθήσεται καὶ ἑκατέρα τῶν αβ καὶ βγ περιφερειῶν. ἐκτεθείσης γὰρ τῆς αὐτῆς καταγραφῆς, ἐπεζεύχθω ἡ αδ, καὶ ἤχθω ἀπὸ τοῦ δ κάθετος, ἐπὶ τὴν αεγ, ἡ δζ. ὅτι μὲν οὖν τῆς αγ περιφερείας δοθείσης, ἥτε ὑπὸ αδζ γωνία, τὴν ἡμίσειαν αὐτῆς ὑποτείνουσα, δεδομένη ἔσται, καὶ ὅλον τὸ αδζ τρίγωνον δῆλον. ἐπεὶ δὲ τῆς αγ εὐθείας ὅλης δεδομένης, ὑπόκειται καὶ ὁ τῆς αε πρὸς εγ λόγος, ὁ αὐτὸς ὢν, τῷ τῆς ὑπὸ τὴν διπλῆν τῆς αβ, πρὸς τὴν ὑπὸ τὴν διπλῆν τῆς βγ, ἥτε αε ἔσται δοθεῖσα, καὶ λοιπὴ ἡ ζε. καὶ διὰ τοῦτο καὶ τῆς δζ δεδομένης, δοθήσεται καὶ ἥτε ὑπὸ εδζ γωνία, τοῦ εδζ ὀρθογωνίου, καὶ ὅλη ἡ ὑπὸ αδβ. ὥστε καὶ ἥτε αβ περιφέρεια δοθήσεται, καὶ λοιπὴ ἡ βγ, ὅπερ ἔδει δεῖξαι. πάλιν ἔστω κύκλος ὁ αβγ, περὶ κέντρον τὸ δ, καὶ ἐπὶ τῆς περιφερείας αὐτοῦ εἰλήφθω τρία σημεῖα τὰ α β γ, ὥστε ἑκατέραν τῶν αβ αγ περιφερειῶν, ἐλάσσονα εἶναι ἡμικυκλίου. καὶ ἐπὶ τῶν ἑξῆς δὲ λαμβανομένων περιφερειῶν τὸ ὅμοιον ὑπακουέσθω. καὶ ἐπιζευχθεῖσαι ἥτε δα καὶ ἡ γβ, ἐκβεβλήσθωσαν, καὶ συμπιπτέτωσαν κατὰ τὸ ε σημεῖον, λέγω ὅτι ἐστὶν ὡς ἡ ὑπὸ τὴν διπλῆν τῆς γα περιφερείας, πρὸς τὴν ὑπὸ τὴν διπλῆν τῆς αβ, οὕτως ἡ γε εὐθεῖα, πρὸς τὴν βε. ὁμοίως γὰρ τῷ προτέρῳ λημματίῳ, ἐὰν ἀπὸ τῶν β καὶ γ ἀγάγωμεν καθέτους, ἐπὶ τὴν δα, τήν τε βζ καὶ τὴν γη, ἔσται διὰ τὸ παραλλήλους αὐτὰς εἶναι ὡς ἡ γη πρὸς τὴν βζ, οὕτως ἡ γε πρὸς τὴν εβ. ὥστε καὶ ὡς ἡ ὑπὸ τὴν διπλῆν τῆς γα, πρὸς τὴν ὑπὸ τὴν διπλῆν τῆς αβ, οὕτως ἡ γε πρὸς τὴν εβ, ὅπερ ἔδει δεῖξαι. καὶ ἐνταῦθα δὲ αὐτόθεν παρακολουθεῖ, διότι κἂν ἡ γβ περιφέρεια μόνη δοθῇ, καὶ ὁ λόγος τῆς ὑπὸ τὴν διπλῆν τῆς γα, πρὸς τὴν ὑπὸ τὴν διπλῆν τῆς αβ δοθῇ, καὶ ἡ αβ περιφέρεια δοθήσεται. πάλιν γὰρ ἐπὶ τῆς ὁμοίας καταγραφῆς ἐπιζευχθείσης τῆς δβ, καὶ καθέτου ἀχθείσης, ἐπὶ τὴν βγ τῆς δζ, ἡ μὲν ὑπὸ βδζ γωνία, τὴν ἡμίσειαν ὑποτείνουσα τῆς βγ περιφερείας, ἔσται δεδομένη, καὶ ὅλον ἄρα τὸ βδζ ὀρθογώνιον. ἐπεὶ δὲ, καὶ ὁ τῆς γε πρὸς τὴν εβ λόγος δέδοται, καὶ ἔτι ἡ γβ εὐθεῖα δοθήσεται, καὶ ἥτε εβ, καὶ ἔτι ὅλη ἡ εβζ. ὥστε καὶ ἐπεὶ ἡ δζ δέδοται, δοθήσεται καὶ ἥτε ὑπὸ εδζ γωνία τοῦ αὐτοῦ ὀρθογωνίου, καὶ λοιπὴ ἡ ὑπὸ εδβ. ὥστε καὶ ἡ αβ περιφέρεια ἔσται δεδομένη. τούτων προληφθέντων, γεγράφθωσαν ἐπὶ σφαιρικῆς ἐπιφανείας, μεγίστων κύκλων περιφέρειαι, ὥστε εἰς δύο τὰς αβ καὶ αγ δύο γραφείσας τὰς βε καὶ γδ, τέμνειν ἀλλήλας κατὰ τὸ ζ σημεῖον. ἔστω δὲ ἑκάστη αὐτῶν ἐλάσσων ἡμικυκλίου. τὸ δ᾽ αὐτὸ καὶ ἐπὶ πασῶν τῶν καταγραφῶν ὑπακουέσθω. λέγω δὴ ὅτι ὁ τῆς ὑπὸ τὴν διπλῆν τῆς γε περιφερείας, πρὸς τὴν ὑπὸ τὴν διπλῆν τῆς εα περιφερείας, συνῆπται ἔκ τε τοῦ τῆς ὑπὸ τὴν διπλῆν τῆς γζ, πρὸς τὴν ὑπὸ τὴν διπλῆν τῆς ζδ, καὶ τοῦ τῆς ὑπὸ τὴν διπλῆν τῆς δβ, πρὸς τὴν ὑπὸ τὴν διπλῆν

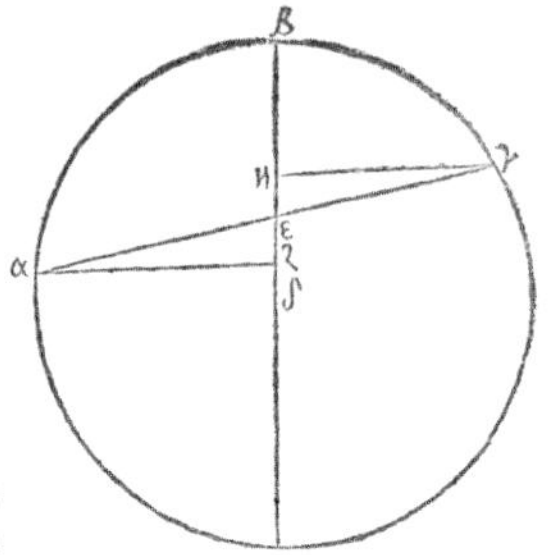

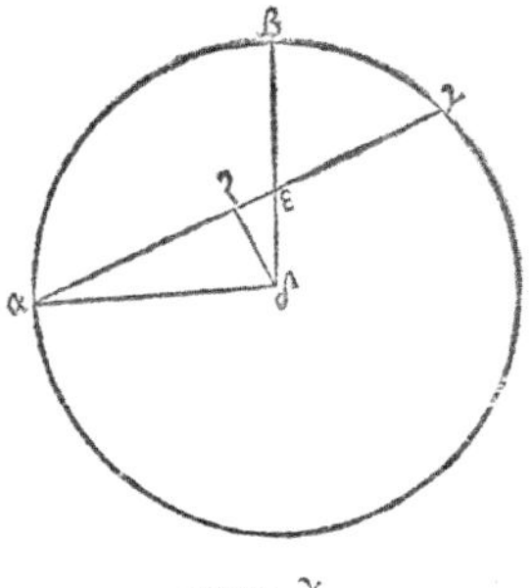

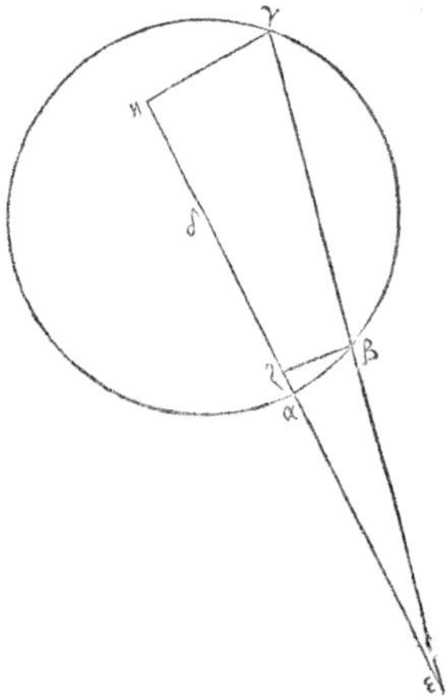

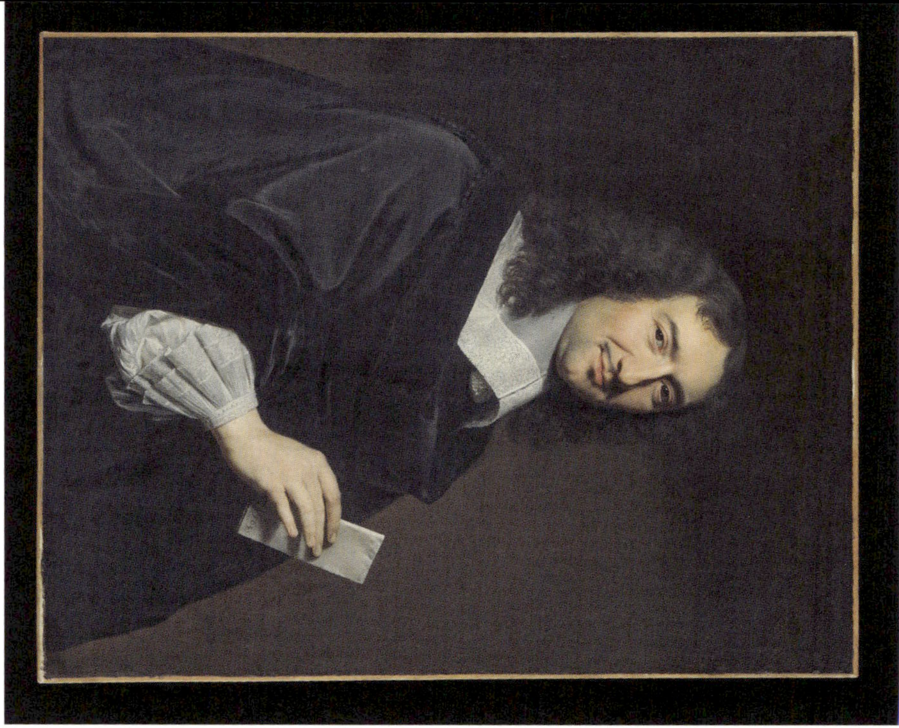

↑ Jean-Baptiste Colbert, an early owner of the Posner copy of the Almagest (1538). Philippe de Champaigne, *Jean-Baptiste Colbert* (1655). By permission of the Metropolitan Museum of Art (Public Domain; Object ID 435886).

CONGRESS, United States.
US Bill of Rights. Philadelphia, 1792.

The first official printing of the US Bill of Rights.

This copy of the US Bill of Rights was part of an edition printed in Philadelphia immediately after the Bill's ratification in December 1791. It represents the first time that the Bill of Rights was accorded force of law and the first time that its articles appeared in print as an integral part of the United States Constitution. Thomas Jefferson, then US Secretary of State, ordered its printing. The edition is a grail of Americana: only five complete copies are known to survive — one at the American Antiquarian Society, one at the Library of Congress, one in the Maryland State Archives, one in the Posner Memorial Collection, and another privately held copy. Henry Posner Senior acquired this copy in 1963.

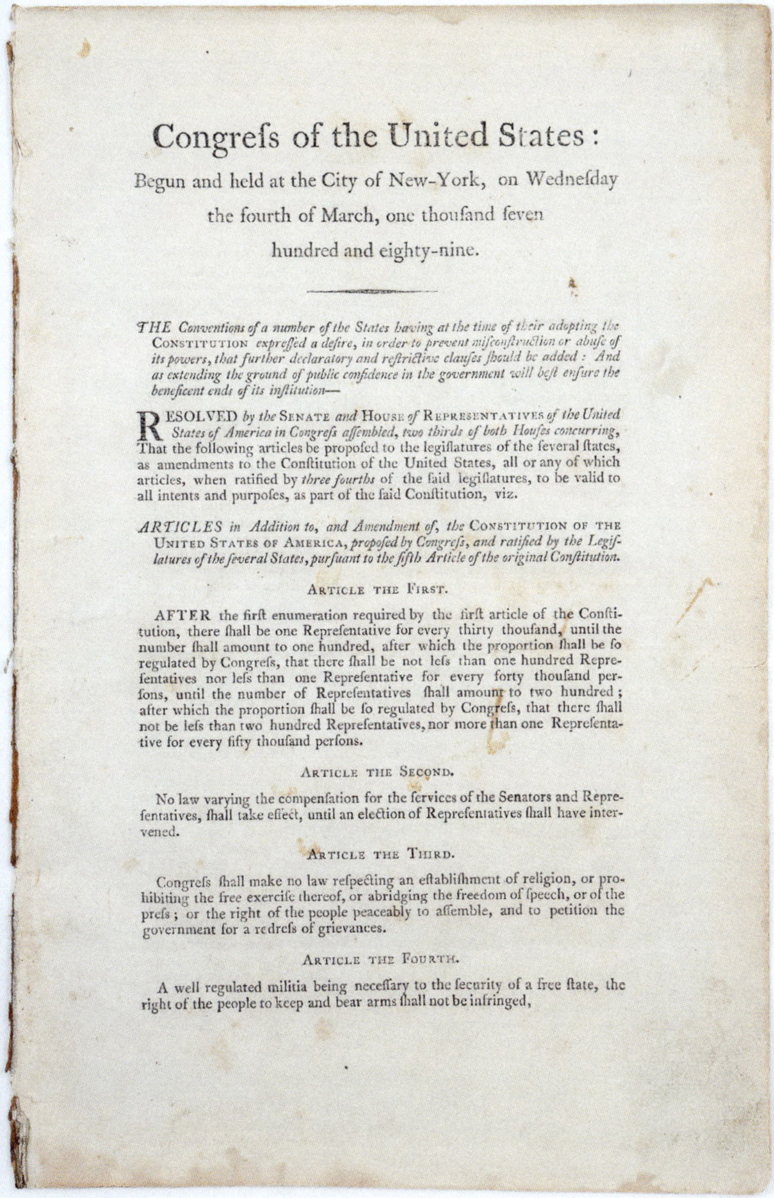

Congreſs of the United States:

Begun and held at the City of New-York, on Wedneſday the fourth of March, one thouſand ſeven hundred and eighty-nine.

THE Conventions of a number of the States having at the time of their adopting the CONSTITUTION *expreſſed a deſire, in order to prevent miſconſtruction or abuſe of its powers, that further declaratory and reſtrictive clauſes ſhould be added: And as extending the ground of public confidence in the government will beſt enſure the beneficent ends of its inſtitution—*

RESOLVED *by the* SENATE *and* HOUSE *of* REPRESENTATIVES *of the United States of America in Congreſs aſſembled, two thirds of both Houſes concurring,* That the following articles be propoſed to the legiſlatures of the ſeveral ſtates, as amendments to the Conſtitution of the United States, all or any of which articles, when ratified by *three fourths* of the ſaid legiſlatures, to be valid to all intents and purpoſes, as part of the ſaid Conſtitution, viz.

ARTICLES in Addition to, and Amendment of, the CONSTITUTION OF THE UNITED STATES OF AMERICA, *propoſed by Congreſs, and ratified by the Legiſlatures of the ſeveral States, purſuant to the fifth Article of the original Conſtitution.*

ARTICLE THE FIRST.

AFTER the firſt enumeration required by the firſt article of the Conſtitution, there ſhall be one Repreſentative for every thirty thouſand, until the number ſhall amount to one hundred, after which the proportion ſhall be ſo regulated by Congreſs, that there ſhall be not leſs than one hundred Repreſentatives nor leſs than one Repreſentative for every forty thouſand perſons, until the number of Repreſentatives ſhall amount to two hundred; after which the proportion ſhall be ſo regulated by Congreſs, that there ſhall not be leſs than two hundred Repreſentatives, nor more than one Repreſentative for every fifty thouſand perſons.

ARTICLE THE SECOND.

No law varying the compenſation for the ſervices of the Senators and Repreſentatives, ſhall take effect, until an election of Repreſentatives ſhall have intervened.

ARTICLE THE THIRD.

Congreſs ſhall make no law reſpecting an eſtabliſhment of religion, or prohibiting the free exerciſe thereof, or abridging the freedom of ſpeech, or of the preſs; or the right of the people peaceably to aſſemble, and to petition the government for a redreſs of grievances.

ARTICLE THE FOURTH.

A well regulated militia being neceſſary to the ſecurity of a free ſtate, the right of the people to keep and bear arms ſhall not be infringed,

GALILEI, Galileo (1564–1642).
Discorsi e dimostrazioni matematiche intorno a due nuove scienze attenenti alla meccanica e i movimenti locali. Leiden, 1638.

The foundation of mathematical physics and mechanics.

Known for his discovery of sunspots, Saturn's rings, and his telescopic observations of the moon, Galileo also pioneered a mathematical language for describing the world. The *Discorsi* forms a series of demonstrations and thought experiments showing that physical phenomena — from the acceleration of falling objects to the force exerted by a fulcrum and lever — can be described in numbers. Isaac Newton claimed to have derived his first two laws of motion — those of inertia and momentum — from this book.

The *Discorsi* was published while Galileo was under house arrest following his conviction for writing in support of a heliocentric model of the cosmos. Effectively banned from publishing his work in Italy, the *Discorsi* was smuggled north by its Dutch publisher, Louis Elsevier, and printed in Leiden.

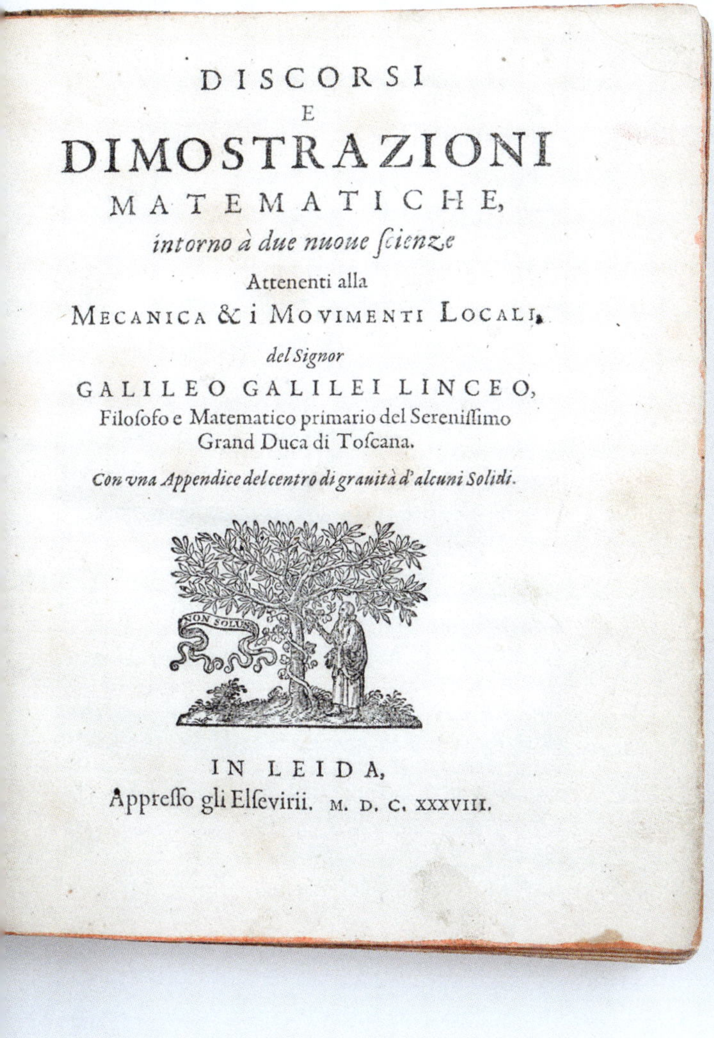

DISCORSI
E
DIMOSTRAZIONI
MATEMATICHE,
intorno à due nuoue scienze
Attenenti alla
MECANICA & i MOVIMENTI LOCALI,
del Signor
GALILEO GALILEI LINCEO,
Filosofo e Matematico primario del Serenissimo
Grand Duca di Toscana.
Con vna Appendice del centro di grauità d'alcuni Solidi.

NON SOLUS

IN LEIDA,
Appresso gli Elsevirii. M. D. C. XXXVIII.

mettendo in considerazione altri momenti che quelli della semplice forza in B, e della resistenza in D, quasi che l'istessa Leua fusse immateriale, e senza grauità. Mà se noi metteremo in conto la grauità

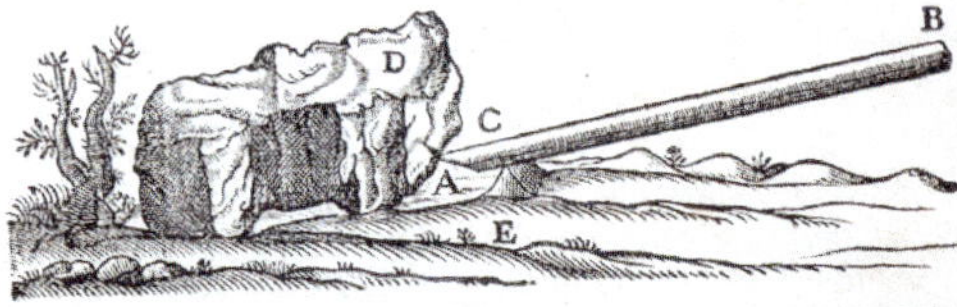

ancora dello strumento stesso della Leua, la quale sarà talor di legno, e tal volta anche di ferro; è manifesto che alla forza in B aggiunto il peso della Leua altererà la proporzione, la quale conuerrà pronunziare sotto altri termini. E però prima che passar più oltre è necessario, che noi conuenghiamo in por distinzione trà queste due maniere di considerare, chiamando, vn prendere assolutamente quello, quando intenderemo lo strumento preso in astratto, cioè separato dalla grauità della propria materia: mà congiugnendo con le figure semplici, & assolute la materia con la grauità ancora, nomineremo le figure congiunte con la materia momento, ò forza composta.

Sagr. E forza ch' io rompa il proposito, che haueuo di non dar' occasione di digredire, mà non potrei con attenzione applicarmi al rimanente, se non mi fusse rimosso certo scrupolo, che mi nasce; & è questo, che mi pare che V. S. faccia comparazione della forza posta in B con la total grauità del sasso D, della qual grauità mi pare che vna parte, e forse forse la maggiore si appoggi sopra 'l piano dell' Orizonte; si che

Salu. Hò inteso benissimo. V. S. non soggiunga altro; mà solamente auuerta, che io non hò nominata la grauità totale del sasso, mà hò parlato del momento, che egli tiene, & esercita sopra 'l punto A estremo termine della Leua B A, il quale è sempre minore dell' intero peso del sasso; & è variabile secondo la figura della pietra, e secondo che ella vien più, ò meno sollenata.

Sagr.

E per vn breue esempio di questo che dico disegnai già la figura di vn' osso allungato solamente tre volte, & ingrossato con tal proporzione, che potesse nel suo animale grande far l'uffizio proporzio-

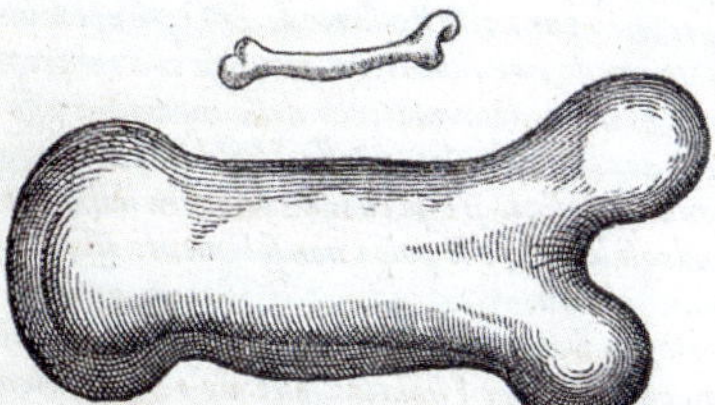

nato à quel dell' osso minore nell' animal più piccolo, e le figure son queste: doue vedete sproporzionata figura, che diuiene quella dell' osso ingrandito. Dal che è manifesto, che chi volesse mantener in vn vastissimo Gigante le proporzioni, che hanno le membra in vn huomo ordinario, bisognerebbe ò trouar materia molto più dura, e resistente per formarne l'ossa, ò vero ammettere, che la robustezza sua fusse à proporzione assai più fiacca, che ne gli huomini di statura mediocre; altrimente crescendogli à smisurata altezza si vedrebbono dal proprio peso opprimere, e cadere. Doue che all' incontro si vede nel diminuire i corpi non si diminuir con la medesima proporzione le forze, anzi ne i minori crescer la gagliardia con proporzion maggiore. Onde io credo che vn piccolo cane porterebbe addosso due, ò trè cani eguali à se, mà non penso già che vn cauallo portasse ne anco vn solo cauallo à se stesso eguale.

Simp. *Mà se così è, grand' occasione mi danno da dubitare le moli immense, che vediamo ne i pesci, che tal Balena, per quanto intendo, sarà grande per dieci Elefanti, e pur si sostengono.*

Salu. *Il vostro dubbio S. Sim. mi fà accorgere d'una condizione da me non auuertita prima, potente essa ancora à far che Giganti,*

NEWTON, Isaac (1643–1727).
Opticks: Or, a Treatise of the Reflextions, Refractions, Inflexions and Colours of Light.
London, 1704.

The discovery of the nature of light and infinitesimal calculus.

In *Opticks* Newton showed that light is composed of a spectrum of elementary colors. Unlike Newton's *Principia Mathematica* (1687) — a first edition copy of which is held in the Posner Memorial Collection — *Opticks* was written in English (see "NEWTON," page 493). This marked a turn away from Latin and toward the vernacular in scientific writing, a trend that accelerated in the eighteenth and nineteenth centuries. Printed at the end of this volume are Newton's treatises on infinitesimal calculus, which he developed independently of the German mathematician Gottfried Wilhelm von Leibniz (1646–1716).

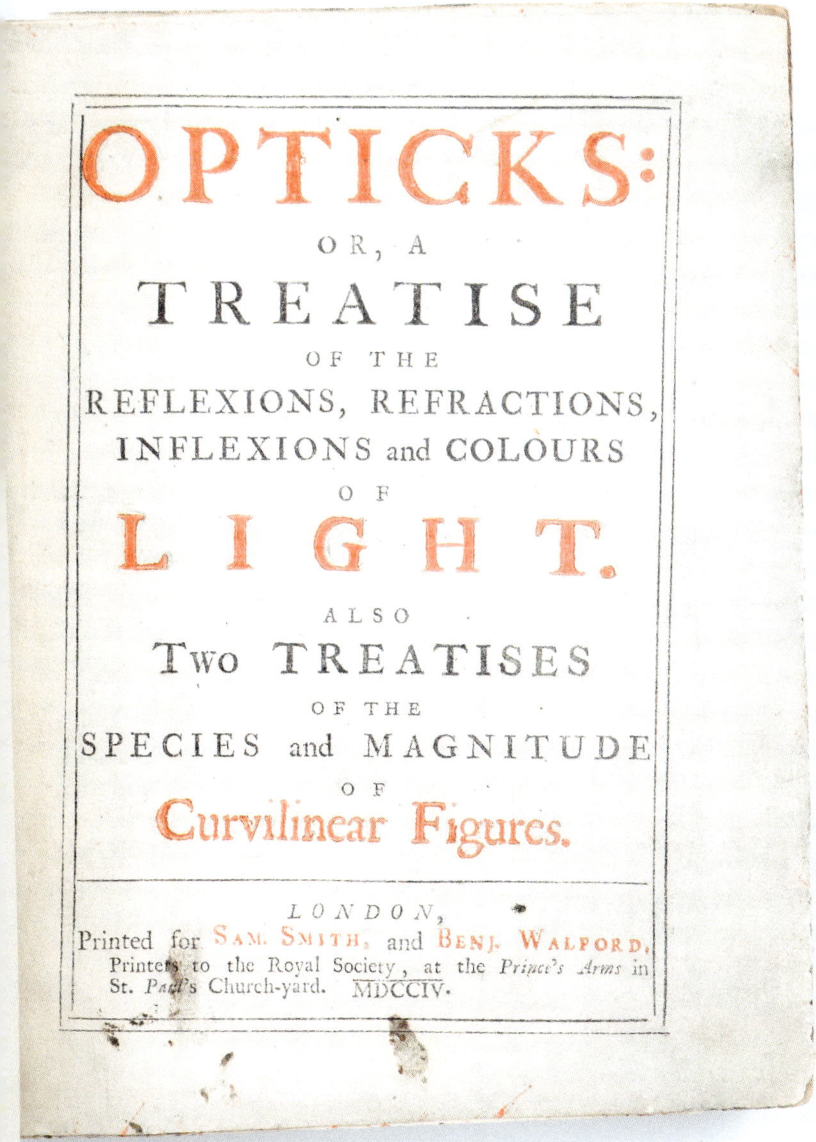
OPTICKS:

OR, A

TREATISE

OF THE

REFLEXIONS, REFRACTIONS, INFLEXIONS and COLOURS

OF

LIGHT.

ALSO

Two TREATISES

OF THE

SPECIES and MAGNITUDE

OF

Curvilinear Figures.

LONDON,

Printed for SAM. SMITH, and BENJ. WALFORD, Printers to the Royal Society, at the *Prince's Arms* in St. *Paul's* Church-yard. MDCCIV.

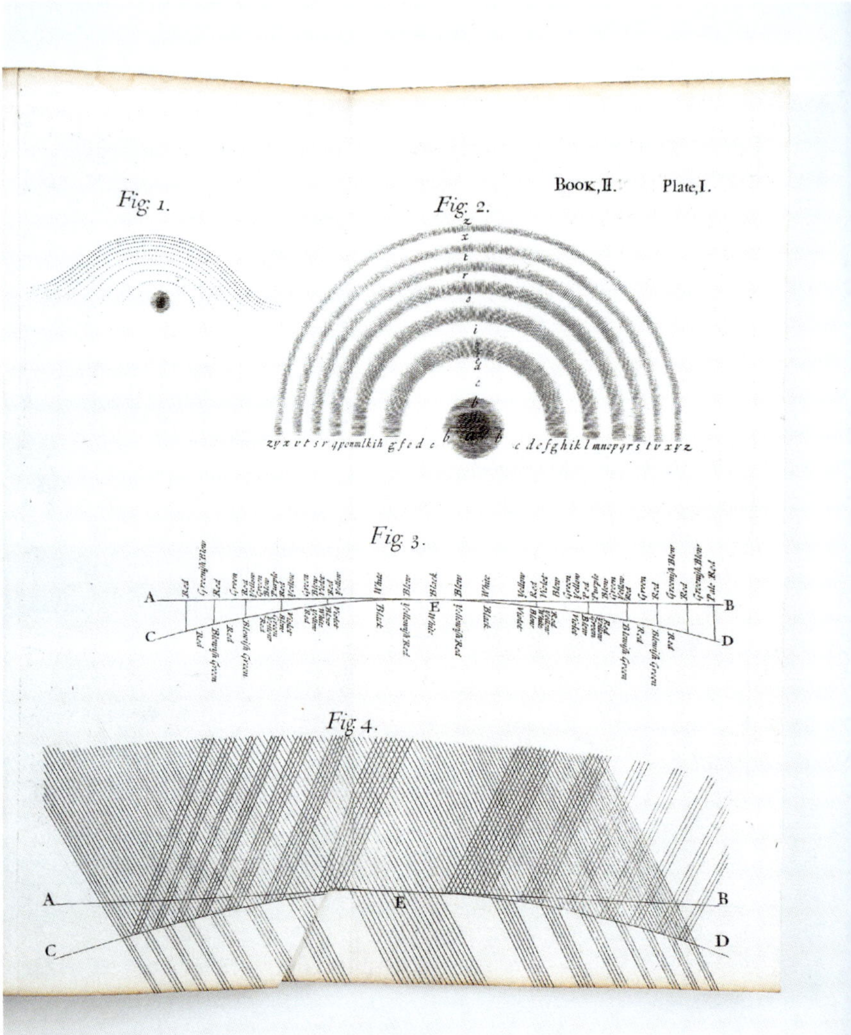
Book, II.
Plate, I.
Fig: 1.
Fig: 2.
Fig: 3.
Fig 4.
A
B
C
D
E

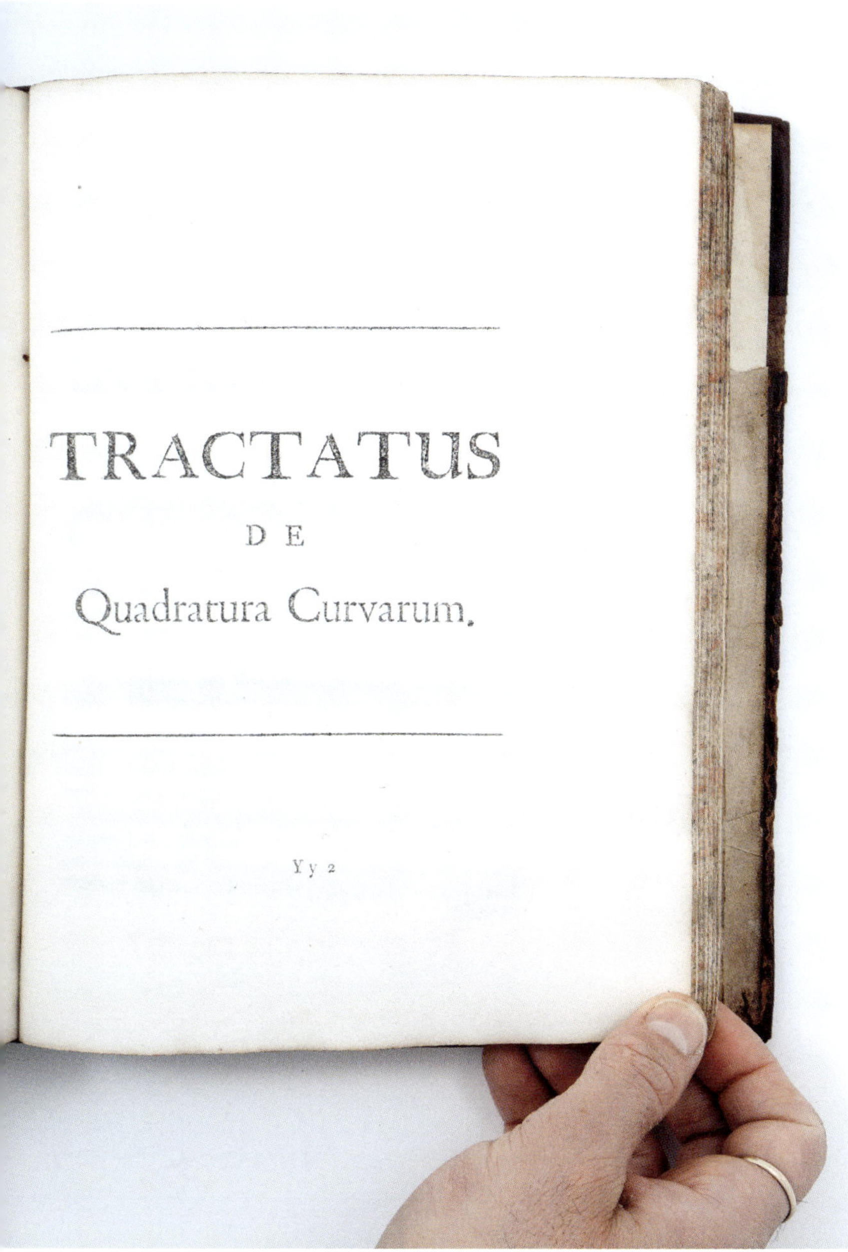
TRACTATUS
DE
Quadratura Curvarum.
Yy 2

JENNER, Edward (1749–1823).
An Inquiry into the Causes and Effects of the Variolæ Vaccinæ. London, 1798.

The invention of vaccines and the origin of modern immunology.

In 1796 British physician Edward Jenner injected the cowpox virus into the arm of one of his patients. The patient suffered a slight headache and a loss of appetite, but on the following day was in good health. More surprisingly, the patient was found to be immune to the smallpox virus: two months later, when Jenner introduced the far deadlier variola virus under the patient's skin, "no sensible effect was produced." This was the invention of modern vaccination, a word Jenner coined: *vacca* is Latin for cow, a nod to the pastoral origins of the first vaccine.

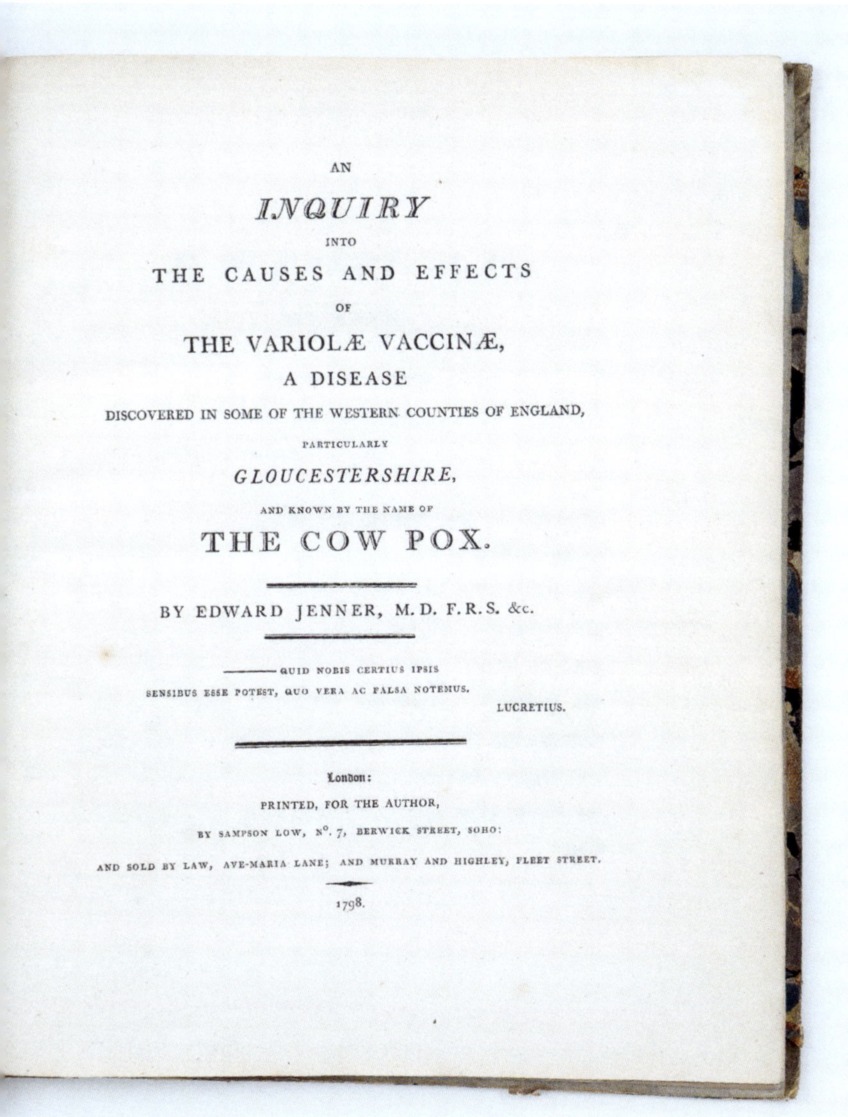

AN

INQUIRY

INTO

THE CAUSES AND EFFECTS

OF

THE VARIOLÆ VACCINÆ,

A DISEASE

DISCOVERED IN SOME OF THE WESTERN COUNTIES OF ENGLAND,

PARTICULARLY

GLOUCESTERSHIRE,

AND KNOWN BY THE NAME OF

THE COW POX.

BY EDWARD JENNER, M. D. F. R. S. &c.

——— QUID NOBIS CERTIUS IPSIS
SENSIBUS ESSE POTEST, QUO VERA AC FALSA NOTEMUS.

LUCRETIUS.

London:

PRINTED, FOR THE AUTHOR,

BY SAMPSON LOW, N°. 7, BERWICK STREET, SOHO:

AND SOLD BY LAW, AVE-MARIA LANE; AND MURRAY AND HIGHLEY, FLEET STREET.

1798.

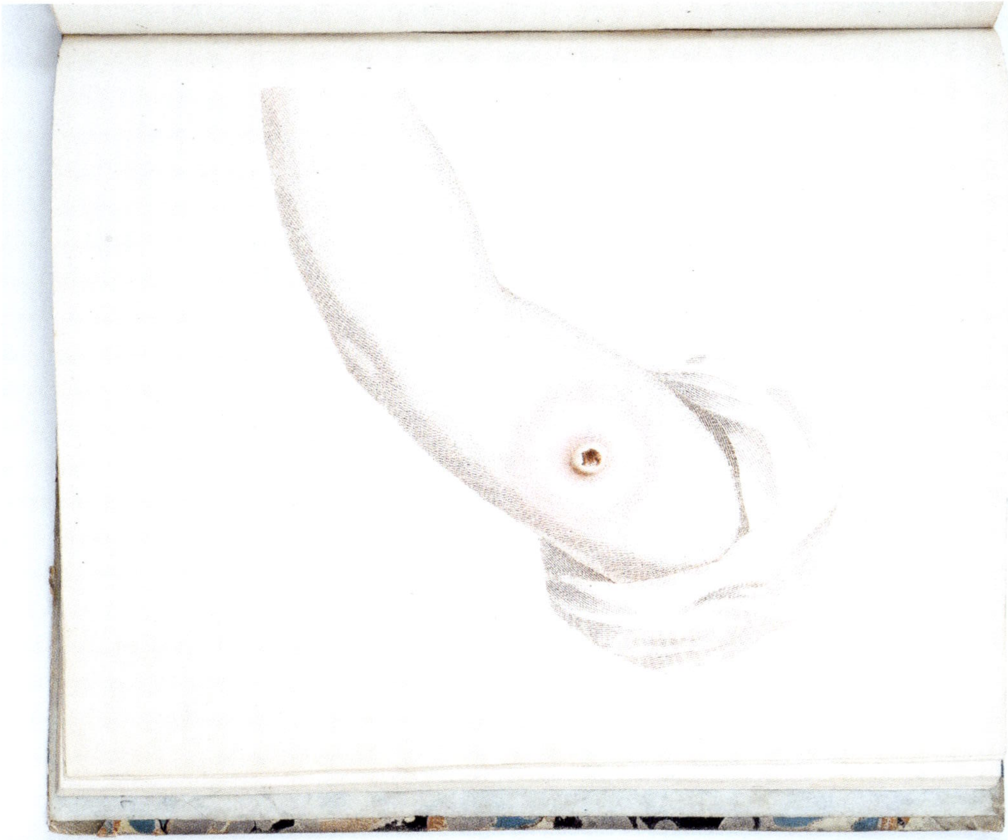

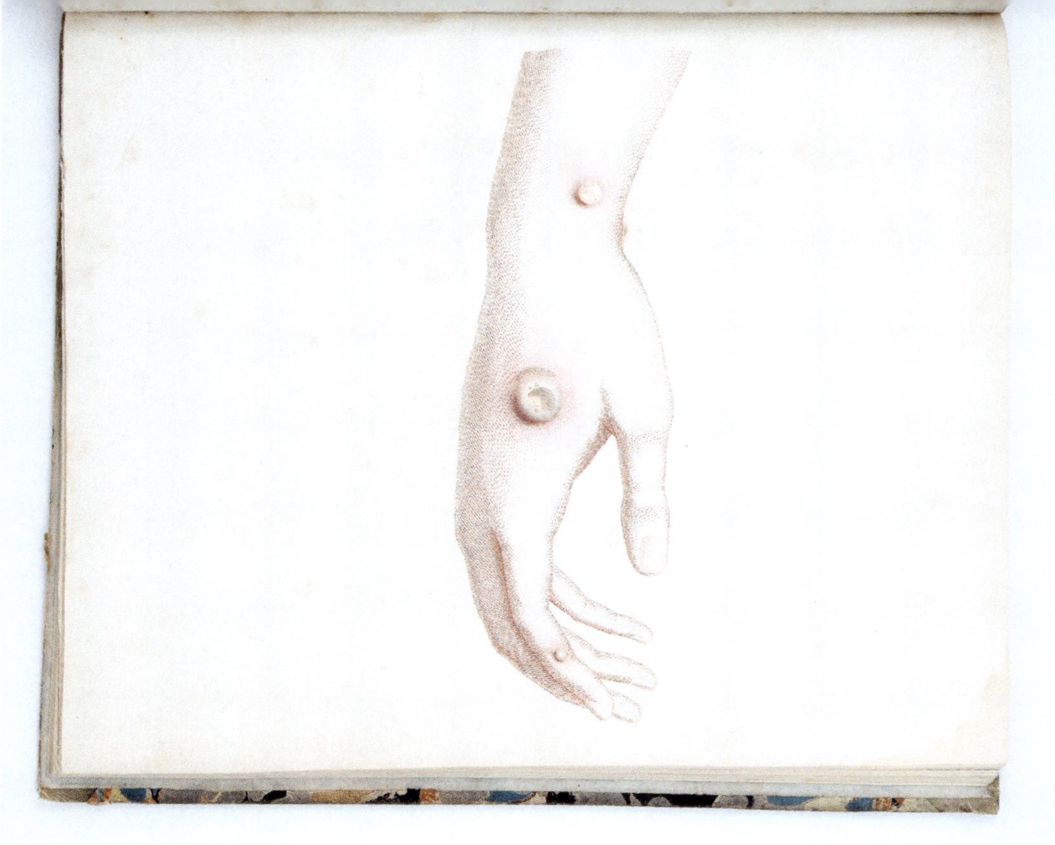

Stab-Stitch Holes, Deckles, & Broken Types: Some Notes on the First Official Printing of the US Bill of Rights (1792)[1]

The US Bill of Rights was printed in Philadelphia by Francis Childs and John Swain immediately after ratification in December 1791. This "first official" printing of the Bill of Rights (figure 1 shows the title page) is undated but is usually assigned to January or February 1792. The edition is considered a grail of Americana. Only five complete copies are known to survive: one at the Maryland State Archives, another at the American Antiquarian Society, another at the Library of Congress, and another in the Posner Memorial Collection in Special Collections. The fifth complete copy — labelled the "Samuel Johnston copy" after its first owner (1733–1816) — was sold at Christie's, 19 December 2002 (lot 225), and is currently untraced.[2] A sixth copy that survives as a single-sheet fragment recently appeared at Sotheby's, 21 July 2024 (lot 1006).[3] The Posner copy was deposited with Carnegie Mellon University Libraries in 1978 in memory of Henry Posner Senior, who purchased it from antiquarian bookseller H.P. Kraus in 1963. Before that, it was sold at Parke-Bernet, 25 November 1952

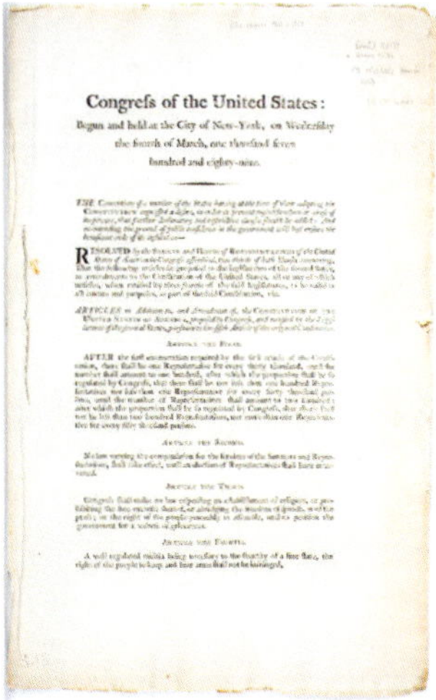

Congrefs of the United States:
Begun and held at the City of New-York, on Wednefday the fourth of March, one thoufand feven hundred and eighty-nine.

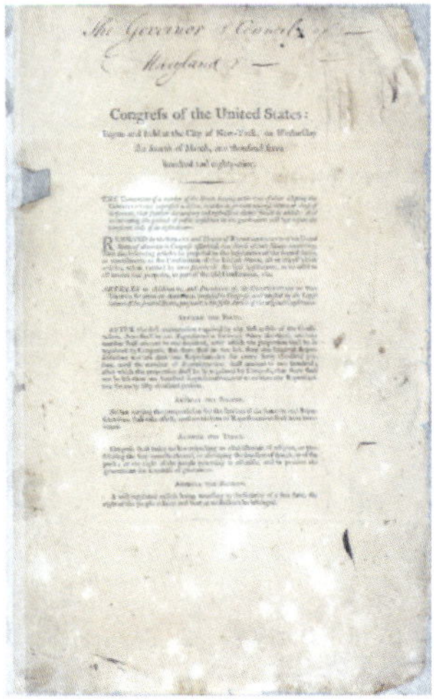

Congrefs of the United States:

Figure 1A & 1B ↓ The MSA copy and the AAS copy, showing differing margins and sheet dimensions. By permission of the Maryland State Archives and the American Antiquarian Society.

(lot 50) and was previously in the collection of Otto Oren Fisher (1881–1961). The Posner copy is perhaps the rarest single item in Special Collections.

The historical significance and legislative origins of the 1792 Childs-Swain printing of the US Bill of Rights are well studied: it represents the first time that the Bill of Rights was accorded force of law and the first time that its articles appeared in print as an integral part of the United States Constitution (see Kaminski et al., 2020 and Hrdlicka 2020). Instead of revisiting the historical background of the Bill of Rights, this essay examines the material and typographic evidence in five of the six surviving copies. These copies (one examined physically, four digitally) show that the edition was produced in two forms, or "issues." In examining this evidence, this essay offers a demonstration of the kind of scientific analysis that work with rare books often involves. It shows, too, that even an object as well studied and celebrated as the first official printing of the US Bill of Rights still has secrets to tell.

Figure 1 collates the first page of the American Antiquarian (AAS) copy and the Maryland State Archive (MSA) copy of the Bill of Rights and shows that the margins of the MSA copy are significantly larger than those in the AAS copy. The lower margin in the MSA copy, mottled by time and splotches of ink, is especially spacious. The difference in size is not the result of trimming: the sheets of the AAS copy retain their irregular "deckled edges" — a result of the papermaker's slurry of macerated linen rags seeping through the frames ("deckles") of the molds used to manufacture sheets of paper in the eighteenth century. The outer *fore* edge as well as the top and bottom edges of each leaf in the MSA and AAS copies display this characteristic "deckling," indicating that they approximate their original dimensions.[4]

No bibliography of early American imprints or published description of the 1792 Childs-Swain Bill of Rights notes that the edition was printed on two stocks of paper that differ appreciably in size.[5] These stocks of paper constitute a bibliographical point that distinguishes two issues of the edition: one "large-paper" issue attested to by the MSA and Sotheby's copies, and another "small-paper" issue attested to by the AAS, Library of Congress (LoC), Samuel Johnston, and Posner copies.

Another point that distinguishes the two issues of the 1792 Bill of Rights is the attestation found at the end of the document. This printed statement was added to certify the document's printing — a kind of legal endorsement authenticating the accuracy of the text. The attestation appears on the final leaf of the edition, below Vermont's notice of ratification (figure 2). In both the MSA copy and the Sotheby's copy, there appears a single line of printed text: "Deposited among the Rolls in the Office of the Secretary of State." Beneath this appears the manuscript signature of

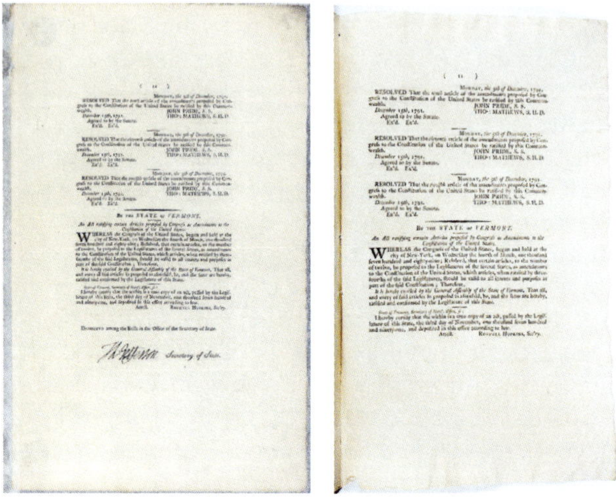

Figure 2A & 2B ↘ The MSA copy bears Jefferson's signature and the typographic attestation. The AAS copy lacks both the signature and the attestation. By permission of the Maryland State Archives and the American Antiquarian Society.

Thomas Jefferson next to his official title, *Secretary of State*, printed in cursive types.[6] This same statement and its accompanying attestation are absent in the small-paper copies I examined (Posner, AAS, and LoC).

The Office of the Secretary of State recorded payment of $30 for 135 copies of the "ratification of amendments to Constitution" to the printers Francis Childs and John Swain in July 1792.[7] The paper for these copies cost an additional $3.12. It seems likely that Jefferson instructed Childs and Swain to print at least twenty-six copies on larger, finer paper — two copies for each of the US states, which were then only thirteen in number. Jefferson enclosed these copies in a letter sent on 1 March 1792 to the states' governors: "I have the honor to send you herein enclosed two copies, duly authenticated, of [...] the ratifications, by three fourths of the Legislatures of the several States, of certain articles in addition to and amendment of the Constitution of the United States."[8] Jefferson's note indicating that both copies were "duly authenticated" suggests that each bore — like the MSA and Sotheby's copies — Jefferson's signature; it also suggests that all "authenticated" copies were printed on the same, larger stock of paper. The MSA copy, superscribed in manuscript on its first page to "The Governor & Council of Maryland," is one of these official, "authenticated" copies. The fragmentary nature of the Sotheby's copy obscures the identity of its original recipient, but a brief note penned on the verso of its final leaf suggests that it, too, was also one of two copies sent to a particular governor and state — "Duplicate of ratification of Certain Articles added to the Constitution." It seems likely that this duplicate copy was discarded as unnecessary, explaining its eventual appearance at auction.

All six surviving copies of the 1792 Bill of Rights also contain evidence that they were stab-stitched after printing. Stab-stitching was a temporary and rudimentary method of binding together the folded sheets of an edition. The thread used in this method was often later replaced by a more durable binding or enclosure. But as Zachary Lesser and Aaron Pratt remind us, the holes left by stab-stitch threads, even after their removal and replacement, "provide an important piece of bibliographic evidence for how a book was first sold" or otherwise distributed (Lesser 2021, 73; Pratt 2015).

Both the LoC and AAS copies retain their original binding cord — a length of string threaded through three holes "stabbed" along the pamphlet's folded edge and knotted at the back of the final leaf. Although the Posner copy was at some point rebound in a leather binding (its spine preserves fragments of this binding from which it was removed before Henry Posner Senior purchased it in 1963), close inspection reveals that it, too, was once stab-sewn: near its folded edge are three holes, their placement similar to those in the AAS copy. Although the MSA copy was recently conserved — each leaf encased in a translucent cocoon of archival paper — it also bears evidence of stab-stitching: stab-stitch holes remain visible on the copy's first leaf, indicating that Jefferson's "authenticated" copies were also sewn together before they were sent to the states.

Meanwhile the LoC copy and the Samuel Johnston copy are both preserved in blue paper wrappers — a sort of eighteenth-century dust jacket tacked to each copy by the same stab-stitch thread that held their sheets together (figure 3). While both the Posner and AAS copies lack this paper wrapper, the relatively clean appearance of their first pages suggests that they were at some point protected by a similar covering. In contrast, the outer leaves of the MSA copy are stained, torn, and discolored, implying that they were never wrapped or otherwise bound.

It seems, then, that Jefferson's instructions to Childs and Swain included notes on how the two issues of the 1792 Bill of Rights — both large- and small-paper copies — were to be handled after printing. Intended for enclosure in his letter to the states' governors, the larger "authenticated" copies may not have required stab-sewn covers, while the smaller unauthenticated copies may have been intended for sale or public distribution and therefore warranted the protection of paper wrappers.

This analysis is borne out by bibliographical evidence in other printings of legislative acts issued during Jefferson's office as Secretary of State, including, for example, "An Act Establishing a Mint, and Regulating the Coins of the United States." Printed by Francis Childs and John Swain and dated 2 April 1792 (two months after the printing of the Childs-Swain Bill of Rights), a copy of this act was

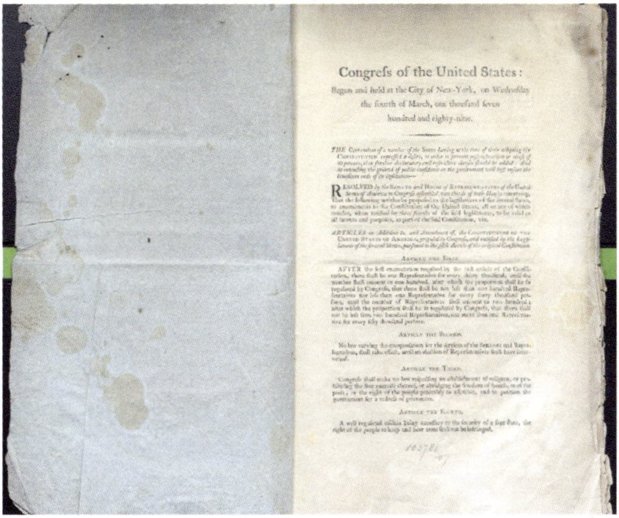

Figure 3 ↑ The first page of the LoC copy, showing the copy's original blue-paper wrapper. By permission of the Library of Congress, Rare Book Division.

sold at Swann Galleries in 2014.[9] The copy is signed by Jefferson and bears Jefferson's title ("Secretary of State") in the same cursive types used by Childs and Swain in the 1792 printing of the Bill of Rights. This copy also bears a statement recording that the act had been "deposited among the Rolls in the office of the Secretary of State," mirroring the printed attestation found in the MSA and Sotheby's copies. The 1792 Childs-Swain Bill of Rights, therefore, seems to have followed an established practice — one by which signed copies of legislative acts were printed on special stocks of paper and in sufficient quantities to distribute to each of the thirteen states.

Perhaps the most striking evidence found in the surviving copies of the 1792 Bill of Rights is a subtle variant in the text. On the second page of the AAS, LoC, and Posner copies, the printed attestations of John Beckley and Samuel A. Otis display two pieces of damaged type. Both the *B* in Beckley and the *O* in Otis are broken (figure 4). These same letters are

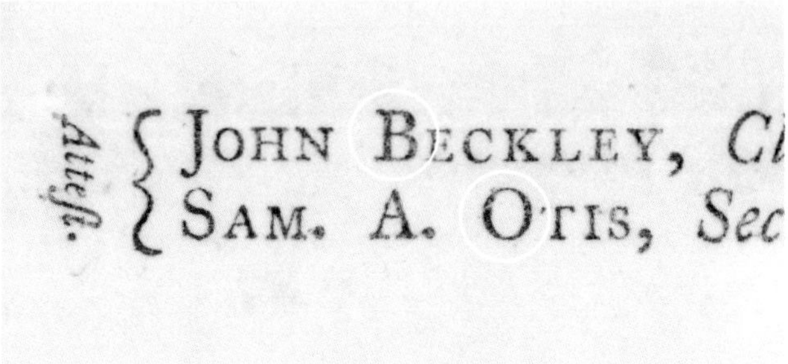

Figure 4 ↓ The undamaged B and O as they appear in the MSA copy. By permission of the Maryland State Archives.

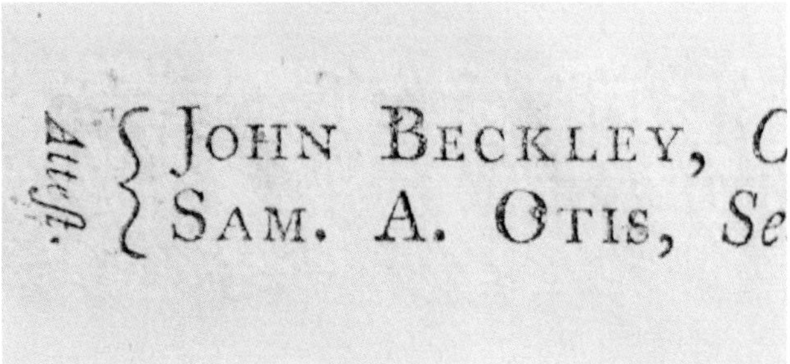

Figure 5 ↘ The damaged B and O as they appear in the Posner copy. By permission of Posner Memorial Collection, Carnegie Mellon University Libraries, Special Collections (Call number KF4743.5 1792).

undamaged in the MSA copy (figure 5). Pieces of type — or “sorts” — were cast in metal, each carrying a reversed image of a single letter. These metal sorts were placed on the press and inked to impart their image, again reversed, onto paper. Made of a relatively malleable lead alloy, type sorts were often damaged during presswork, crushed or bent by the heavy action of the press or spoiled by debris that settled on the surface of the types during printing. This small accident of the press — a fractured *B* and *O* — thus offers a window onto the past, revealing that the MSA copy — and likely the entire run of Jefferson’s large-paper, “authenticated” issue — was printed *before* the small-paper AAS, LoC and Posner copies; before, that is, the *B* in Beckley and the *O* in Otis were broken. Minute similarities in the types in copies of each issue meanwhile demonstrate that both issues were printed from the same setting of type.

Historical books and documents have much to tell us — as this essay has shown. And bibliographically significant evidence can be as inconspicuous as a hole through paper or a minutely damaged uppercase *O*. It's this idea of the unexpected — the hidden-in-plain-sightedness of bibliographical evidence — that I aim to share with CMU students when they visit Special Collections. I ask them to consider the material fact of the objects in front of them. Far from simple vessels of history or objects marking milestones in science or culture, books carry traces of the people and materials that contributed to their making. A visit to Special Collections is a cultural experience not unlike visiting a museum — an encounter defined as much by reverence as it is by obscure historical fact. But I also invite my students to consider and interpret things as seemingly unremarkable and unprepossessing as holes and stains on paper. I want them to feel included in the process of analysis and discovery involved in my day-to-day work. The beauty of this approach is that it assumes no cultural or even linguistic expertise: books printed in obscure languages and strange scripts become newly accessible, and students who might otherwise be intimidated by the reverential hush of Special Collections are instead invited into the fold. When the alchemy of this encounter works, it works brilliantly. And there are many discoveries yet to be made.

1 A version of this essay has been accepted for publication in *The Library* (Lemley 2025). I would like to thank the editors for their permission to reprint the essay in this form.

2 Christie's, "Printed Books and Manuscripts" (Live Auction 1060), 19 December 2002, lot 225: https://www.christies.com/en/lot/lot-4043658

3 The fragment makes up the final sheet (signed "C") of the edition and bears Thomas Jefferson's manuscript signature on the last leaf recto. The fragment was unsold. Sotheby's, "Fine Books and Manuscripts, Including Americana (Part 1)," 21 July 2024, lot 1006: https://www.sothebys.com/en/buy/auction/2022/fine-books-and-manuscripts-including-americana-part-1/the-united-states-constitution-and-the-bill-of-3

4 The large-paper Sotheby's copy measures 412 × 249 mm (uncut), indicating an original sheet size of approximately 500 × 412 mm. The sheets of the Sotheby's copy show posthorn watermarks and are countermarked "HS Sandy Run" — marks identifying paper stocks manufactured by Henry Schütz outside of Philadelphia. Gravell locates paper bearing these same marks in Thomas Jefferson's papers at the Library of Congress and notes that Schütz supplied paper to George Washington during his presidency. Gravell dates the marks to 1792. The leaf dimensions of the small-paper AAS copy are 339 × 208 mm (uncut). The Posner copy, rebound and trimmed, measures 313 × 200 mm. See Thomas L. Gravell, American Watermarks, 1690-1835 (New Castle, DE: Oak Knoll Books, 2002), 94 [430, 433 (308), 434 (306), 435 (307)], PM-184. The small-paper issue evidently was made up of assorted paper stocks: the second sheet ("signed B") in the AAS copy bears a clover or trefoil watermark over a ring (not in Gravell); the second sheet (signed "B") of the Posner copy bears a damaged and unidentifiable mark (not in Gravell); and the final sheet (signed "C") of the Posner copy bears a Britannia watermark (not in Gravell).

5 The auction record for the Sotheby's copy notes that "the paper size of the present fragmentary copy is much larger than the [Samuel] Johnston copy (recorded as 13 $\frac{7}{16}$ x 8 $\frac{5}{16}$ in.) and was clearly intended for a special purpose." Sotheby's, "Fine Books and Manuscripts, Including Americana (Part 1)," 21 July 2024, lot 1006: https://www.sothebys.com/en/buy/auction/2022/fine-books-and-manuscripts-including-americana-part-1/the-united-states-constitution-and-the-bill-of-3

6 The auction record for the Sotheby's copy erroneously claims that none of the other surviving copies are signed by Jefferson. Both the MSA copy and the Sotheby's copy bear Jefferson's signature. Sotheby's, "Fine Books and Manuscripts, Including Americana (Part 1)," 21 July 2024, lot 1006: https://www.sothebys.com/en/buy/auction/2022/fine-books-and-manuscripts-including-americana-part-1/the-united-states-constitution-and-the-bill-of-3

7 "VII. Contingent Expenses of the Department of State, 1790–1793, 14 January 1790–31 December 1793," Founders Online, National Archives, https://founders.archives.gov/documents/Jefferson/01-17-02-0099-0008. [Original source: The Papers of Thomas Jefferson, vol. 17, 6 July–3 November 1790, ed. Julian P. Boyd. Princeton: Princeton University Press, 1965, pp. 359–379.]

8 Thomas Jefferson, "Circular to the Governors of the States, 1 March 1792," Founders Online, National Archives, https://founders.archives.gov/documents/Jefferson/01-27-02-0774. [Original source: The Papers of Thomas Jefferson, vol. 27, 1 September–31 December 1793, ed. John Catanzariti. Princeton: Princeton University Press, 1997, p. 815.]

9 Swann Galleries, "Printed & Manuscript Americana" (Sale 2344), 8 April 2014: https://www.swanngalleries.com/3dcat/2344/files/assets/basic-html/page69.html

Vaccination's Dead Metaphors: Jenner's Inquiry & the Invention of Vaccines[1]

In the summer of 1796 the English physician Edward Jenner removed pus from the sores of a cowpox-infected dairymaid. He then injected the infectious matter into a pair of incisions made into the arm of James Phipps, age eight. As Jenner tells it in his 1798 treatise, *An Inquiry into the Causes and Effects of the Variolae Vaccinae* — a first edition copy of which is held in Special Collections — this crude injection soon ran its course: on the seventh day, the boy suffered a slight headache and loss of appetite, but on the following day, he was in good health. More surprisingly, the boy was found to be immune to smallpox: two months later, when Jenner introduced the far deadlier variola virus under the boy's skin, "no sensible effect was produced" (34). The local inflammation and fevered delirium that normally followed inoculation were absent, a sign of Phipps's enduring immunity.

Jenner wasn't aware of the cause of the new inoculant's success — the cowpox (*variola vaccinia*) and smallpox (*variola major*) viruses are genetically related,

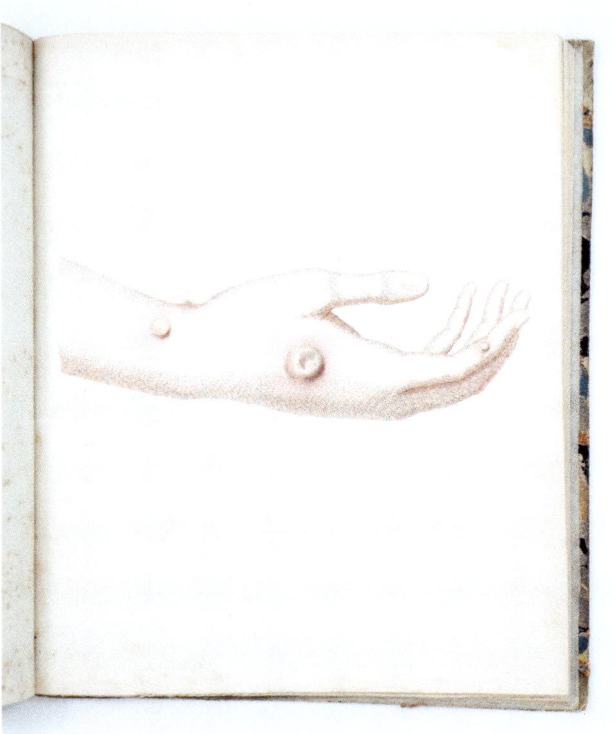

Figure 1 ↓ Artist unknown, The cowpox afflicted hand of Sarah Nelmes (1798), hand-colored engraving on paper, 244 × 207 mm. By permission of Posner Memorial Collection, Carnegie Mellon University Libraries, Special Collections (Call number RC183 .A2 J4).

distinct but similar enough to ensure that contact with the first induces immunity to the second — but he was familiar with its effect: "Morbid matter of various kinds, when absorbed into the system, may produce effects in some degree similar," he wrote, "but what renders the Cow-pox virus so extremely singular, is, that the person who has been thus affected is for ever after secure from the infection of the Small Pox [...]" (6).

True to its name, Jenner's *Inquiry* forms a catalogue of notes that sets out Jenner's findings in tidily numbered cases. "In support of so extraordinary a fact," Jenner explains in the work's preface after laying out his hypothesis, "I shall lay before my Reader a great number of instances" (7). Of course, Jenner's "instances" are human beings, and the *Inquiry* is populated by a faceless cross section of Jenner's rural society, from an aging undergardener to a "respectable Gentlewoman" identified only as "Mrs. H —" (14).

Insofar as the *Inquiry's* experiments emerged organically from Jenner's provincial community, he owed much to the homespun learning of his neighbors. As Eula Biss remarks in her 2014 book *On Immunity*, Jenner's coinage of a new Latinate disease (*varoliae vaccinae*, "smallpox of the cow") belies the humble origins of vaccination: "Its roots are in folk medicine," she writes, "and its first practitioners were farmers" (Biss 2014, 51). This folk knowledge is still with us, distilled into aphorism: "as smooth as a milkmaid's skin." This phrase hints at centuries of unlettered medical learning and observation: when the early modern world suffered from smallpox's staggering thirty percent mortality rate, the dairymaid was commonly held to be uniquely resistant to the virus's disfiguring traces. What Jenner's *Inquiry* showed is that the fabled immunity (if not the complexion) of milkmaids could be artificially imparted.

Many of Jenner's case notes are paired with disquietingly beautiful illustrations that depict pox sores and pustules in various states of suppuration. One such image shows the lithe hand of Sarah Nelmes, a milkmaid who had contracted cowpox through a small scratch on her right hand while tending her bovine charges. Jenner's written observations on Nelmes's hand couch anatomy in the winsome language of fairy tale, hinting at a poetics of vaccination: "She received the infection on a part of the hand which had been previously in a slight degree injured by a scratch from a thorn" (31). From a cow's infected udder through an accidental cut on an unprotected

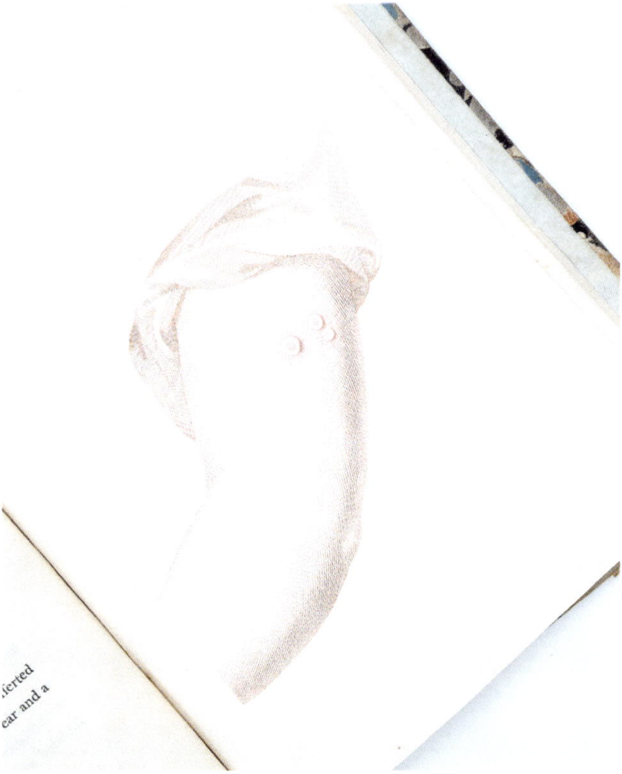

Figure 2 ↓ Artist unknown, The vaccinated arm of Hannah Excell, seven years old (1798), hand-colored engraving on paper, 244 × 207 mm. By permission of Posner Memorial Collection, Carnegie Mellon University Libraries, Special Collections (Call number RC183 .A2 J4).

finger, Jenner traces the disease's progress through brambles, fields, and barnyards. To speak of herd immunity, then, is to use a dead metaphor with roots in eighteenth-century pastoral. Vaccination (*vacca*, Latin for "cow") and the almost magical protection it confers originated in a herd of cattle.

Bucolic images imbue the language of Western medicine, and Jenner's cow-cure is no exception. The phrase "herd immunity" itself, though of later date than Jenner's vaccine, also relies on agrarian metaphors: herd immunity works by a kind of collective insulation, where the weaker few are protected by the vaccinated or antibody-bearing many — much like a herd of ruminants shielding its young and weak from predators and privation. But if *vaccine* is a product of the pastoral, *inoculation* — an earlier method of immunizing against a specific disease — stems from the georgic, borrowed from the labors of arborists in orchards. "To inoculate" means to graft a plant onto new rootstock, an image that recalls both the insertion of foreign matter into an unwilling (or at least indifferent) host and the immunological response that follows. Not by chance, Jenner ran a vaccination clinic from a thatch-roof cottage built among the plantings and paths of his suburban garden in Berkeley, Gloucestershire. He named this structure the Temple of Vaccinia and offered from its single room his new vaccine to the poor of the district. Vaccination has its own strange mythos, then, with Jenner as a kind of benevolent Lancer of the Glen.

Though we've come a long way from Jenner's horticultural clinic, his experiments are relevant again — all the more so, given that one of the vaccines for the COVID-19 virus was developed at the Jenner Institute in Oxford, United Kingdom. Jenner's name has become a word with cultural weight and authority, and we expect an institute bearing his name to provide therapeutic deliverance. Despite Jenner's reputation among epidemiologists and medical historians, though, the history of vaccination resists being pigeonholed into categories that structure the history of science. Vaccines are not, as we might expect, entirely the invention of Western ingenuity, institutes, or formal experimentation. A form of immunization against smallpox known as variolation was practiced in China, Africa, and West Asia as early as the sixteenth century. This procedure involved inserting a sample of live smallpox virus (again in the form of carefully harvested pus) under a patient's skin. This method of deliberate infection was hazardous, sometimes leading to the death of the inoculated or new outbreaks of the disease. Insufflation, another method of immunization practiced in early modern China, involved forcefully blowing pulverized smallpox scabs into a patient's nasal cavity — with searing results, no doubt. Jenner's method was not new then, but it was groundbreaking, both for its harmlessness and its quotidian source: a cow named Blossom.

Jenner's illustrations were similarly groundbreaking. The sores that adorn Jenner's engravings of afflicted arms and hands are colored in mucid tones of ecru and green, each pitted ulcer haloed by a daubed corona of faded pink. Jenner describes these dermic signs as "superficial suppurations" that "put on a circular form, with their edges more elevated than their centre" (4). Obscured by the immediacy of these illustrations is the fact that they are composites: "the pustule on the fore finger," Jenner notes of the illustration of Nelmes' hand, "shews the disease in an earlier stage. It did not actually appear on the hand of this young woman, but was taken from that of another, and is annexed for the purpose of representing the malady after it has newly appeared" (31).

Each of the illustrations accompanying Jenner's case studies was first engraved in copperplate and printed using an intaglio press in a sanguine shade of ink. But the sores themselves were finished by hand in watercolors, making every copy of Jenner's *Inquiry* a unique artifact and work of art. The copy held in Special Collections contains illustrations of remarkable freshness and intensity. The effect of the engravings' hybrid medium is startling, conveying a realism usually unattainable in printed anatomical illustrations of the period. The illustrations therefore offer welcome clarity in the face of Jenner's vacillating prose: the sores are sometimes "palish blue, or rather of a colour somewhat approaching to livid," Jenner says (3).

Ekphrasis fails where images enthrall. Jenner's grisly illustrations of misbehaving skin are as much instruments of comfort, then, as they are graphic records of therapeutic intervention: the accurate portrayal of the disease's defining attribute — the sores, or "pocks" that give poxes their name — makes Jenner's text useful for visual diagnosis and subsequent treatment.

Nineteenth-century doctors and amateur vaccinators could — and did — consult Jenner's *Inquiry* to guide the successful vaccination of their own patients. But Jenner's images also encourage his readers to view the sores they depict as positive symptoms, specifically as signs indicating immunity and recovery rather than impending crisis. They thus offered beauty and quietus of a clinical kind, anticipating a genre of medical image we now recognize as detached, austere, diagnostic — the pictorial stuff of WebMD and anatomy textbooks. We seek out these depictions of disease not from morbid fascination with illness and disfigurement — or at least not only — but in an effort to humanize both. Harvard historian of art Sarah Lewis observes that the rapid spread of COVID-19 was uniquely disorienting because early news reports lacked images of the dead, diseased, and dying. What's more, there is scant visual evidence of the virus's effects; it is a disease that injures from within, slowly draining its victims of breath. We were instead fed data and models that predicted the virus's grim progress through a population of anonymous victims. When we did catch glimpses of the COVID-dead, they were shrouded in sealed body bags, stacked in makeshift morgues parked unceremoniously in the loading docks of hospitals.

Disease maps operate at a still further remove. Broadcast with the evening news, these maps depicted regional micro-epidemics, like virulent clouds passing over the contours of familiar geography. An animated graphic released on social media by the British Consulate in the US showed Earth from space, our pale blue dot grotesquely transformed into the COVID-19 virus, its corona of monstrous spikes extruding through erstwhile atmosphere.

While these infographics and diagrams hint at death, they mask its realities — that the victims of viral disease are known to many of us, for instance. By contrast, forthright images of disease, Sarah Lewis writes, "can help us more deeply understand the severity of the situation as we work to curb the virus. [...] They help us humanize clinical statistics, to make them comprehensible. They step unto the breach" (Lewis 2020). With the terror of smallpox faded from our collective memory, it is easy to forget that Jenner's illustrations would have served this function for an audience trained to view pox sores as signatures of death or disfigurement. This explains their strange and unsettling beauty. They are pockmarked harbingers of hope.

Anxious to see these reassuring signs, some of Jenner's patients worried about the new vaccine's potency. An earlier form of inoculation, called variolation, was often the cause of hand-wringing when it failed to incite an attenuated form of smallpox: "There are many, who from some peculiarity in the habit resist the common effects of variolous matter inserted into the skin, and who are in consequence haunted through life with the distressing idea of being insecure from subsequent infection" (70). Jenner proposed his new method as a reassuring alternative: "A ready mode of dissipating anxiety originating from such a cause must now appear obvious" (70). Jenner's cowpox vaccine, unlike other forms of inoculation, never failed to give coincident rise to oracular pustules and a fevered crisis. Its potency was authenticated and effected by physical trial.

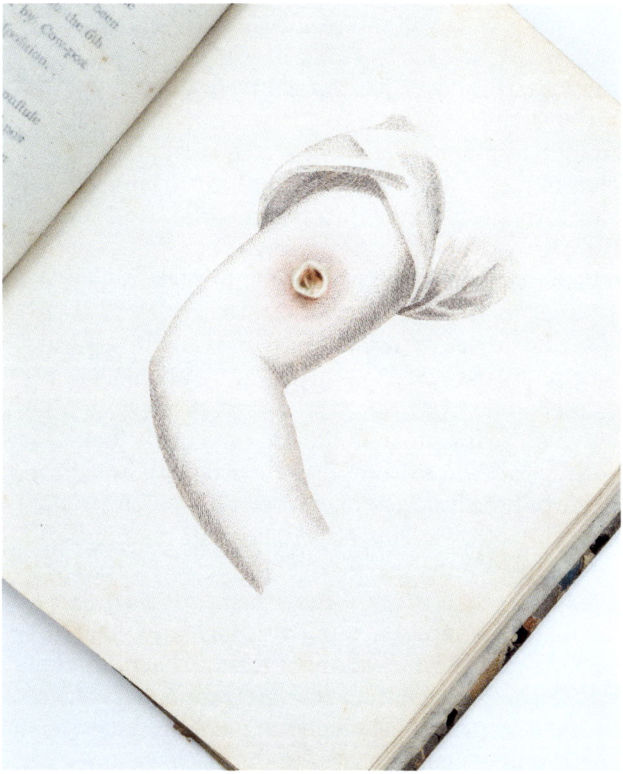

Figure 3 ↑ Artist unknown, The vaccinated arm of John Baker, five years old (1798), hand-colored engraving on paper, 244 × 207 mm. By permission of Posner Memorial Collection, Carnegie Mellon University Libraries, Special Collections (Call number RC183 .A2 J4).

Modern vaccines, like other modern technologies, are instead expected to work surreptitiously: they bestow immunity (or ought to) without dermic eruptions or foreboding symptoms. Perhaps this expectation accounts, in part, for the obstinacy of antivax sentiment. There is something innate in us that distrusts treatments free from noxious stimulus or disfiguring effect. The idea of drinking bleach as a curative, for instance, holds our fascination for its unsophisticated violence. Cytotoxicity is the therapeutic preserve of the desperate: we poison ourselves, hoping that the disease falters and abates before we do.

Jenner viewed vaccination as a way to restore balance to humankind's native immunity. In his preface to the *Inquiry*, he ascribes the eighteenth century's proliferation of diseases to "[t]he deviation of Man from the state in which he was originally placed by Nature" (1). The precarity of our health, Jenner's words suggest, is a product of our withdrawal from the natural world and its perceived benefits, benefits which vaccination might reinject through fragile skin. This seems an odd doctrine, accustomed as we are to lists of pharmaceutical ingredients whose botanical origins are obscured behind chemical nomenclature. While Jenner's position was colored by a bucolic idyll in which he imagined milkmaids and cattle took willing part — Blossom, the brown dairy cow believed to harbor the first case of cowpox used in Jenner's experiments, was later flayed, her hide mounted and displayed in the halls of St. George's Hospital, London — his attitude is remarkably modern. Set against recent and alarming disinformation that seeks to cast vaccines as unnatural and harmful, the optimism of vaccination's earliest proponent comes as some surprise: "I shall myself continue to prosecute this inquiry," Jenner writes on the *Inquiry's* final page, "encouraged by the hope of its becoming essentially beneficial to mankind" (75).

Encountering these words 225 years after their publication reminds us of the fragility of wellness and the often fraught course by which progress in medicine is won: the last recorded case of smallpox was diagnosed in 1977, and it wasn't until 1980 that the World Health Organization announced the disease's final, fateful eradication.

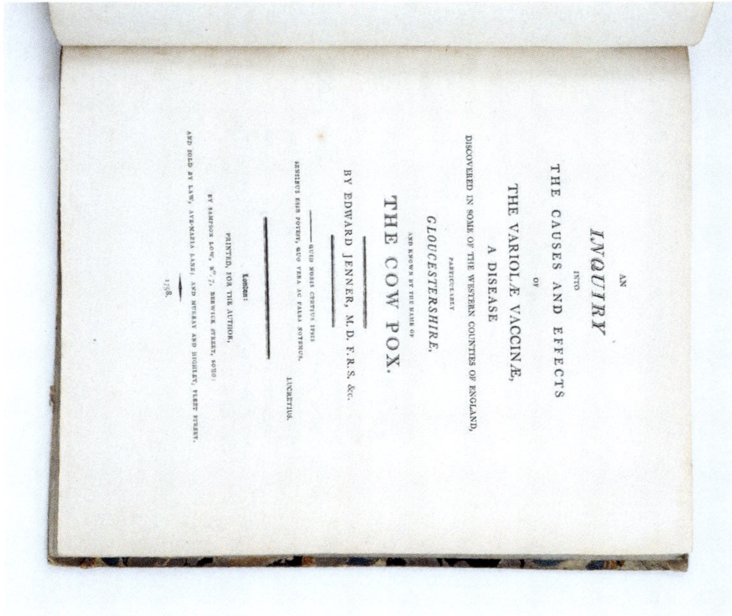
AN
INQUIRY
INTO
THE CAUSES AND EFFECTS
OF
THE VARIOLÆ VACCINÆ,
A DISEASE
DISCOVERED IN SOME OF THE WESTERN COUNTIES OF ENGLAND,
PARTICULARLY
GLOUCESTERSHIRE,
AND KNOWN BY THE NAME OF
THE COW POX.

BY EDWARD JENNER, M. D. F. R. S. &c.

——QUID NOBIS CERTIUS IPSIS
SENSIBUS ESSE POTEST, QUO VERA AC FALSA NOTEMUS.
LUCRETIUS.

London:
PRINTED, FOR THE AUTHOR,
BY SAMPSON LOW, N°. 7, BERWICK STREET, SOHO:
AND SOLD BY LAW, AVE-MARIA LANE; AND MURRAY AND HIGHLEY, FLEET STREET.
1798.

Figure 4 ↑ Title page, *An Inquiry into the Causes and Effects of the Variolae Vaccinae* (London 1798). By permission of Posner Memorial Collection, Carnegie Mellon University Libraries, Special Collections (Call number RC183 .A2 J4).

1 An early version of this essay appeared in the online journal *Hyperallergic* (Lemley 2021).

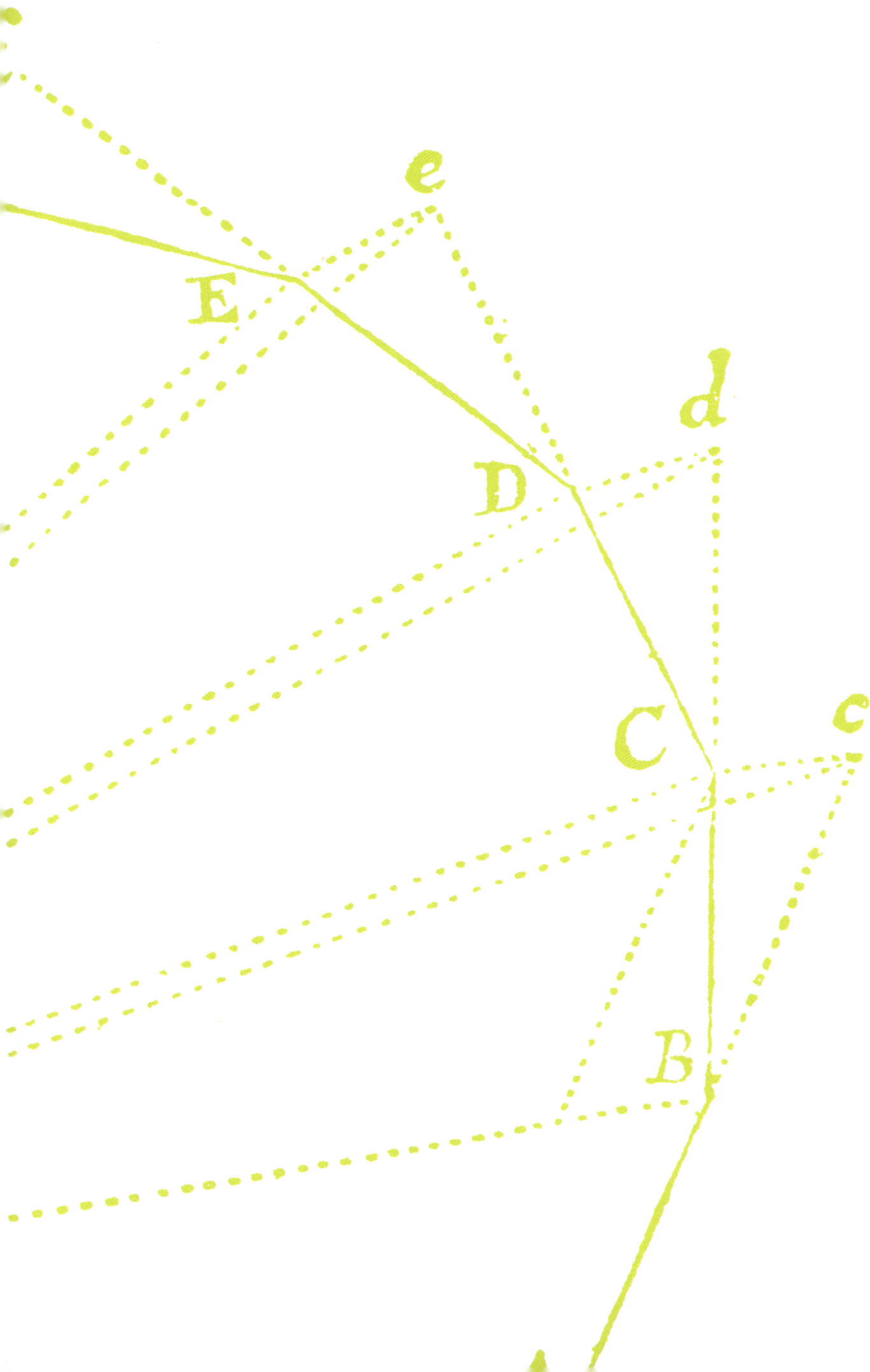

e
E
d
D
C
c
B

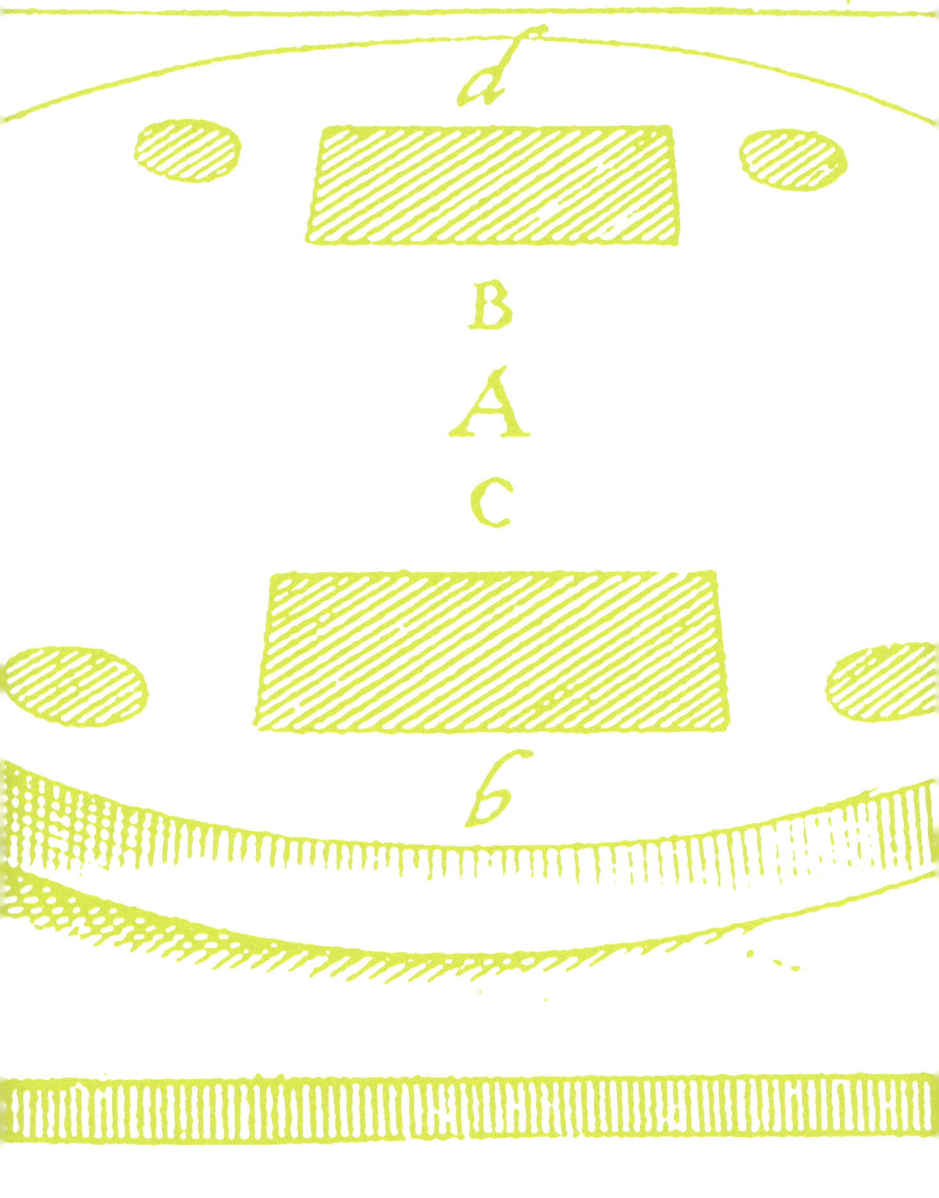
d
B
A
C
b

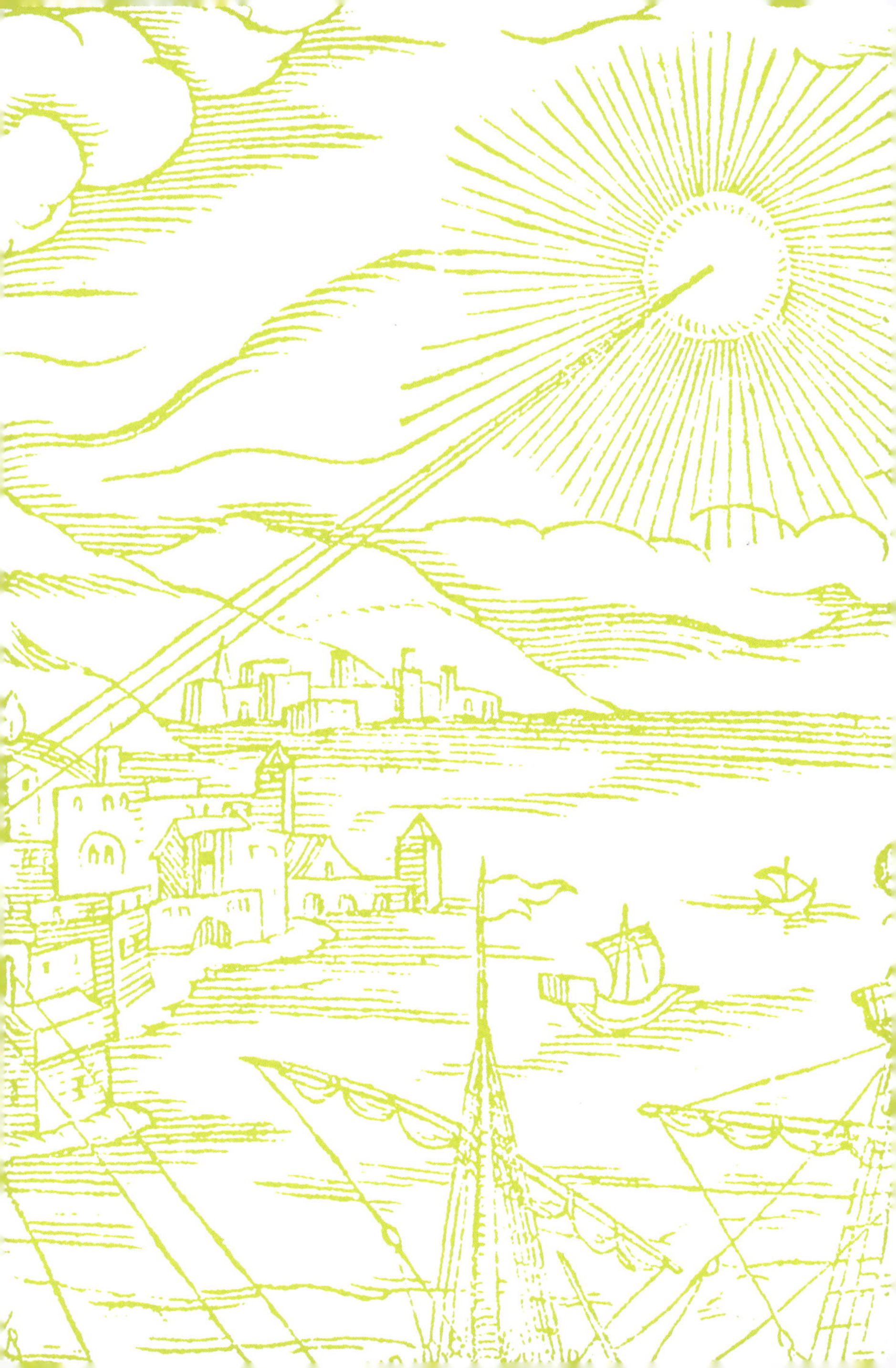

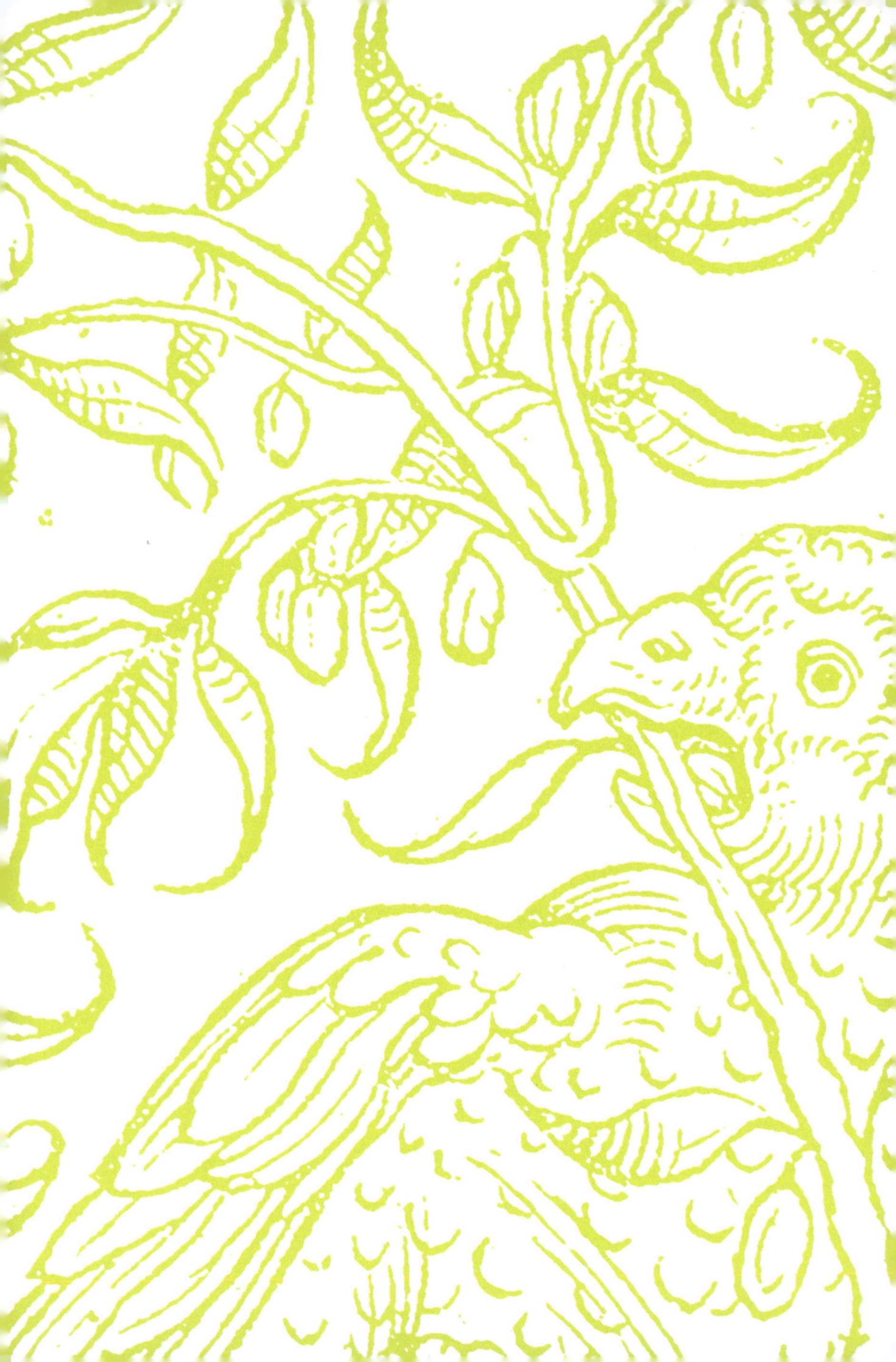

B

alſo called, * *the Princ*

ruleth in the darkneſs of

And in conſequence her

in oppoſition to the faithf

called the *Children of*

PART III

Ciphers & Encryption

Cryptology — the science and mathematics of building and breaking cipher systems — has a long history. Gathered here are books and objects that illustrate this history, from the first Western treatise on ciphers, printed in Venice in 1568, to a World War II-era Enigma machine.

DELLA PORTA, Giambattista (1535–1615).
De furtivis literarum notis vulgo, de ziferis libri IIII.
London, 1591; Naples, 1563.

A paper cipher machine.

Giambattista della Porta's *De furtivis literarum* ("On Secret Writing") contains a number of cipher discs, or volvelles (from Latin *volvere*, "to turn"). These carefully mounted dials of paper rotate on pieces of thread to encrypt or decrypt any message. The underlying mechanics of the Enigma machine (two of which are held in Special Collections) are more advanced, but no different: rotating discs establish fleeting links between plaintext and ciphertext.

Special Collections holds copies of the first Naples edition (1563) and a later London edition (1591).

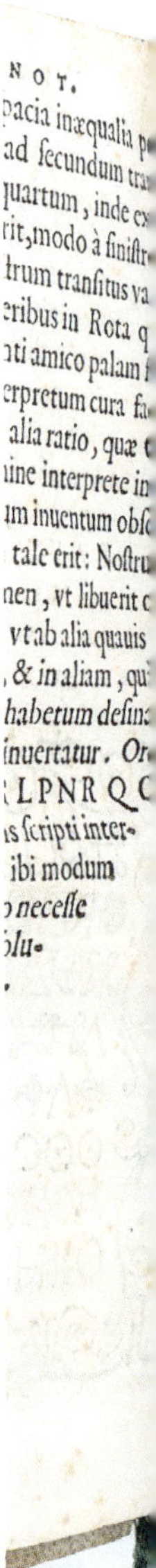
N O T.
pacia inæqualia p
ad ſecundum tra
quartum , inde ex
rit,modo à ſiniſtr
trum tranſitus va
eribus in Rota q
ati amico palam
erpretum cura fa
alia ratio , quæ
iine interprete in
um inuentum obſ
tale erit : Noſtru
nen , vt libuerit c
vt ab alia quauis
, & in aliam , qu
habetum deſin
inuertatur . Or
L P N R Q C
as ſcripti inter-
ibi modum
neceſſe
olu-

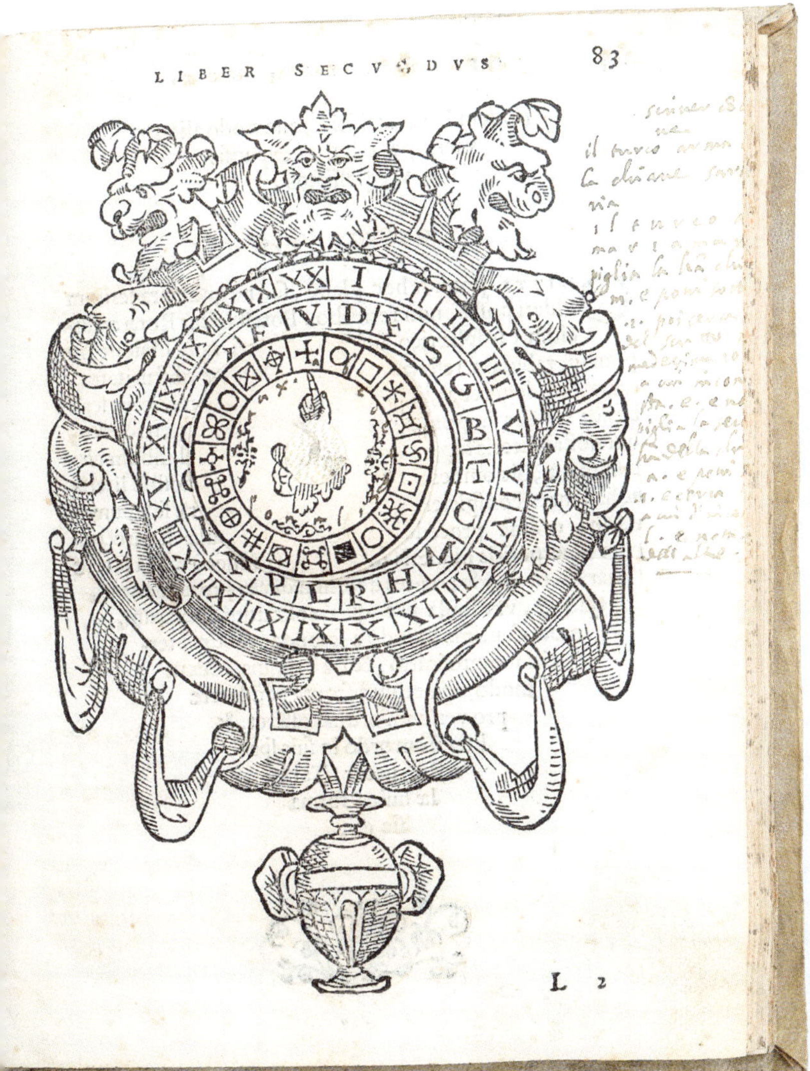
LIBER SECVNDVS 83

L 2

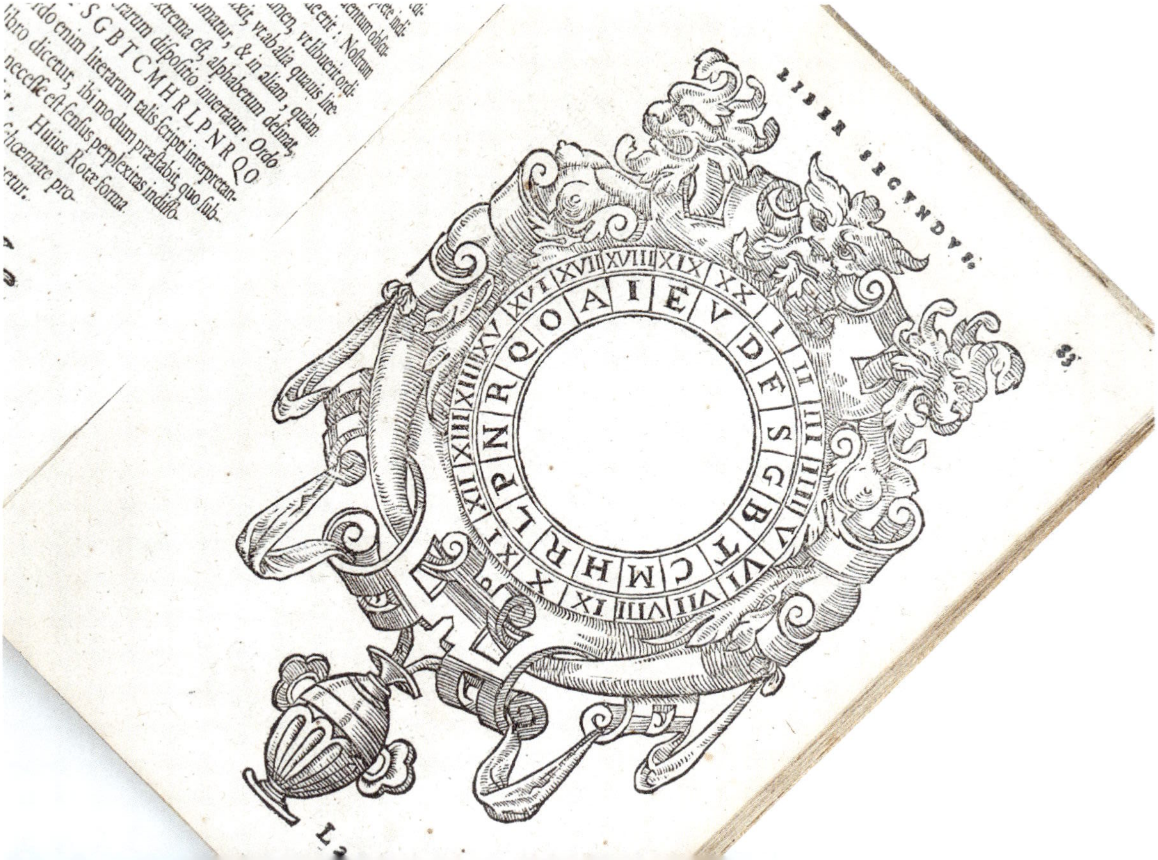
LIBER SECVNDVS.
83.

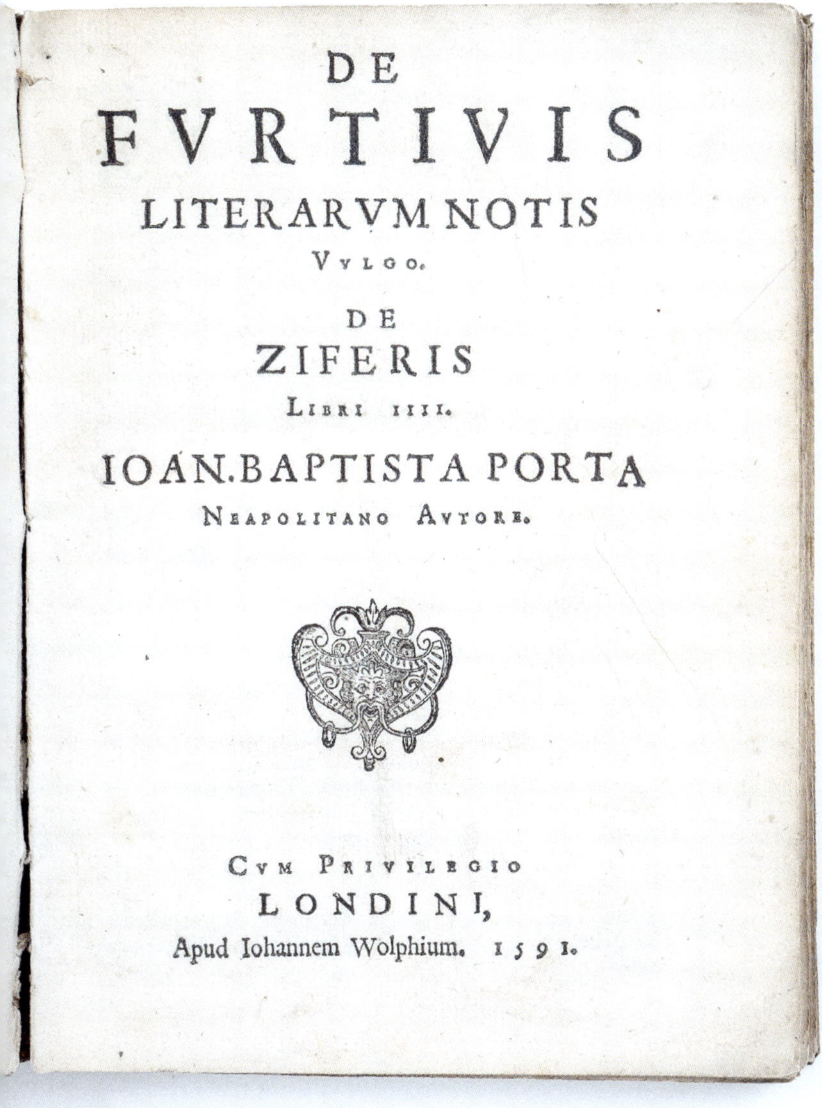

DE
FVRTIVIS
LITERARVM NOTIS
VVLGO.
DE
ZIFERIS
LIBRI IIII.
IOAN. BAPTISTA PORTA
NEAPOLITANO AVTORE.

CVM PRIVILEGIO
LONDINI,
Apud Iohannem Wolphium. 1591.

ALBERTI, Leon Battista (1404–1472).
Opuscoli morali. Venice, 1568.

The first Western work on cryptology.

Alberti was a Renaissance polymath: architect, diplomat, mathematician, and athlete. He also invented polyalphabetic ciphers. These are cryptographic systems that use multiple alphabets to disguise messages, and whose complexity resists cryptanalysis, or code breaking. Unlike simple substitution ciphers (where one plaintext letter always substitutes for the same ciphertext letter), polyalphabetic ciphers change periodically, allowing individual letters to be enciphered differently within a single message — a concept fundamental to modern encryption techniques.

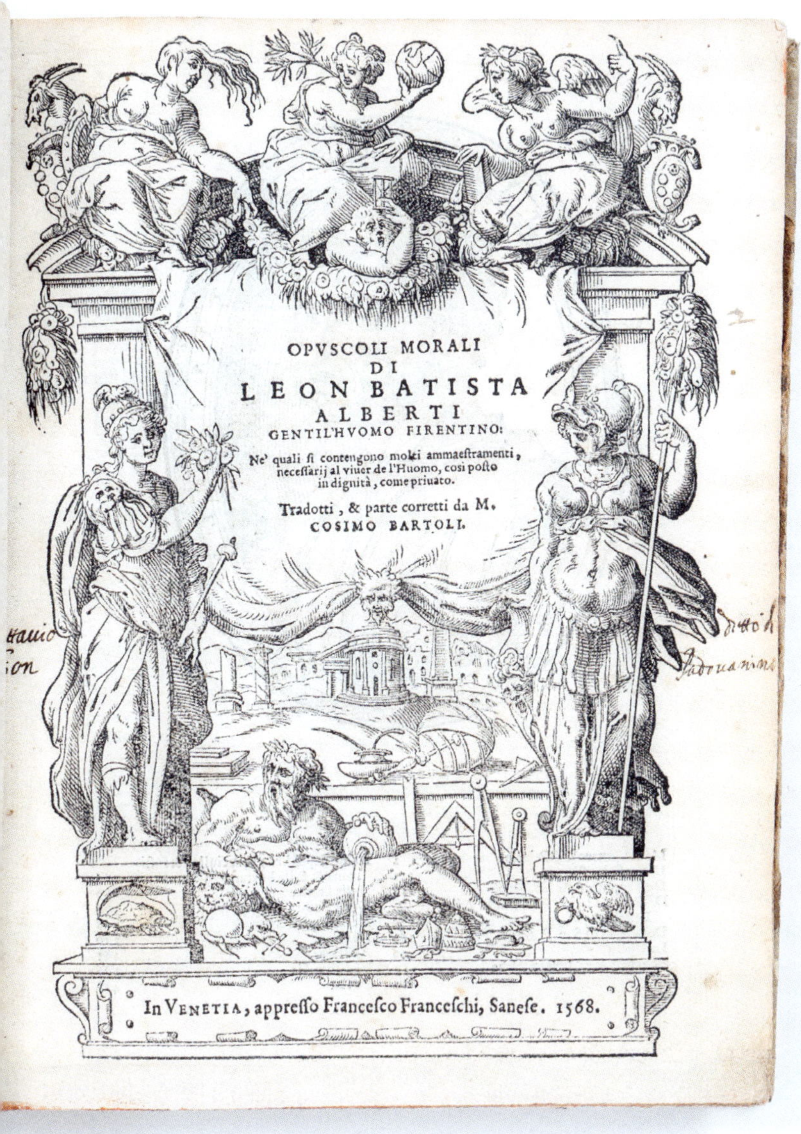
OPVSCOLI MORALI
DI
LEON BATISTA
ALBERTI
GENTIL'HVOMO FIRENTINO:

Ne' quali ſi contengono molti ammaeſtramenti, neceſſarij al viuer de l'Huomo, coſi poſto in dignità, come priuato.

Tradotti, & parte corretti da M.
COSIMO BARTOLI.

In VENETIA, appreſſo Franceſco Franceſchi, Saneſe. 1568.

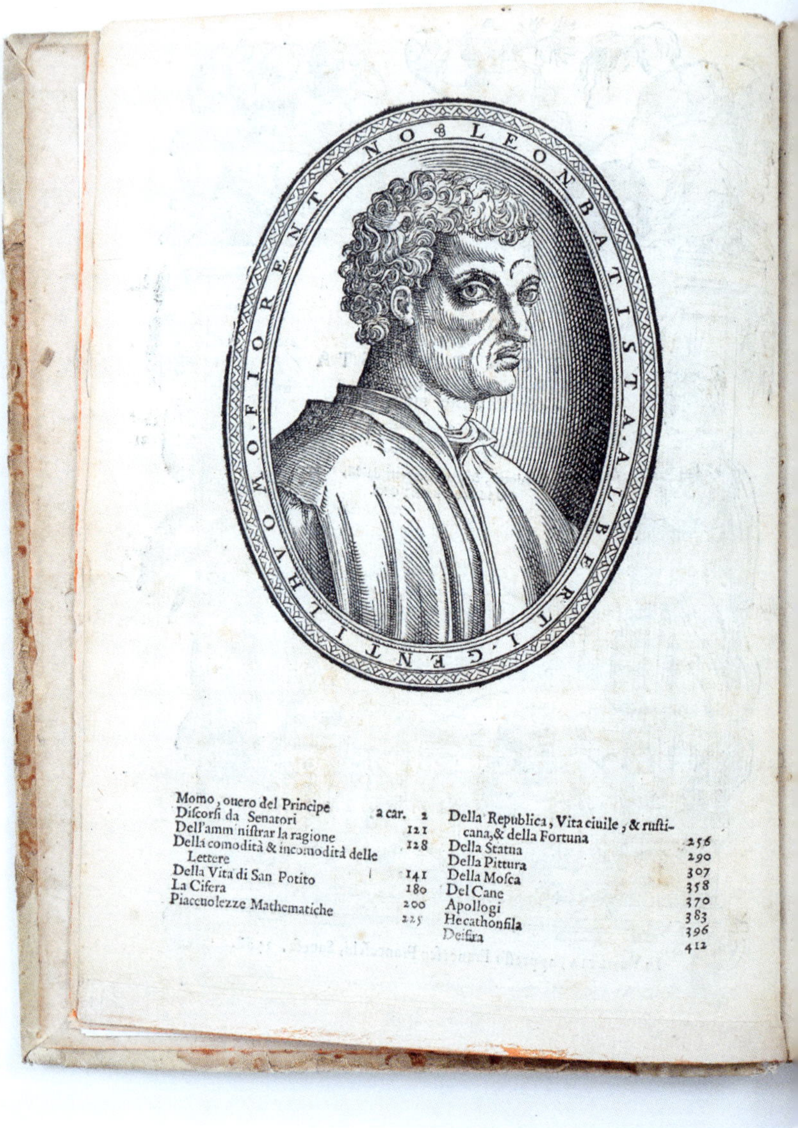

Momo, ouero del Principe a car. 2
Discorsi da Senatori 121
Dell'amministrar la ragione 128
Della comodità & incomodità delle Lettere
Della Vita di San Potito 141
La Cifera 180
Piaceuolezze Mathematiche 200
225
Della Republica, Vita ciuile, & rustica, & della Fortuna 256
Della Statua 290
Della Pittura 307
Della Mosca 358
Del Cane 370
Apollogi 383
Hecathonfila 396
Deifira 412

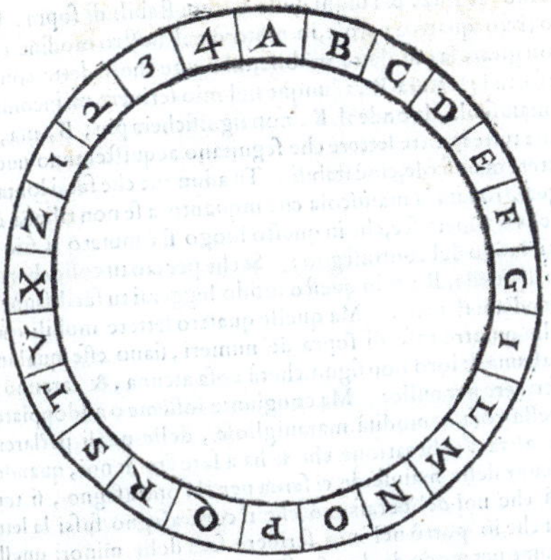

in questa maniera di cosi fatti duoi cerchi , chiamo
Lo instrumento composto in questa maniera di cosi fatti duoi cerchi , chiamo
io il Modine o uuoi Modello, & di questi Modini bisogna hauerne dua, uno che
resti appresso di te, & l'altro che porti seco quel tale al quale andando egli lonta
no tu harai a scriuere, & siano amenduoi questi Modini , & quanto al numero
delle lettere, & alla positura, & allo ordine similissimi, in maniera che non sie-
no punto differenti in cosa alcuna . Fatte queste cose bisogna che noi delibe-
riamo in fra di noi quale habbi ad esser quella lettera, che ci serua per contra-
segno, conciosia che il contrasegno serue come quasi per una chiaue da aprirci
la entrata per potere penetrare negli intimi segreti . Ma questo contrasegno
è di due sorti, perche uno è quello che si piglia dalle Maiuscole stabili, & l'altro
è quello che si piglia dalle lettere piccole & mobili, & l'uno & l'altro a uoglia
nostra . Diremo prima del contrasegno Mobile, sia per modo di dire risolu-
tosi fra noi che il contrasegno della Tauoletta mobile sia, K, io uolterò la Tauo
letta del modine, in modo che, hauendo a scriuere, uorrò che per modo desem-
pio il , K , uengha sotto la Maiuscola , B , & che la seguente corrisponda a quel
la che segue. Quando adunque io ti scriuerrò , la prima lettera che io scriuer
rò , inanzi a tutte l'altre , sarà la maiuscola B , sotto la quale io che harò a scri-
uere , harò posto il contrasegno K , & questo dimostrerà in che modo io mi sia
seruito del Modine, in esso scriuere. Et uolēdo tu che sarai lontano leggere quel
che io ti harò scritto, ti bisognerà girare la tua tauoletta mobile del modine me
desimamente in maniera, che sotto la maiuscola , B, caschi il contrasegno, K,
di qui finalmēte ti auuerrà che tutte le altre lettere minori trouate nello scritto,

O 3 ti di

MONTAGU, Mary Wortley (1689–1762). *Letters of the Right Honourable Lady M[ar]y W[ortle]y M[ontagu]e.* London, 1763.

Floral ciphers at the Ottoman court.

British writer and aristocrat Lady Mary Wortley Montagu recorded her observations of Turkish culture in letters. The letters resemble anthropological field notes, though refracted through Montagu's wit and curiosity. One of these letters describes a method for disguising messages in flower arrangements — a kind of floristic cryptography in use at the Ottoman court:

> There is no colour, no flower, no weed, no fruit, herb, pebble, or feather, that has not a verse belonging to it; and you may quarrel, reproach, or send letters of passion, friendship, or civility, or even of news, without ever inking your fingers.

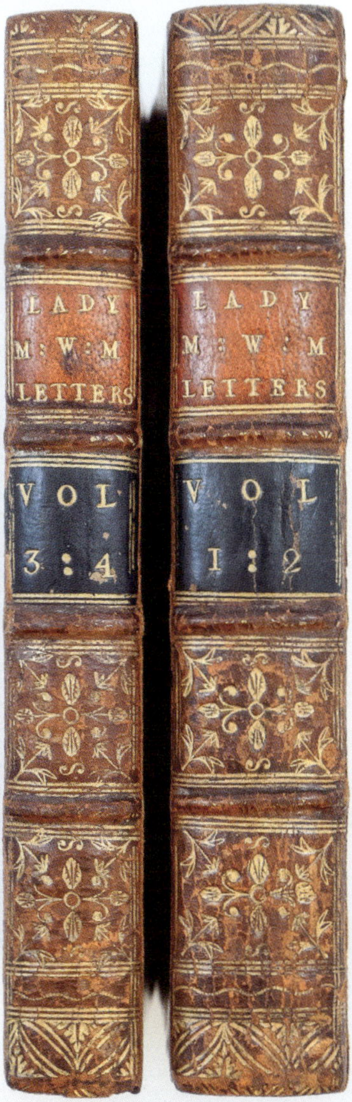
LADY
M:W:M
LETTERS
VOL
3:4
LADY
M:W:M
LETTERS
VOL
1:2

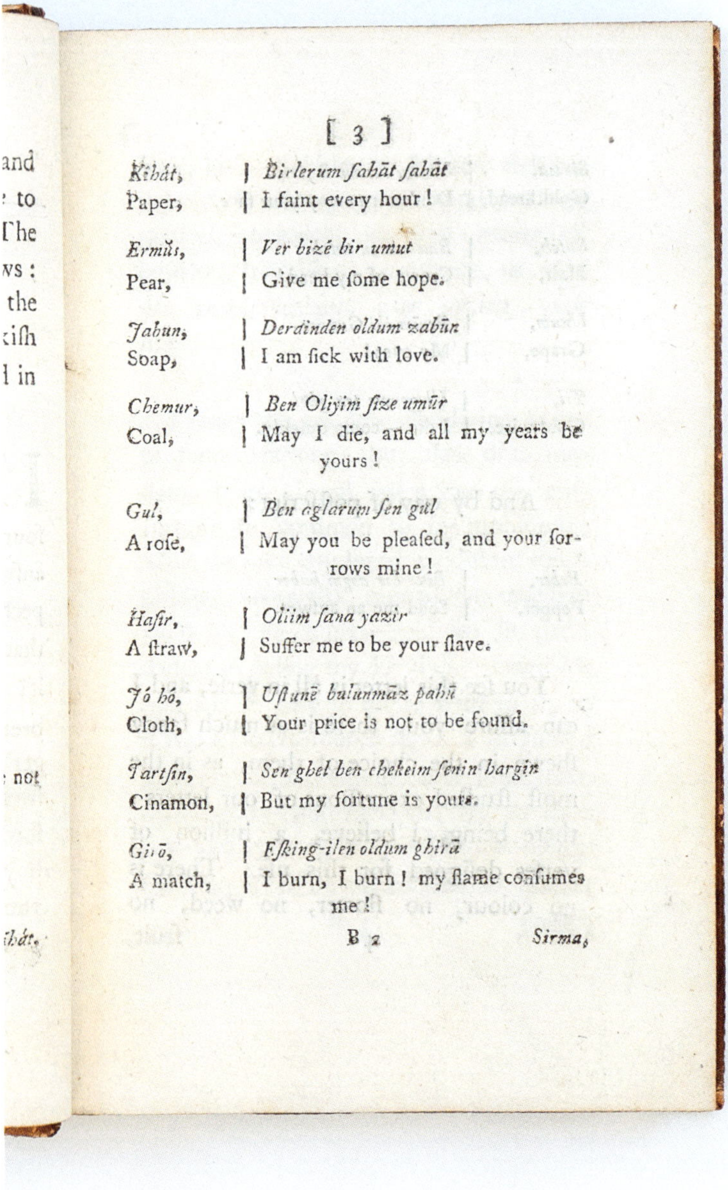

[3]

Kihát,	*Birlerum ſahāt ſahāt*
Paper,	I faint every hour!
Ermūs,	*Ver bizé bir umut*
Pear,	Give me ſome hope.
Jabun,	*Derdinden oldum zabūn*
Soap,	I am ſick with love.
Chemur,	*Ben Oliyim ſize umūr*
Coal,	May I die, and all my years be yours!
Gul,	*Ben aglarum ſen gul*
A roſe,	May you be pleaſed, and your ſorrows mine!
Haſir,	*Oliim ſana yazir*
A ſtraw,	Suffer me to be your ſlave.
Jó hó,	*Uſtunē bulunmāz pahū*
Cloth,	Your price is not to be found.
Tartſin,	*Sen ghel ben chekeim ſenin hargin*
Cinamon,	But my fortune is yours.
Girō,	*Eſking-ilen oldum ghirā*
A match,	I burn, I burn! my flame conſumes me!

B 2 *Sirma,*

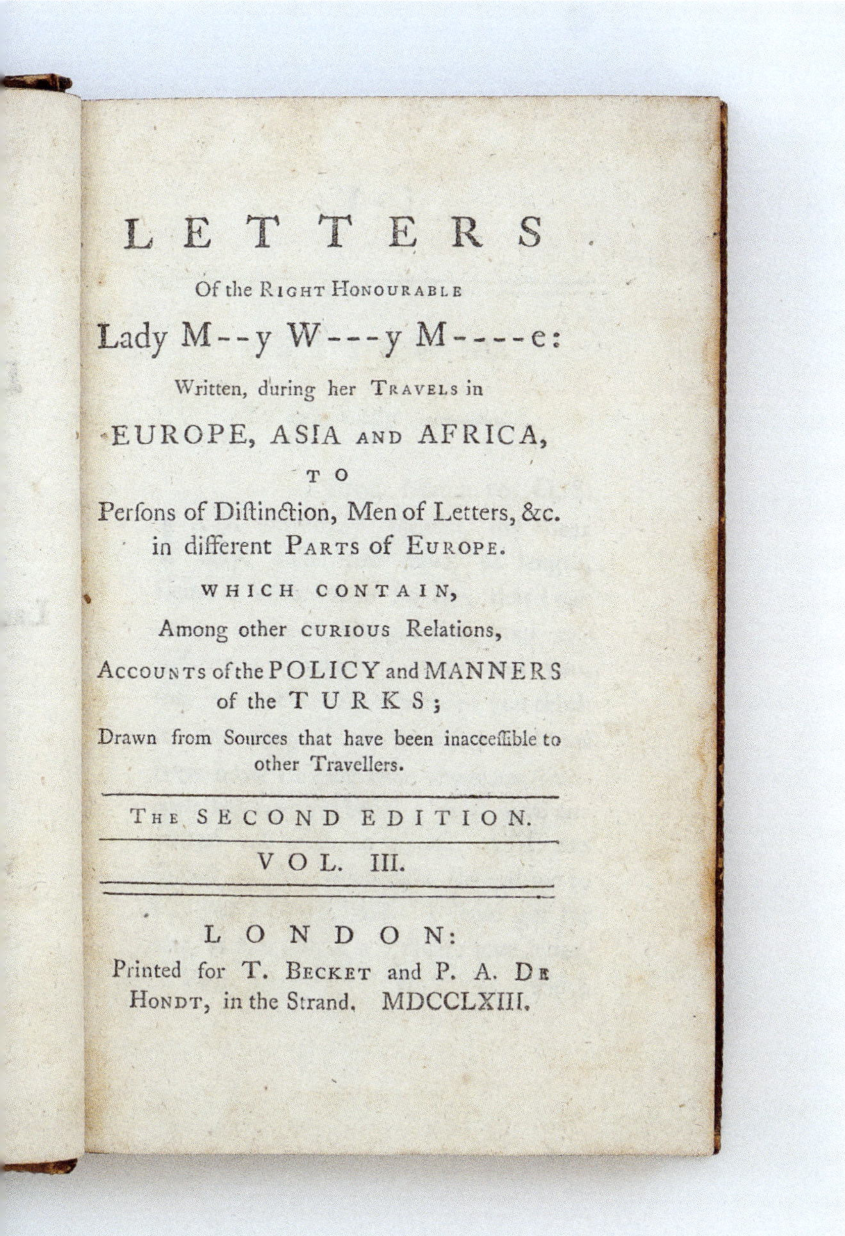

LETTERS

Of the RIGHT HONOURABLE

Lady M--y W---y M----e:

Written, during her TRAVELS in

EUROPE, ASIA AND AFRICA,

TO

Perſons of Diſtinction, Men of Letters, &c. in different PARTS of EUROPE.

WHICH CONTAIN,

Among other CURIOUS Relations,

ACCOUNTS of the POLICY and MANNERS of the TURKS;

Drawn from Sources that have been inacceſſible to other Travellers.

THE SECOND EDITION.

VOL. III.

LONDON:

Printed for T. BECKET and P. A. DE HONDT, in the Strand. MDCCLXIII.

ENIGMA, Model M3.
Germany, circa 1937–1941.

A complex electromechanical device with a dark history.

The Enigma M3 encrypted and decrypted dispatches for the Nazi armed forces during World War II. The story of its decipherment — accomplished in two phases, first by Marian Rejewski of Polish intelligence and then by Alan Turing and a team of cryptanalysts at Bletchley Park — is well known. The battery-operated Enigma M3 enciphers messages using an array of three wheels that advance with each stroke of its typewriter-like keys. The resulting jumble of ciphertext is nearly unbreakable, but proved vulnerable to the common code-breaking techniques of frequency analysis and crib matching. These long-standing deciphering techniques were made more effective with the help of early computers.

Special Collections also holds a rarer four-rotor Enigma M4 (see page 295).

↓ The TRAUB-McCORDUCK COLLECTION of The HISTORY OF COMPUTING

CMU's Enigma M3 came to Special Collections in 2018 with the Traub-McCorduck Collection of the History of Computing. Assembled by Joseph F. Traub (1932–2015; head of CMU School of Computer Science, 1971–79), the collection illustrates the evolution of machine calculation and encryption from the Renaissance to the twentieth century. Pamela McCorduck (1940–2021), an authority on the early history of AI and Traub's wife, donated the collection to CMU Libraries (see "TRAUB-McCORDUCK COLLECTION" page 293).

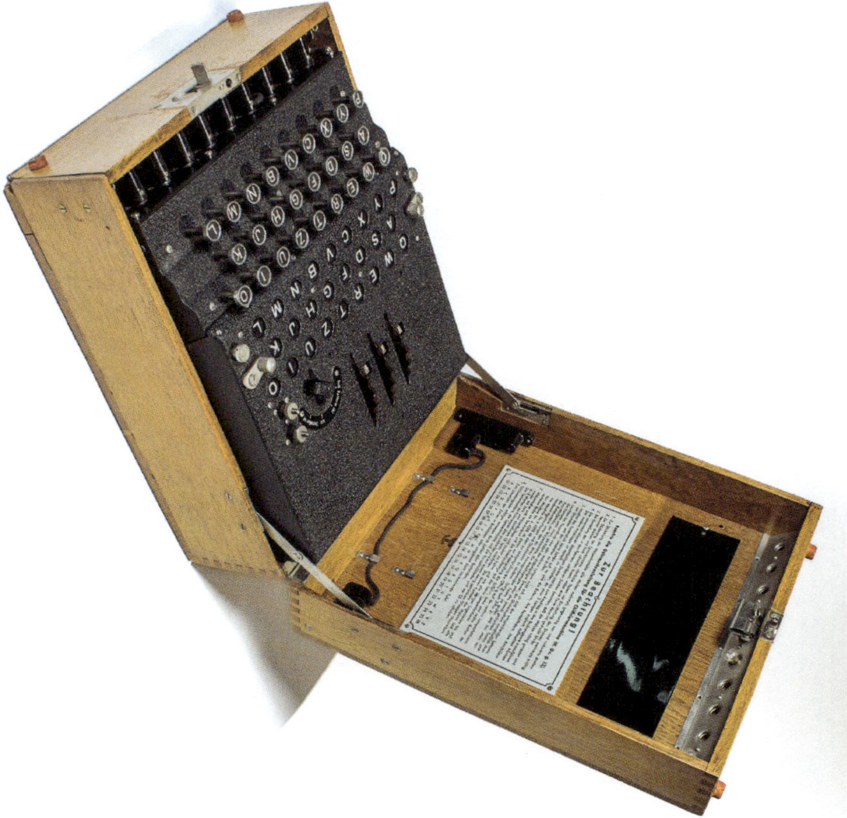
Zur Beachtung!

Essay

On Cipher Discs & Early Encryption

Cipher discs are paper-and-string encrypting machines made up of two or more rotating dials called volvelles (from the Latin *volvere*, "to turn"). Embellished with alphabets, numbers, or other symbols, these ciphering wheels encode or decode any message: when the cells of the volvelle's inner dial align with those in its outer dial, the facing pairs of symbols or letters determine the cipher's plaintext-to-ciphertext substitutions. In the cipher disc shown in figure 1, for instance, *g* replaces *A*, *k* replaces *B*, and *l* replaces *C*, etc. Though made of simple materials — paper, ink, paste, vellum strips, and string — and constructed on basic principles, cipher discs are a foundational tool in the history of cryptology.

The cipher disc reached a high point of artistry in the Renaissance, a period as enamored of learning and beauty as it was of "concealment, hidden revelation, and secrecy," as Jonathan Sawday reminds us (Sawday 2007, 103). Its invention is often credited to that

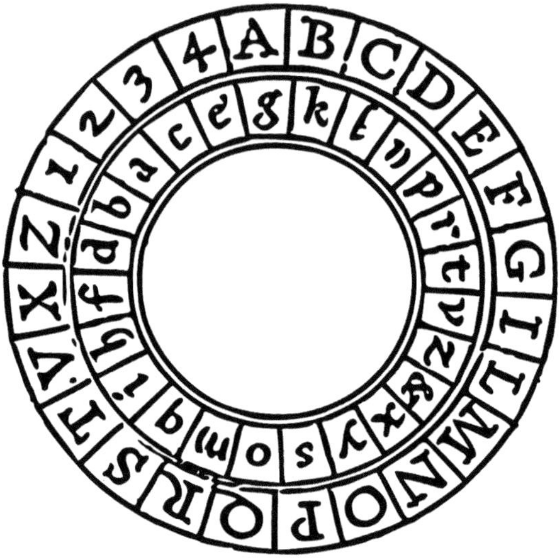

Figure 1 ↓ A Renaissance cipher disc.
By permission of Augusto Buonafalce, CC BY-SA 3.0
<https://creativecommons.org/licenses/by-sa/3.0>,
via Wikimedia Commons

consummate Renaissance figure, Leon Battista Alberti (1404–1472), mathematician, poet, athlete, and architect, but similar instruments probably date to antiquity: the Caesar Cipher — a "shift" cipher named for Julius Caesar, its earliest documented user — can employ a cipher disc to track its cryptic substitutions, for instance (Kahn 1996, 83 and 136; Grafton 2002; Kahn 1996, 125–130; Akkerman and Langman 2024, 71–75).

Alberti's writing on ciphers and cipher discs appears in a short treatise written in 1466 or 1467 titled *De Cifra* ("On Ciphers"). This was the first Western work on cryptography and, as Arielle Saiber has argued, "one of the earliest treatises on coding" (2017, 22). First circulated in manuscript among Alberti's friends and patrons, *De Cifra* was eventually printed in Venice in 1568 as part of a posthumously published collection of essays titled *Opuscoli Morali*. The *Opuscoli* also contains Alberti's brief and experimental works on mathematical perspective, family life, architecture, and morality — hence the collection's name: "moral essays," *opuscoli* meaning "little works" (figure 2). Revealing Alberti's intellectual range and drift, the *Opuscoli* also includes a short treatise on the house fly. Here, though, *De Cifra*, and its pioneering contributions to the mathematics and theory of encryption, offer a fascinating point of departure for Special Collections' holdings of books and artifacts from the history of cryptography and encryption.

Near the beginning of the treatise, Alberti lists the advantages of ciphers, noting that their primary function is to shield strategic information from prying eyes. This posed a problem for intelligence-hungry diplomats and spies, who relied on spycraft and code breaking to reveal "the schemes and machinations of enemies." But, Alberti adds, "in my opinion it is even more advantageous to be able to communicate our intentions to another, no matter how far away, in a way that no other mortal except the intended recipient... is able to read them" (Williams et al. 2010, 170).

In this passage Alberti sets up a useful distinction: the first branch of cryptology, cryptanalysis (from *analýein* [ἀναλύειν], the Greek verb meaning to untie, unknot, or unbind), involves uncovering secrets that one's enemies would prefer to keep hidden. The second, and according to Alberti, more important, branch is cryptography: the art of designing ciphers of sufficient complexity to frustrate code breaking. While Alberti treats both cipher-making and cipher-breaking in *De Cifra*, his first interest is in the design of new ciphers that evade the cryptanalyst's probing art. For what use is a cipher, Alberti implies, if it fails to conceal?

Alberti's solution was a technology of his own invention: polyalphabetic ciphers. These are cryptographic

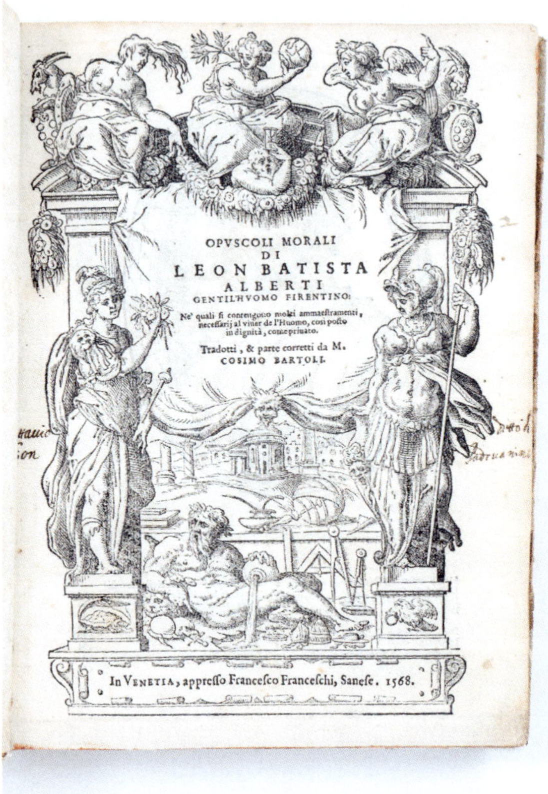
OPVSCOLI MORALI
DI
LEON BATISTA
ALBERTI
GENTILHVOMO FIRENTINO:
Ne' quali si contengono molti ammaestramenti, necessarij al viuer de l'Huomo, così posto in dignità, come priuato.
Tradotti, & parte corretti da M.
COSIMO BARTOLI.
In VENETIA, appresso Francesco Franceschi, Sanese. 1568.

Figure 2 → The title page of the first edition of the *Opuscoli Morali*, which contains Alberti's pioneering treatise on ciphers: *De Cifra*. By permission of Carnegie Mellon University Libraries, Special Collections (Call number PQ4562 .A6 1568).

systems that employ multiple alphabets to disguise messages. Unlike simple substitution ciphers that replace each occurrence of a plaintext letter with the same ciphertext letter, polyalphabetic ciphers periodically change, allowing the same letter to be enciphered differently within the same message. Five hundred years later, this remains a concept fundamental to algorithmic methods of encryption.

But as Alberti well knew, complex ciphers are difficult to use — the human brain is poorly adapted to tracking the thousands of permutations involved in even basic polyalphabetic systems. To address this, Alberti devised a cipher instrument: a metal volvelle made up of two concentric brass plates, each marked with an alphabet around its perimeter. The bulk of Alberti's treatise is given over to a description of this device, forming a series of notes that read like a step-by-step instruction manual on the disc's fabrication and use. While Alberti's straightforward account of the instrument seems to work against his stated interest in secrecy (at the end of the treatise, Alberti requests that "this little work of mine [be] kept among our friends, not in the public domain" [Williams et al. 2010, 183]), the result is a startlingly vivid description of an early cryptographic technology:

> make two circles cut from two bronze sheets, one larger, called fixed [*fermo*], the other smaller, which we call moveable [*mobile*]. The diameter of the fixed circle is larger by the ninth part. I divide the entire circumference into twenty-four coequal parts; these parts of the circle are called houses [i.e., cells]. In the individual houses of the larger circle I write in red ocher upper-case letters following the usual alphabetic order, that is, first A, second B, third C, and so on with the rest. The H and the K are left out because they are used so rarely. These upper-case letters will be twenty in number... and will occupy twenty houses.
>
> In the smaller circle there will then be houses similar in number to those of the larger disk and with corresponding lines dividing them into houses [in which] will be written a letter in black ink... not in alphabetical order as in the fixed

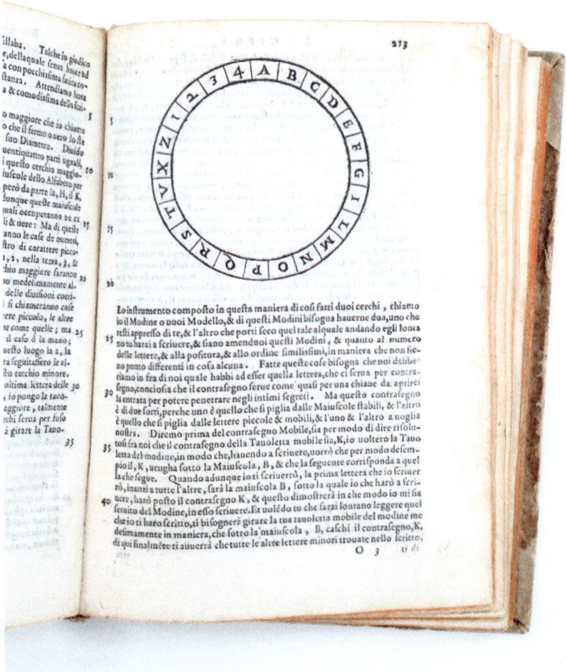

Figure 3 ↓ Alberti's cipher disc without its central volvelle. By permission of Carnegie Mellon University Libraries, Special Collections (Call number PQ4562 .A6 1568).

circle, but strewn by chance, say that the first of these mobile letters is a, the second g, the third q, and so forth with the rest, until all twenty-four houses are filled. There are that many Latin alphabetic characters, the last of which is '&'. This done, we place the mobile circle on top of the larger fixed circle such that a single pin pierces the centre of both acting as the axis, so that the mobile circle rotates (Williams et al. 2010, 180–181).

In a gesture to his work on mathematical games, Alberti named his metal cipher instrument the "formula," and advised making at least two: one for the sender and another for the recipient. Agreeing on an initial position, or "index," for their discs, two (or more) correspondents could disguise their communications in a code made legible only by means of their carefully calibrated devices.

An illustration of Alberti's cipher wheel appears alongside its description in *De Cifra*, but only its outer, stationary disc is shown (figure 3). This was intentional. Rather than provide a completed disc, Alberti invites the reader to craft one of their own and mount it onto the page. Having done so, Alberti's printed illustration would function as a paper replica of his brass prototype. Although made of flimsier stuff, this paper surrogate was no less powerful; whether made of paper or metal, Alberti assures us, the device will "serve as a sort of key that opens a door onto even the most intimate of secrets" (*serve come quasi per una chiave da aprirci la entrata per potere penetrare negli intimi segreti*).

A century or so later, Alberti's cipher disc inspired more intricate forms. In 1563, the Italian scientist Giambattista Della Porta published a book-length study of cryptography titled *De Furtivis Literarum*. Unlike Alberti's *De Cifra* — which only provided instructions on making his "formula" — Della Porta's work is a cipher machine in the shape of a book: its pages carry several printed cipher discs with concentric, rotating rings filled with arcane symbols that seem to conceal divinatory secrets and obscure arts (figure 4).

Threaded onto a piece of robust string, the central dial of Della Porta's cipher disc rotates clockwise within two static dials: the first contains a scrambled and modified twenty-letter alphabet (it's missing letters J, K, U, W, X, Y, and Z and has two Rs), while the second contains Roman numerals from one to

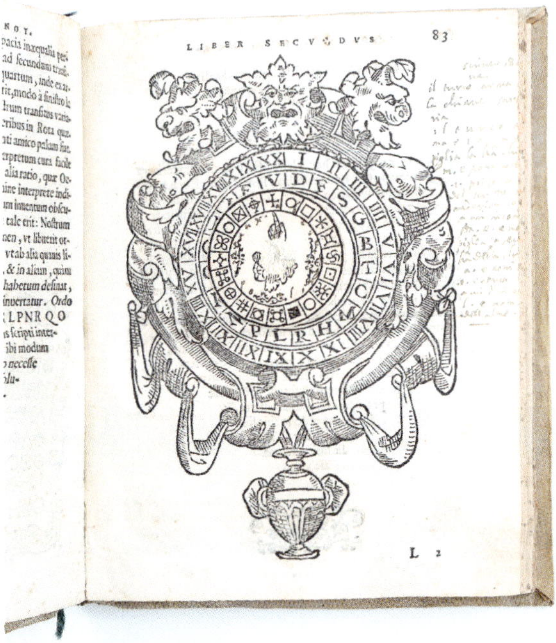

Figure 4 ← One of Della Porta's still functional cipher discs in the 1563 Naples editions held in Special Collections. By permission of Carnegie Mellon University Libraries, Special Collections (Call number Z103.5 .P67 1563).

twenty. The hand inside the inner volvelle points to the disc's dial position. Each position (I–XX) determines which of Della Porta's strange symbols replace a letter in the plaintext, and vice versa. As the pointing hand makes its way around the outer disc — advancing one step for each symbol or letter in the message — the process of enciphering or deciphering is simple: when enciphering, each letter in the plain-text is replaced with its aligned, corresponding symbol; and when deciphering, each symbol in the ciphertext is replaced with its aligned, corresponding letter. This amounts to an Alberti-inspired polyalphabetic substitution cipher, though in Della Porta's case the substitution involves pseudo-hieroglyphic signs instead of the familiar characters of the Latin alphabet.

The publication history of Della Porta's book itself forms a kind of cipher. Originally printed in Naples in 1563, a second, unauthorized edition of the book was printed in London in 1591. The two editions are nearly indistinguishable, suggesting that the publisher of the London edition, John Wolf, sought to disguise his edition as the first, likely in an attempt to conceal his plagiarism: in many surviving copies of Wolf's edition, the imprint on the title page is false, reading "Naples, 1563" instead of "London, 1591." Special Collections holds a copy of the original Naples printing as well as a copy of Wolf's 1591 edition (figures 5 and 6). Despite their resemblances, these copies differ in one significant way: the cipher discs in Special Collections' copy of Wolf's edition are blank, left unassembled by its earliest readers, while those in the Naples edition are present, even retaining their sixteenth-century axial threads. This reflects the random nature of historical survival: when Della Porta's book was printed, a set of readymade central dials was provided on a separate leaf of paper bound at the end of the volume (figure 7). Handwritten numerals (1–3) directed the dials' placement on Della Porta's first, second, or third cipher disc. It was left to the original buyer (or a later reader) to carefully cut out these dials and mount them with needle, thread, and paste to the cipher discs' vacant centers. While an early owner of the copy of the Naples edition held in Special Collections followed these instructions, no early reader of the copy of the London edition bothered to do so.

Besides its set of carefully assembled and functional cipher discs, Special Collections' copy of the Naples edition of *De Furtivis Literarum* also contains extensive evidence of use: across its pages, one early reader has amended Della Porta's text, experimented with the book's cipher systems, and made observations on cryptography in the margins (figure 8). This interaction with the text amounts to a uniquely cryptographic

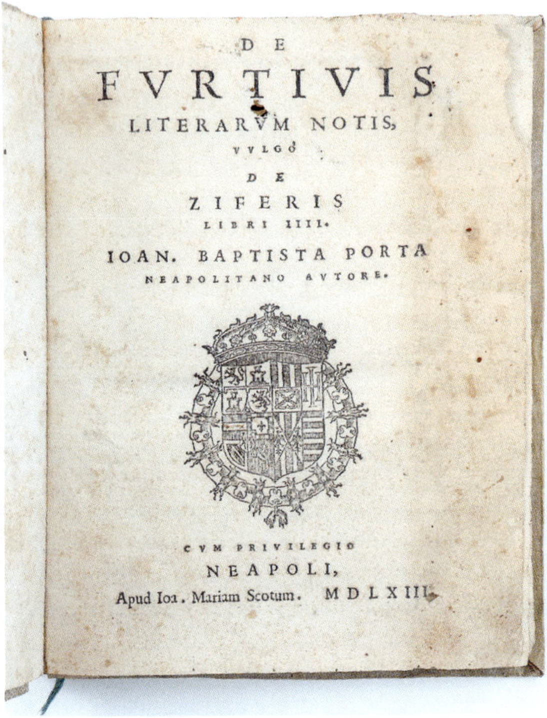
DE FVRTIVIS LITERARVM NOTIS, VVLGÓ DE ZIFERIS LIBRI IIII.

IOAN. BAPTISTA PORTA NEAPOLITANO AVTORE.

CVM PRIVILEGIO

NEAPOLI,

Apud Ioa. Mariam Scotum. MDLXIII.

Figure 5 → The title page of Della Porta's *De Furtivis Literarum* (Naples, 1563). By permission of Carnegie Mellon University Libraries, Special Collections (Call number Z103.5 .P67 1563).

example of Renaissance readers' propensity to "use" their books directly on the page (see Sherman 2009). In this case, the annotations that survive in Special Collections' Neapolitan edition of Della Porta's book show an amateur cryptographer honing their cryptographic skills. In contrast, Special Collections' copy of the London edition offers up a comparative void: its margins are blank, and its pages show no sign of readerly labor. Special Collections' copy of Alberti's *De Cifra* is likewise devoid of annotation. The dissimilarities that set apart Special Collections' copies of Della Porta's and Alberti's books extend to the tone and intellectual framework of the works themselves. Whereas the *Opuscoli* enacts a detached,

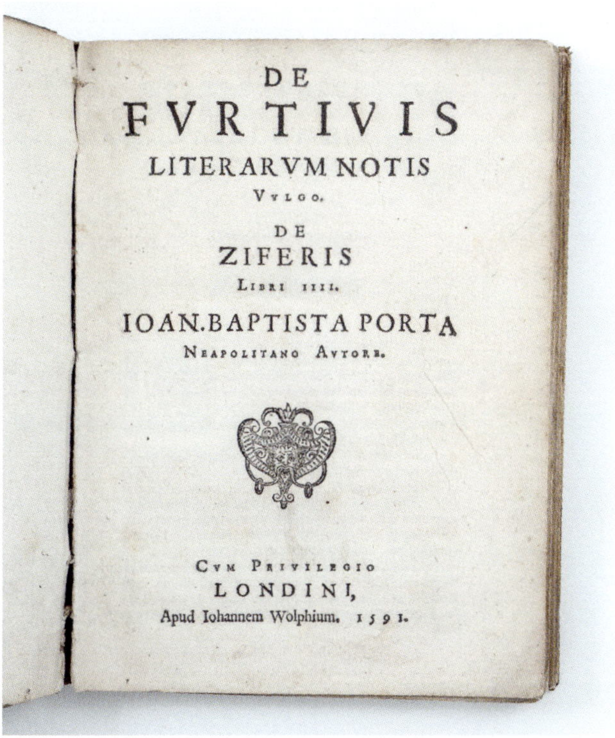

Figure 6 ↑ The title page of Della Porta's *De Furtivis Literarum* (London, 1591). By permission of Carnegie Mellon University Libraries, Special Collections (Call number Z103.5 .P67 1591).

rational, and technical account of cryptology, *De Furtivis Literarum* leans into the field's reputation for the esoteric and the occult. Cryptology was for Alberti an instrument of state power, but Della Porta claims a wider remit, enlisting cryptography as part of an elaborate system of secrets and coded learning. Even Della Porta's title imputes a certain mystique to cryptography: *De Furtivis Literarum* translates as "on secret writing" or "on secret letters." Della Porta's cipher wheels, meanwhile, are flanked by sphinxes — guardians and riddlers of Greek myth who make apt emblems for the cryptologic arts (figure 9). In a similarly emblematic mode, Della Porta's invented alphabet of obscure characters recalls the "lines,

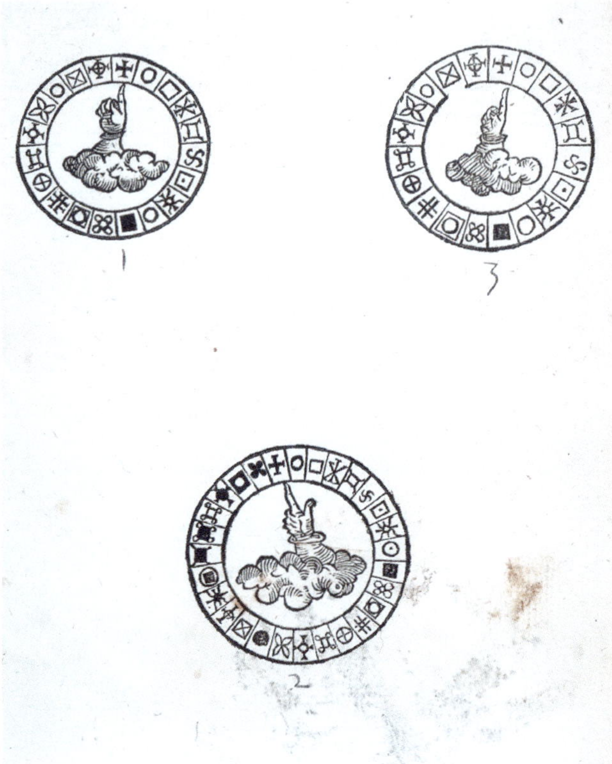

Figure 7 → The central, unassembled index volvelles for Della Porta's cipher discs in Special Collections' copy of the 1591 London edition. By permission of Carnegie Mellon University Libraries, Special Collections (Call number Z103.5 .P67 1591).

circles, signs, letters, and characters" Faustus uses to summon infernal spirits in Marlowe's play (Marlowe 2014, 34). Not by chance, perhaps, Della Porta was placed under surveillance by the papal curia and came to be known as a magus and "professor of secrets." He would eventually help found the Accademia dei Segreti (the Academy of Secrets), a learned society whose elite members were charged with investigating unexplained phenomena of nature.

This preoccupation with mysteries and the occult explains, in part, the strangeness of Della Porta's ciphers. But this strangeness had a methodological edge: while for Alberti, ciphers entailed substituting letters for letters, Della Porta's pseudo-hieroglyphs showed that encryption was not limited to alphabetic pairs. Alighting on this same idea in the 1570s, the English philosopher Francis Bacon (1561–1626) devised a cipher based on only two characters: A and B. With this "biliteral," or two-lettered system, Bacon realized that all twenty-six letters of the Latin alphabet could be encoded in strings of As and Bs, as follows:

A = aaaaa
B = aaaab
C = aaaba
D = aaabb
E = aabaa
F = aabab
G = aabba
H = aabbb
I/J = abaaa
K = abaab
L = ababa
M = ababb
N = abbaa
O = abbab
P = abbba
Q = abbbb
R = baaaa
S = baaab
T = baaba
U/V = baabb
W = babaa
X = babab
Y = babba
Z = babbb

This was a startlingly early form of binary code (see "LEIBNIZ," page 365). But more than that, it had major implications for the history and theory of encryption. In developing his biliteral cipher, Bacon realized that A and B were effectively arbitrary. Any pair of signifiers — two colors, a pair of musical notes, or a doubled gesture of the hand — could provide the basis for a binary system of ciphered communication. In addition to anticipating binary code, then, Bacon also anticipated Morse code, which represents individual letters as long or short pulses of light,

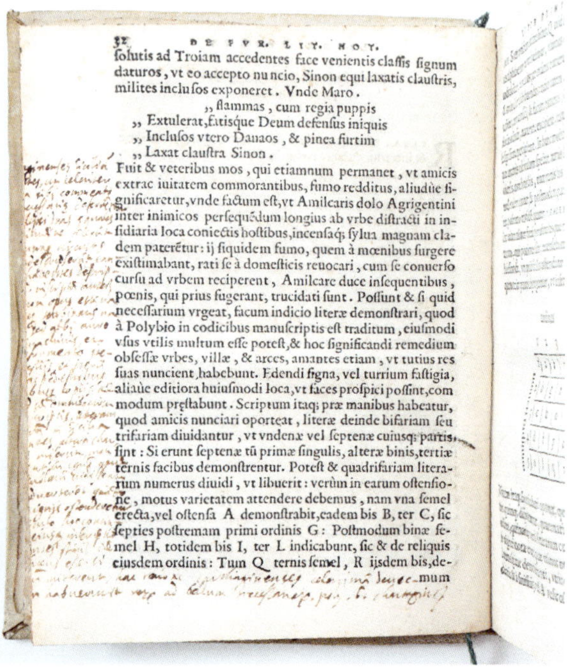
32 DE FVR. LIT. NOT.

ſolutis ad Troiam accedentes face venientis claſſis ſignum daturos, vt eo accepto nuncio, Sinon equi laxatis clauſtris, milites incluſos exponeret. Vnde Maro.

,, flammas, cum regia puppis
,, Extulerat, fatisque Deum defenſus iniquis
,, Incluſos vtero Danaos, & pinea furtim
,, Laxat clauſtra Sinon.

Fuit & veteribus mos, qui etiamnum permanet, vt amicis extra ciuitatem commorantibus, fumo redditus, aliudùe ſignificaretur, vnde factum eſt, vt Amilcaris dolo Agrigentini inter inimicos perſequẽdum longius ab vrbe diſtracti in inſidiaria loca coniectis hoſtibus, incenſaq; ſylua magnam cladem paterẽtur: ij ſiquidem fumo, quem à mœnibus ſurgere exiſtimabant, rati ſe à domeſticis reuocari, cum ſe conuerſo curſu ad vrbem reciperent, Amilcare duce inſequentibus, pœnis, qui prius fugerant, trucidati ſunt. Poſſunt & ſi quid neceſſarium vrgeat, facum indicio literæ demonſtrari, quod à Polybio in codicibus manuſcriptis eſt traditum, eiuſmodi vſus vtilis multum eſſe poteſt, & hoc ſignificandi remedium obſeſſæ vrbes, villæ, & arces, amantes etiam, vt tutius res ſuas nuncient, habebunt. Edendi ſigna, vel turrium faſtigia, aliaùe editiora huiuſmodi loca, vt faces proſpici poſſint, commodum pręſtabunt. Scriptum itaq; præ manibus habeatur, quod amicis nunciari oporteat, literæ deinde bifariam ſeu trifariam diuidantur, vt vndenæ vel ſeptenæ cuiusq; partis ſint: Si erunt ſeptenæ tũ primæ ſingulis, alteræ binis, tertiæ ternis facibus demonſtrentur. Poteſt & quadrifariam literarum numerus diuidi, vt libuerit: verùm in earum oſtenſione, motus varietatem attendere debemus, nam vna ſemel erecta, vel oſtenſa A demonſtrabit, eadem bis B, ter C, ſic ſepties poſtremam primi ordinis G: Poſtmodum binæ ſemel H, totidem bis I, ter L indicabunt, ſic & de reliquis eiusdem ordinis: Tum Q ternis ſemel, R ijsdem bis, decimum

Figure 8 ↓ Annotations show an early reader working in the margins of Della Porta's *De Furtivis Literarum*. By permission of Carnegie Mellon University Libraries, Special Collections (Call number Z103.5 .P67 1563).

electricity, or sound. Bacon concluded with palpable surprise that "anything could signify anything" ["omnia per omnia"]:

> And by this Art a way is opened, whereby a man may expresse and signifie the intentions of his mind, at any distance of place, by objects which may be presented to the eye, and accommodated to the eare ... as by Bells [and] Trumpets, by Lights and Torches ... and any instruments of like nature (quoted in Sherman 2010).

As Bacon almost predicts here, binary code now forms the mathematical basis for our society; heaps of information are enciphered as mere 1s and

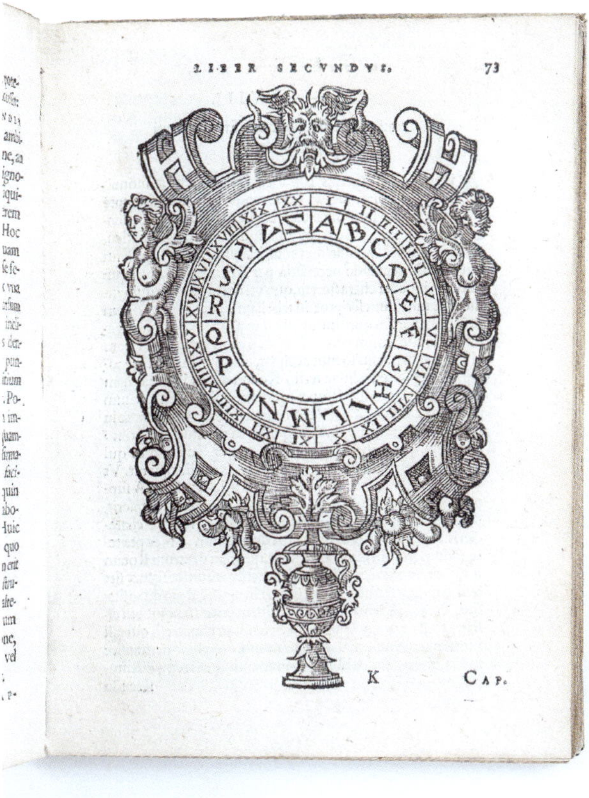

Figure 9 ↑ Sphinxes — riddlers, and guardians of the occult — flank one of Della Porta's cipher discs in the 1591 London edition. By permission of Carnegie Mellon University Libraries, Special Collections (Call number Z103.5 .P67 1591).

0s — everything signified by almost nothing. But stranger signifiers have found expression in less familiar cryptographic systems. The British aristocrat and writer Lady Mary Wortley Montagu (1689–1762) described a method for disguising messages in flower arrangements, a kind of floristic cryptography practiced in eighteenth-century Istanbul that she observed during her husband's appointment as British ambassador to the Ottoman Porte (see "MONTAGU," page 123). Recalling Bacon's own enthusiasm for the cryptic potential of everyday things, Montagu wrote:

> There is no colour, no flower, no weed, no fruit, herb, pebble, or feather, that has not a verse belonging to it; and you may quarrel, reproach, or send letters of passion, friendship, or civility, or even of news, without ever inking your fingers (Montagu 1763, 4–5).

The visual language of early cryptography was clearly a mixed and sweeping one: flowers, musical notes, torches, obscure alphabets, and arcane symbols. While few if any of these systems have survived to influence modern encryption, the cipher disc can claim a long afterlife. The underlying mechanics of the Enigma machine are more advanced, but no different: rotating discs establish fleeting links between plaintext and ciphertext (see "ENIGMA," page 129). What early cipher discs show us, then, is that the history of cryptology is woven through with old ideas, technological iterations, and reinventions.

* * *

Following the 2018 gift of the Traub-McCorduck Collection of the History of Computing — an important group of artifacts and books on early computing and cryptography, including two World War II-era Enigma machines — Special Collections is one of the few institutional collections that tells the kinds of stories that Renaissance cipher discs prompt us to revisit and explore (see "TRAUB-McCORDUCK COLLECTION," page 293).

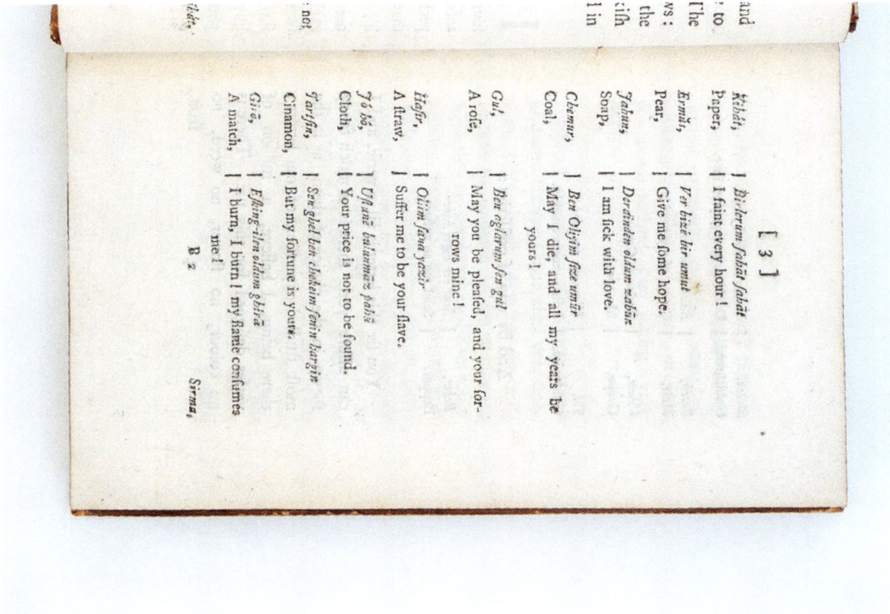

[3]

Kihât, Paper,	*Birlerum ſahât ſahât* I faint every hour!
Ermûs, Pear,	*Ver bizé bir umut* Give me ſome hope.
Jabun, Soap,	*Derdinden oldum zabûn* I am ſick with love.
Chemur, Coal,	*Ben Oliyim ſize umûr* May I die, and all my years be yours!
Gul, A roſe,	*Ben aglarum ſen gul* May you be pleaſed, and your ſorrows mine!
Hasír, A ſtraw,	*Oliim ſana yazir* Suffer me to be your ſlave.
Jó hó, Cloth,	*Uſtuné bulunmaz pahâ* Your price is not to be found.
Tartſin, Cinamon,	*Sen ghel ben chekeim ſenin hargin* But my fortune is yours.
Giro, A match,	*Eſking-ilen oldum ghira* I burn, I burn! my flame conſumes me!

B 2 *Sirma,*

Figure 10 ↑ A table of correspondences showing the ciphered meaning of plants and flowers in Montagu's *Letters* (London, 1763). By permission of Carnegie Mellon University Libraries, Special Collections (Call number PR756 .T72 C65 1763).

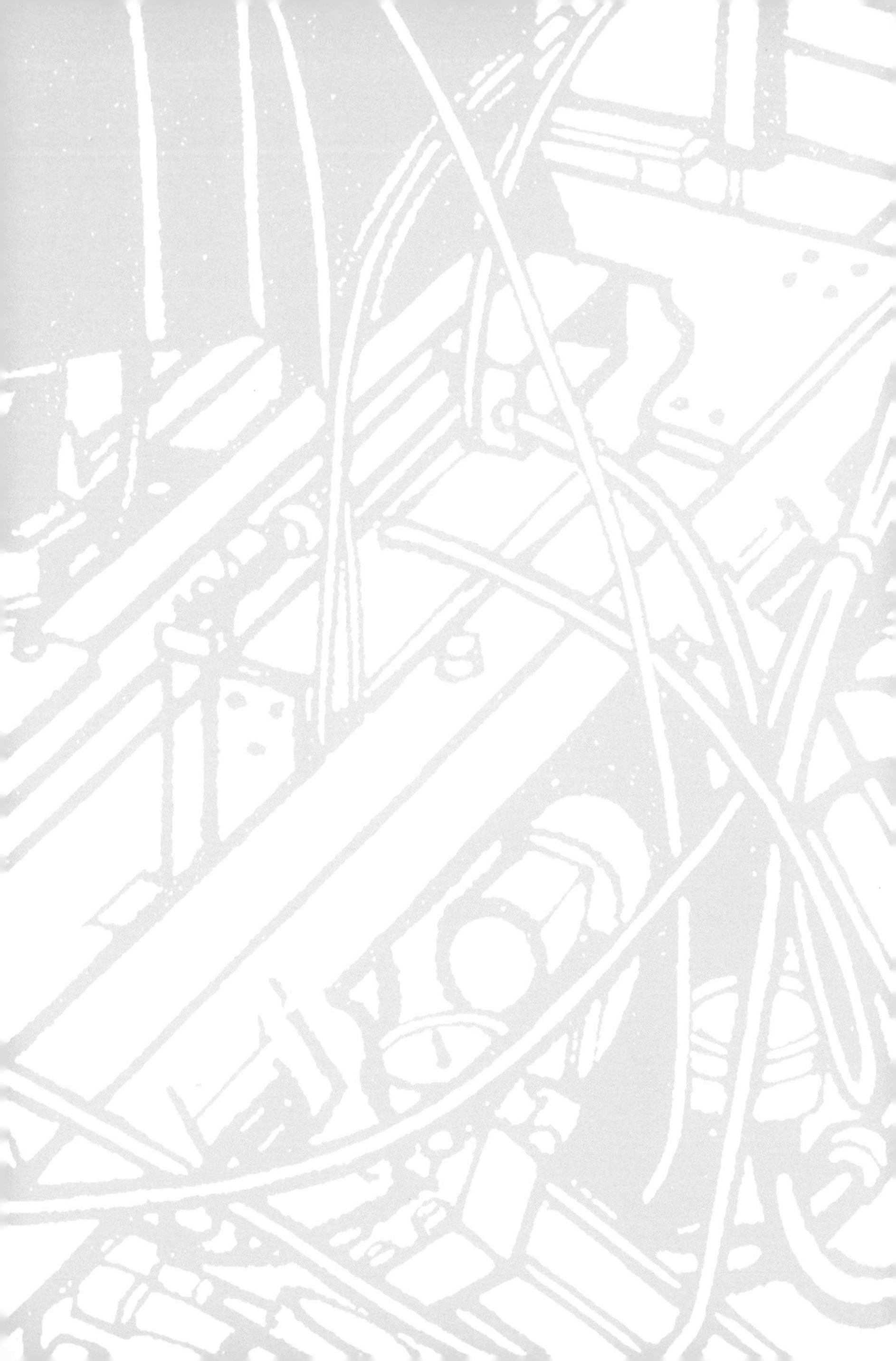

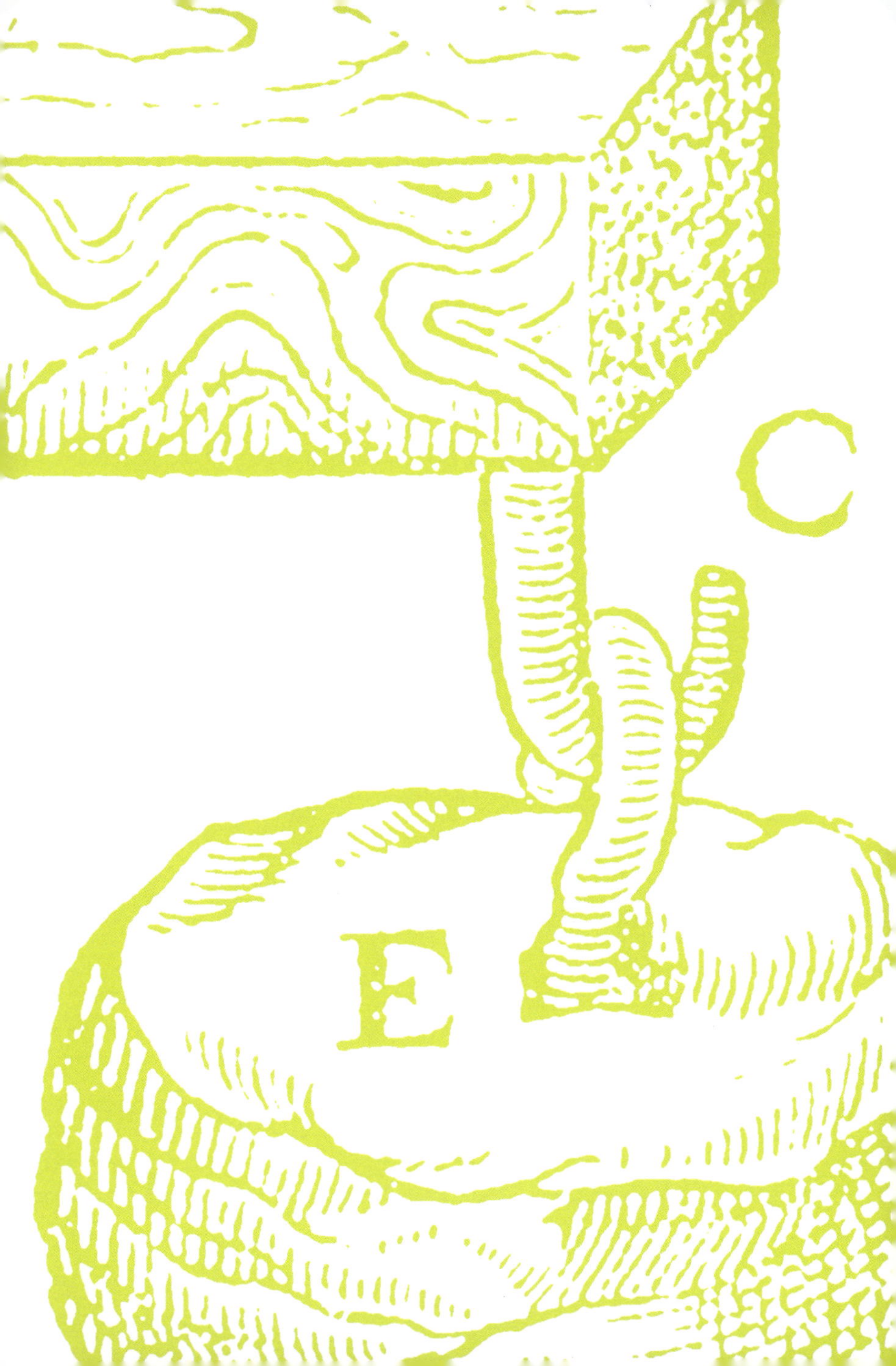
C
E

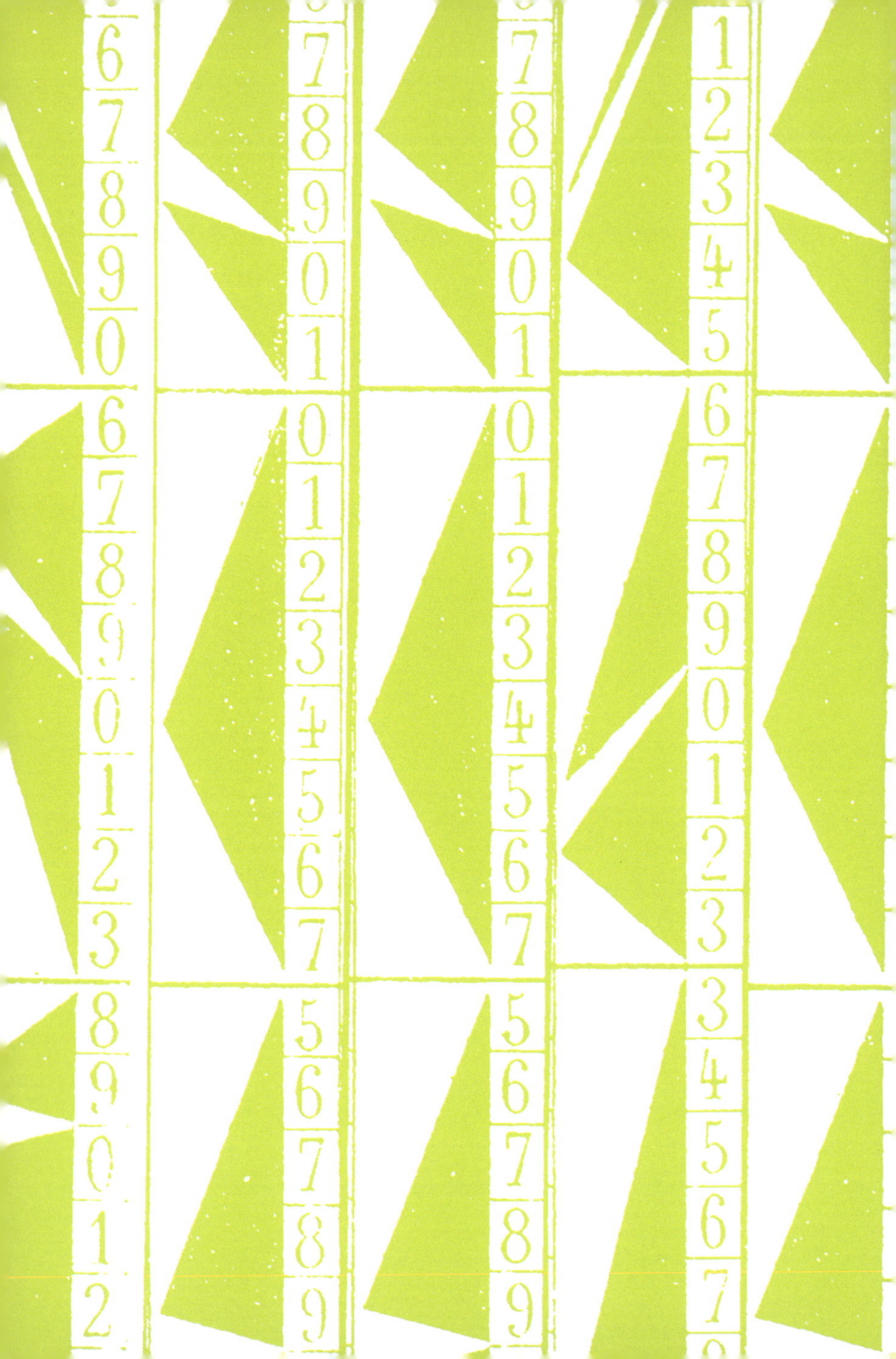

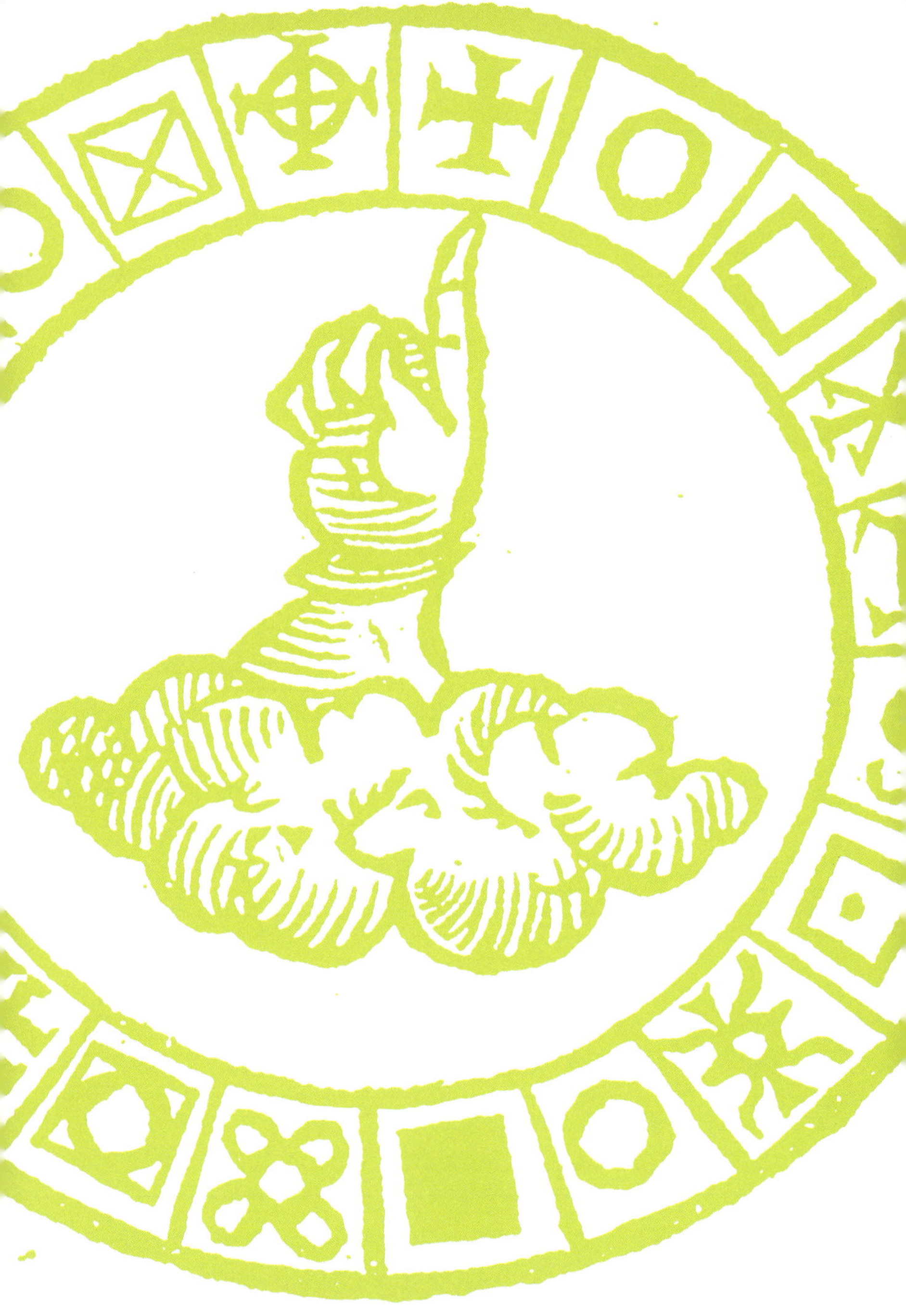

PART III

Robotics & Automata

Robots and other automata have long captured the imagination of human beings. Reflecting the culture of innovation at CMU — a leader in robotics — Special Collections curates the prehistory of the field. Objects in this part include a treatise on mechanical automata printed in nineteenth-century Japan and a Renaissance book on “machines that move themselves.”

HOSOKAWA HANZŌ Yorinao (1749–1796). 機巧圖彙 [Karakuri zui]. Kyoto, 1808.

The first Japanese work on robotics and automata.

This book illustrates a number of Japanese *karakuri*: miniature automatons, clocks, and other mechanized tricks and technologies. The origin of the word *karakuri* is unknown. It may derive from the verbs *karamu* (to twine) and *kuru* (to reel), suggesting the internal pulleys, springs, and weight-driven mechanisms that propelled these devices. Japanese instrument makers engineered *karakuri* of dazzling complexity in the eighteenth century. These devices and Hosokawa's treatise show that the history of robotics and programmable machines crosses cultures and centuries.

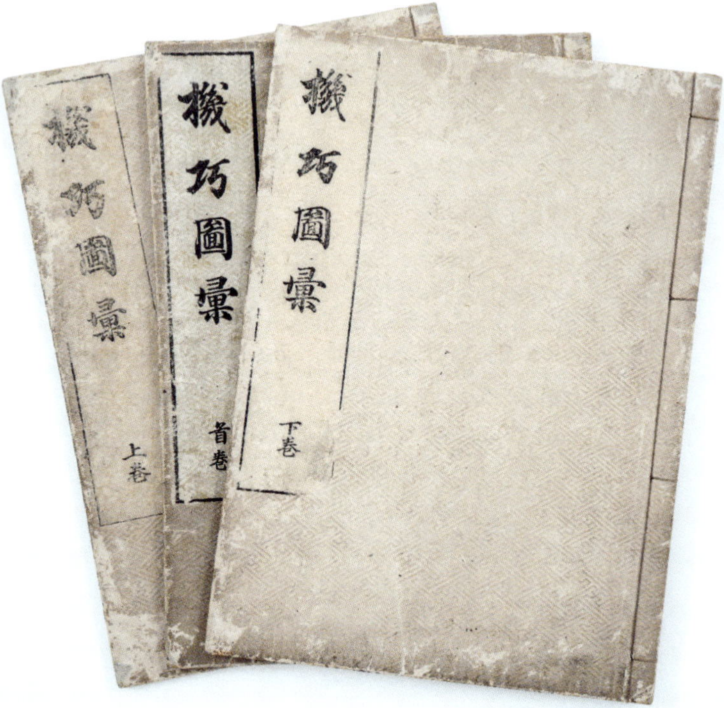
機巧圖彙
上卷
機巧圖彙
首卷
機巧圖彙
下卷

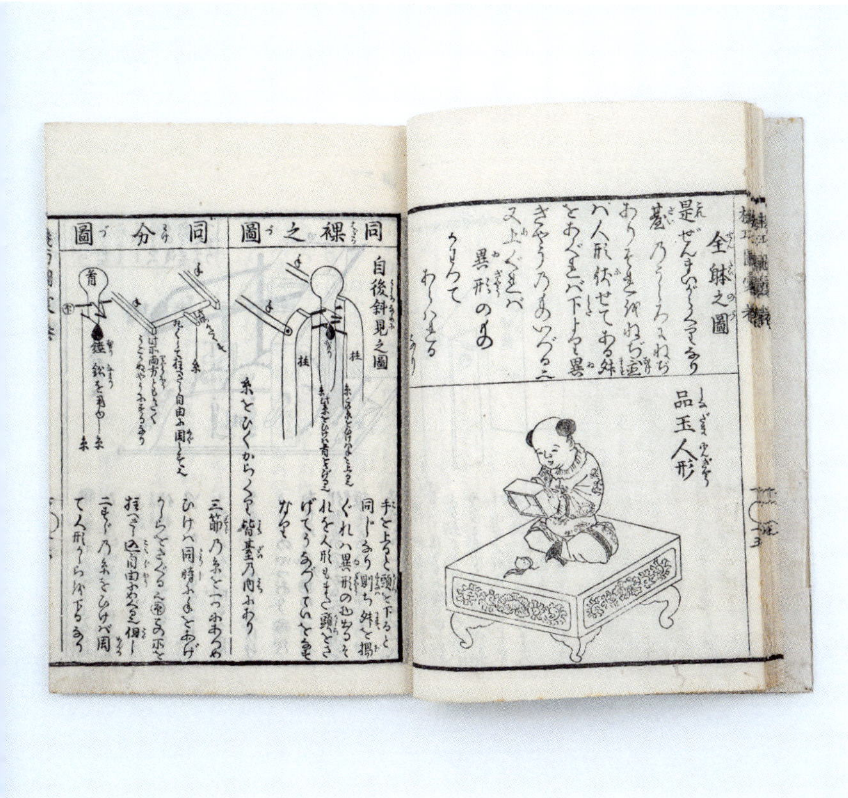
全躰之圖
品玉人形
同裸之圖
同分圖
自後釘見之圖

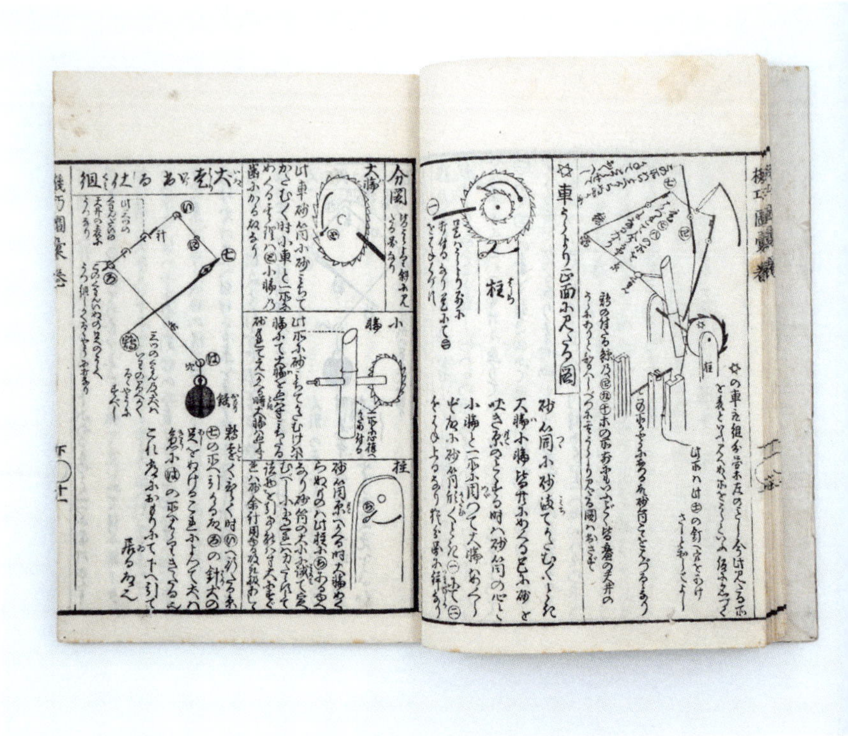

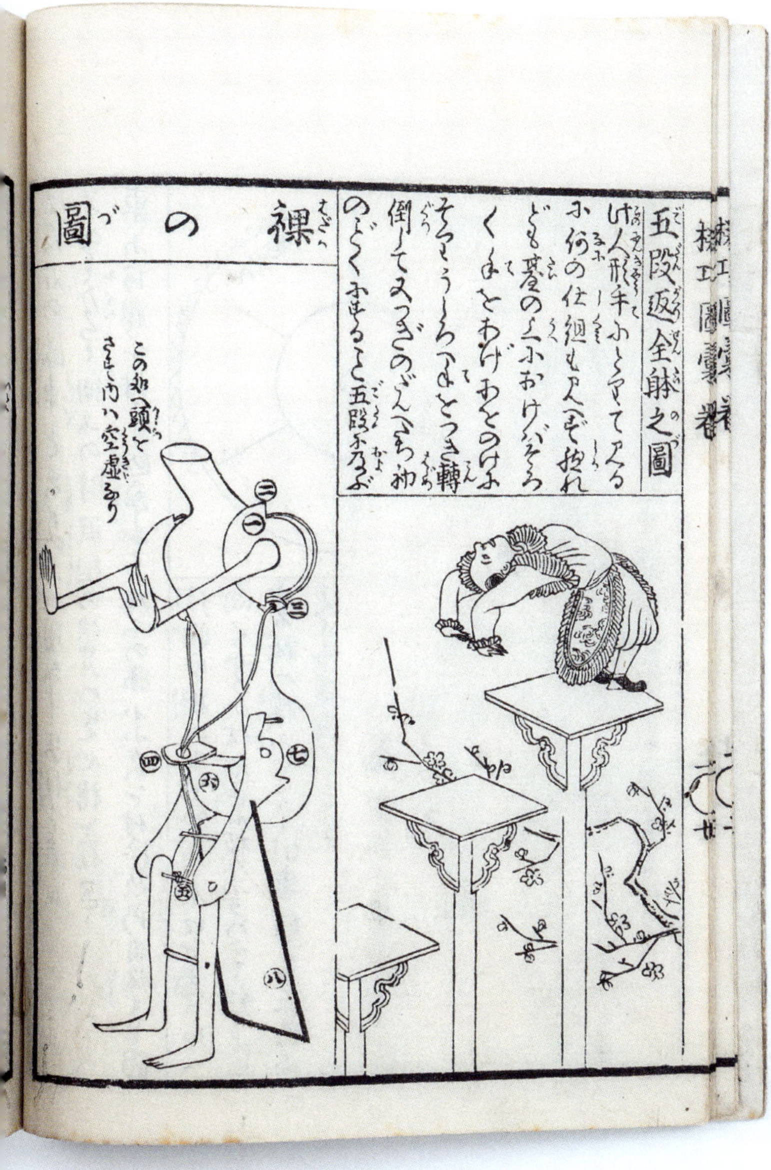
機巧圖彙
五段返全躰之圖
裸の圖

BALDI, Bernardino (1553–1617).
De gli automati, ouero machine se mouenti.
Venice, 1589.

The first printed book on robotics and automata.

This is the first printed book devoted to the subject of robotics. Published in Venice in 1589, it is an Italian translation of Hero of Alexandria's first-century treatise on automata — mechanical devices powered by air, water, and weights to simulate the motions of living beings. Or, in Baldi's memorable phrase, "machines that move themselves." The woodcut shown on page 174 depicts a striking example: a mechanism apparently designed to chop wood, its ax suspended mid-strike by a disembodied arm.

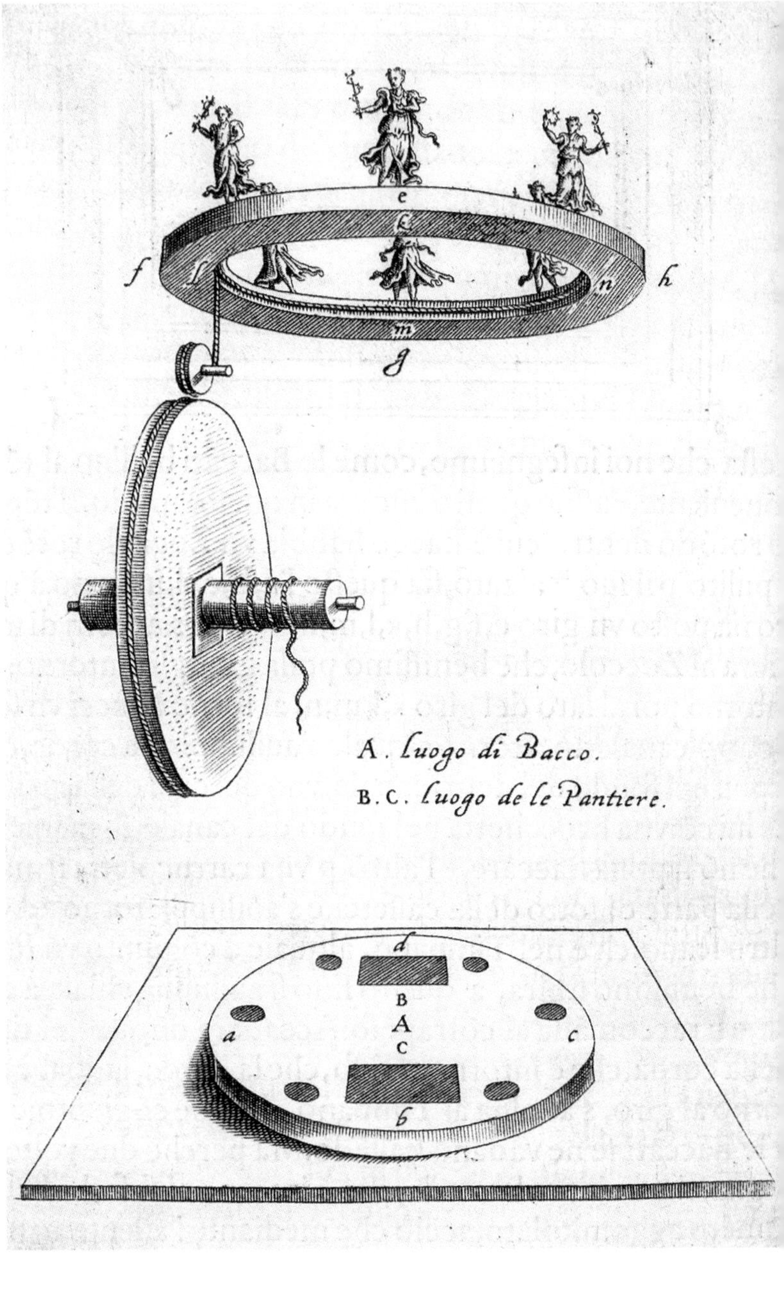
e
f
h
n
m
g
A. luogo di Bacco.
B.C. luogo de le Pantiere.
d
B
A
a
c
C
b

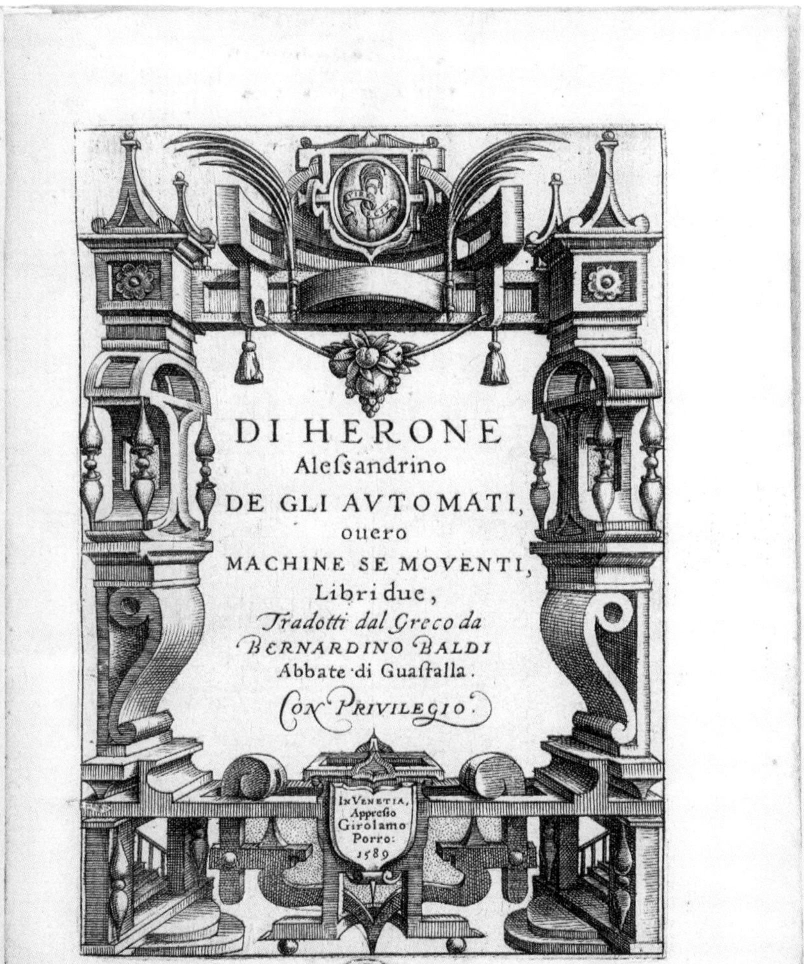

DI HERONE
Alessandrino
DE GLI AVTOMATI,
ouero
MACHINE SE MOVENTI,
Libri due,
Tradotti dal Greco da
BERNARDINO BALDI
Abbate di Guastalla.
CON PRIVILEGIO.

IN VENETIA,
Appresso
Girolamo
Porro:
1589

SUTHERLAND, Ivan Edward (1938–).
A walking robot. Pittsburgh, 1983.

Robotics at CMU.

Turing award-winner Ivan Sutherland developed the hulking Trojan Cockroach — a "walking robot" — at CMU's Robotics Institute in the 1980s. This is a copy of the first printing of his book on the project. CMU's University Archives is home to The Robotics Project, an archive that documents the past, present, and future of robotics at CMU.

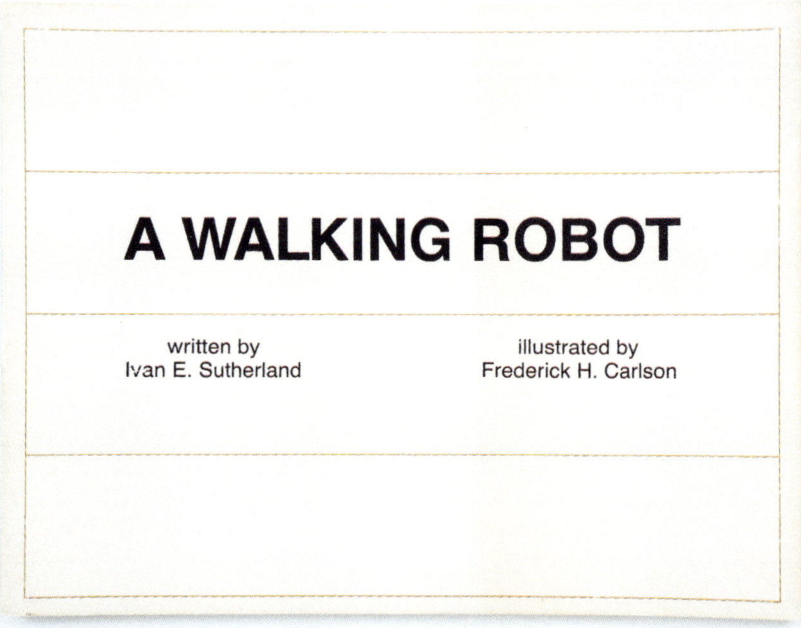
A WALKING ROBOT
written by
Ivan E. Sutherland
illustrated by
Frederick H. Carlson

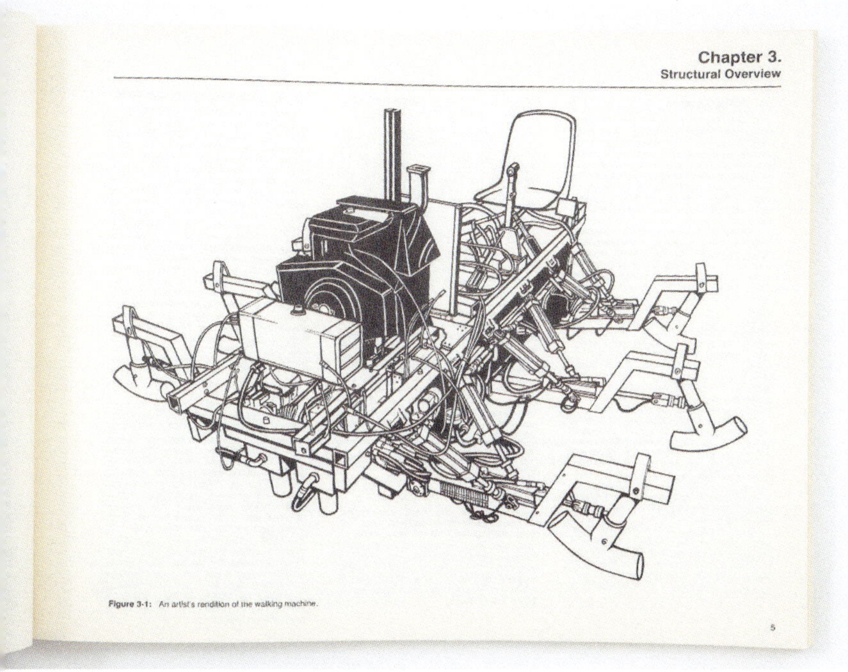

Chapter 3.
Structural Overview

Figure 3-1: An artist's rendition of the walking machine.

5

Essay

On Renaissance Robots, 'Engines of Flesh,' & Artificial Life

"...For what is the heart, but a spring; and the nerves, but so many strings; and the joints, but so many wheels, giving motion to the whole body, such as was intended by the artificer?"

— Thomas Hobbes, *Leviathan* (1651)

The device shown in figure 1 is a strange hybrid of machine and human parts. Made up of an arm mounted to a notched wheel and weight-driven pulley, it resembles a piece of fanciful clockwork: the disembodied limb of a jacquemart, perhaps — the automaton figure who strikes the hour on late-medieval clocks. But this arm holds an axe, not a mallet.

While the device's operation is easy to work out — the gear turns, the arm moves, the axe falls, the action repeats — the diagrammatic quality of the image invites the viewer to deconstruct its mechanism: to follow the pull of weight, the clockwise advance of the gear, the arm's ratcheted swing and drop. And yet there's also something cryptic here. The image ensnares as much as it instructs, suggesting a kind of aliveness that somehow isn't confined to the machine's muscled arm. Instead, the joining of the two — where arm ends and armature begins — makes the image uncanny. It blurs categories that shouldn't overlap, mingling flesh and instrument, anatomy and machine.

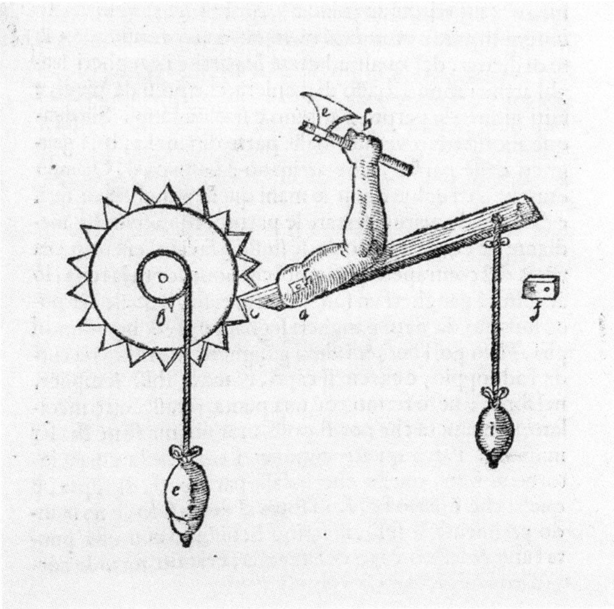

Figure 1 ↓ Bernardino Baldi's axe-wielding arm. By permission of Carnegie Mellon University Libraries, Special Collections (Call number TJ210.2 .H37 1589)

This illustration appears in Bernardino Baldi's Italian translation of *On Automata*, a first-century treatise on self-propelled machines by the Hellenistic engineer and technologist Hero of Alexandria (Grillo 2019). Published in Venice in 1589, Baldi's translation is regarded as the first printed book on robotics, marking the beginning of a rich literary tradition of writing on automated machines. The work reflects Baldi's humanistic training in ancient languages and classical mechanics, in that it offers a meticulous excavation of centuries-old technology reimagined for sixteenth-century use. The book accordingly embodies a fundamental paradox in Renaissance culture: it is simultaneously ancient and modern, backward-looking and cutting edge.

The word "automaton" offers up a similar contradiction. Derived from the Greek αὐτόματον, the word refers to a thing that moves of its own will. In this sense, automata are related to Aristotle's definition of an animal: namely, a beast or being that moves by itself. But so-called autonomous devices offer only a semblance of autonomy. Beneath their lively exteriors are cog-and-wheel mechanisms designed to execute a set of predetermined functions. It is this narrower meaning that we now understand by the word: automatons are precisely will-*less*, their rote actions constrained by mechanical subroutines and human ingenuity. They are more like intricate puppets, in this respect, than proto-robots. But in Baldi's sixteenth-century context, this was far from a settled view; automatons had not yet been fully drained of life.

In the seventeenth century, just as machines were understood to perform a kind of aliveness, human bodies were described in mechanical terms. French philosopher René Descartes (1596–1650) in his *Discourse on the Method* (1637; see "DESCARTES," page 313) famously likened the parts of human anatomy to machinery. He was particularly interested in the function of the eye, comparing it to a lens embedded into the wall of a shadowy camera obscura. Figure 2 illustrates an experiment Descartes designed to demonstrate the eye's optics. It shows a bearded figure in a darkened room peering through a bisected eye. The retinal wall has been removed to reveal the cornea and lens, and lines describe paths of light, refracted through the eye's vitreous humour onto points T, S, and R. Meanwhile, Descartes compared the circulation of blood and bodily humors to a kind of self-contained hydraulic system: vapors and fluids, like those in a heat-driven siphon, rise and fall with the temperature of the body and its parts, while sensations — heat, cold, touch, pain — are channeled through pipe-like nerves to the seat of perception in the brain.

Figure 2 ↓ Descartes' illustration showing an experiment designed to demonstrate the eye's optics. By permission of Posner Memorial Collection, Carnegie Mellon University Libraries, Special Collections (Call number Q155 .D43)

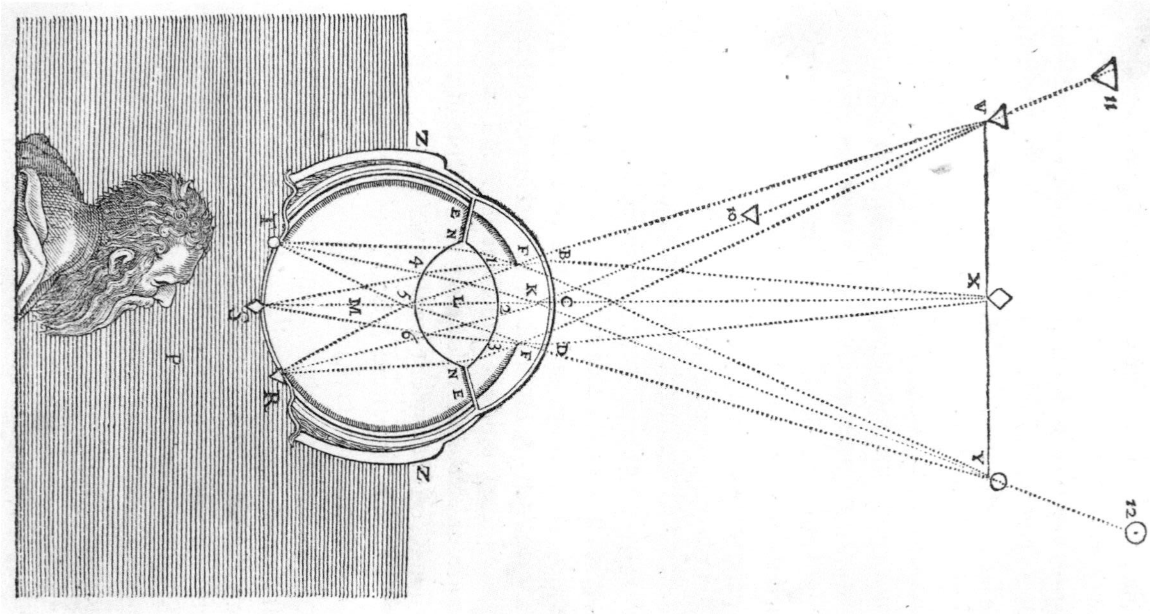

Descartes' mechanistic anatomy left plenty of room for debate: Dutch mathematician and physicist Christiaan Huygens (1629–1695) derided the ideas of those "new philosophers" who imagined the human body as "nothing but Clocks and Engines of Flesh" (quoted in Riskin 2016, 70). Huygens' skeptical view was not commonly held, however; Descartes' philosophy of machine-like humans and humanlike machines remained alive throughout the seventeenth century. Fleshy engines and mechanical bodies animate the work of leading scientists in the period, including Thomas Hobbes (1588–1679; see the epigraph to this essay) and Giovanni Alfonso Borelli (1608–1679), whose groundbreaking treatise on muscles employed Cartesian metaphors to equate muscular contraction with the mechanical action of levers and pulleys.[1] And for Borelli's contemporaries, the analogy cut both ways: automatons, by imitating the motion of living things, seemed to flicker with life.

This paradox of animated inanimate machines suffuses Baldi's book. His woodcut image of the axe-wielding arm almost gives the conundrum a physical form: is the device performing its predetermined and purely mechanical function, or is it straining to exert a darkly vital force? *Of Automata* is full of such "machines that move themselves" (*machine se moventi*) — to use Baldi's evocative translation of the Greek — though many operate in a more lighthearted mode: one of Baldi's (or Hero's) devices is a carousing automaton Dionysus who pours a libation of milk or wine while automated cymbals and drums clatter and crash in the background.

Reprinted in a second edition in 1601, Baldi's book tapped into the period's fascination with automata and other self-propelled machines. But these objects were not exclusive to the Renaissance, nor indeed to Western Europe. Automata and other mechanical tricks have

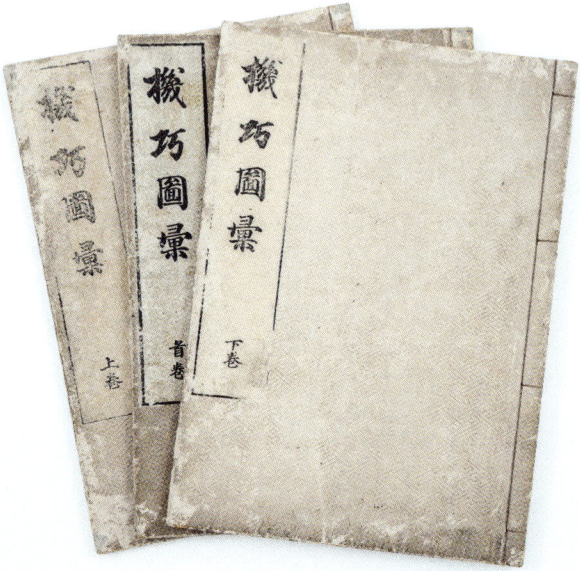

Figure 3 ↓ The copy of Hosokawa's *Karakuri Zui* in Special Collections is bound in three slim volumes. By permission of Carnegie Mellon University Libraries, Special Collections (Call number TJ181 .H87 1808).

captured the imagination of technologists working across cultures and centuries. In 1206, for example, the Islamicate engineer and polymath Ismail al-Jazari described several self-propelled devices in *The Book of Knowledge of Ingenious Mechanical Devices*.[2] These included clocks driven by water and heat, mechanical servants, and automata moved by the subtle working of siphons and pumps. This last category included a fountain shaped like a peacock, "in which flowing water triggers little figures to offer the washer first a dish of perfumed soap powder, then a hand towel" (Riskin, "Frolicsome Engines").

Another book on automata in Special Collections is a work by Japanese engineer Hosokawa Hanzō Yorinao (1749–1796). Printed at the turn of the nineteenth century in Kyoto, the book depicts several Japanese *karakuri* — automatons, clockwork, and other mechanized tricks and technologies — showing the cross-cultural reach of self-moving machines and their long history in Japan. Titled *Karakuri Zui* [Manual on Automatic Mechanical Devices], the work pairs instructive notes on how to build and operate Hosokawa's devices with meticulously drawn woodcut illustrations.

The origin of the word *karakuri* is unknown, though it may derive from the verbs *karamu* (to twine) and *kuru* (to reel), suggesting the internal pulleys, springs, and weight-driven mechanisms that propelled the devices. Many of Hosokawa's inventions, like Baldi's before them, join puppetry and conjuration, including a "whistling little drummer," (Kazuo 2012, 179); a tea-serving automaton that glides forward with eerie poise; a sake bowl with an automaton turtle at its base; and a mechanical cockfighting scene complete with a spring-loaded dog who "frighten[s] the cocks who fly away" (199). One of Hosokawa's most elaborate machines is an automaton fish that swims up an artificial stream. The mechanical tableau stages a Chinese legend about a carp that swims to the top of a waterfall and transforms into a dragon (Kazuo 2012, 169). Mounted on a magnet hidden behind the waterfall's cascade, the fish "climbs automatically [...] transforms into a dragon, creates clouds, and then rises into the sky." Even Hosokawa's clouds are given gaseous form. The book instructs the machine's operator to trap tobacco smoke in a vacuum-sealed box. At the moment of the fish's transformation, the box's lid springs open, releasing a puff of acrid smoke. Thus, as Hosokawa concludes in his accompanying commentary, "when the dragon appears the dark cloud appears with it" (177).

Hosokawa's automata emerged from a moment of intellectual and cultural ferment. Exposure to

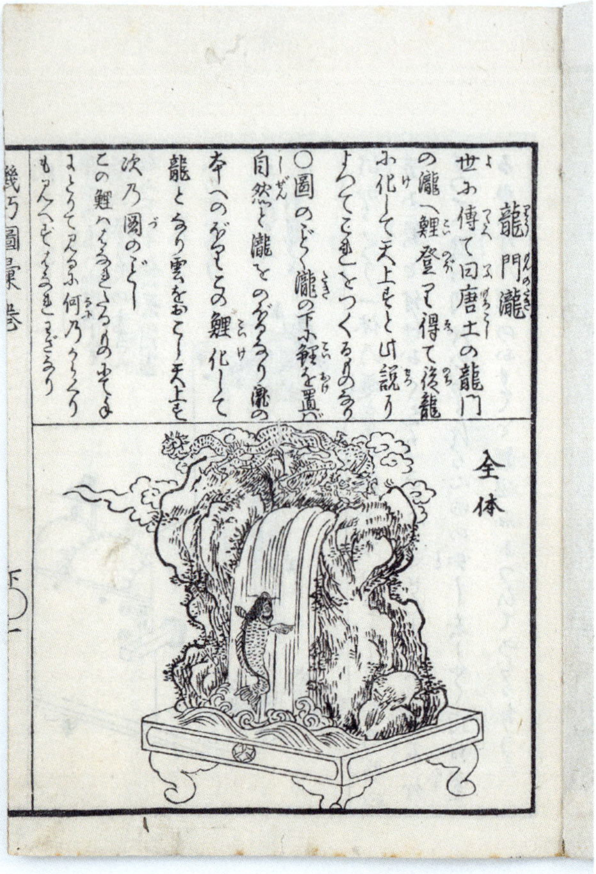

Figure 4 ↓ Hosokawa's metamorphic fish. By permission of Carnegie Mellon University Libraries, Special Collections (Call number TJ181 .H87 1808).

Western science — *rangaku*, or "Dutch learning"— in Edo Japan (1603–1867) led to innovations across the scientific disciplines. Rangaku scholars studied imported European texts, replicating and adapting foreign mechanical, medical, and scientific principles to fit local traditions. While this Western influence is apparent in Hosokawa's book, it appears in transmuted form: Baldi's automata reflect Renaissance humanism's impulse to recover and refine ancient

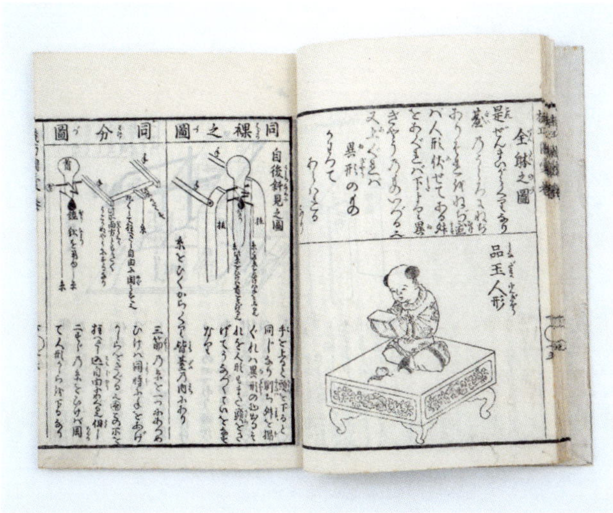

Figure 5 ↑ Karakuri mechanics. By permission of Carnegie Mellon University Libraries, Special Collections (Call number TJ181 .H87 1808).

knowledge, but Hosokawa's work blends indigenous culture and craft with new and foreign technology.

What Baldi's and Hosokawa's books show us is that the history of robotics is a great deal weirder and more wonderful than we might expect. These books suggest that, from the sixteenth to the nineteenth century, there was a period of sustained innovation that unfolded in parallel across cultures and continents. Yet early designs for automata did not always result in functioning prototypes, and the early modern world was not uniform in its embrace of self-moving machines. Rather, by the end of the seventeenth century, skepticism began to gain ground.

Offering a satirical case in point, Jonathan Swift's *Gulliver's Travels* (1726) describes a machine whose rotating parts generate new knowledge. Recounted as part of Lemuel Gulliver's fictional travelogue of his visit to the island of Laputa, the device — a full "twenty foot square" — comprises tiny "bits of wood"

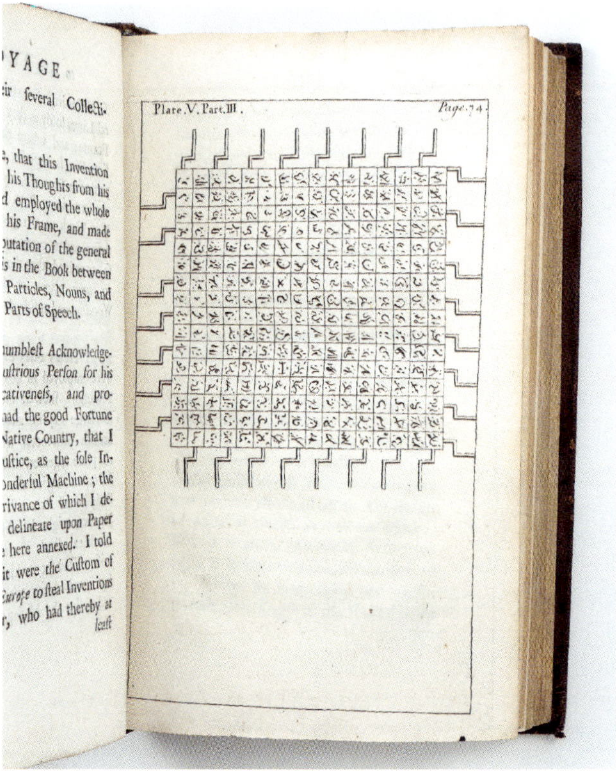

Figure 6 ↓ The Laputan text generator from Swift's *Gulliver's Travels*. By permission of Carnegie Mellon University Libraries, Special Collections (Call number R3724 .G7 1726).

that rotate on "slender wires." Carrying paper labels marked with individual words, these small wooden blocks are turned repeatedly by scholars to produce a nearly infinite number of word permutations and sentence fragments. When the scholars "found three or four words together that might make part of a sentence," they promptly recorded them into a growing number of large volumes. The "Professor," Gulliver's Laputan host tasked with oversight of the machine, suggests that this massive collection of "broken sentences" might one day be pieced together to form a "compleat body of all the arts and sciences" (Swift 1726, vol. 2, 73–75).

Besides delivering a joke at the expense of Britain's machine-obsessed Royal Society, Swift's Laputan ChatGPT seems to predict our own preoccupation with machines that claim to generate new knowledge (I'd argue that Swift's account offers the first intimation of a Large Language Model, or "LLM"). But Swift's tone suggests that he wouldn't have shared in our enthusiasm for these tools. His depiction of science-by-vending-machine instead ridicules mankind's misplaced reliance on technology. But it also marks something of a turning point. Following the seventeenth-century's feverish embrace of automatons and other mechanical conceits, Swift exposes the field's diminishing returns: what at first glance might appear to be a thinking collaborator turns out to be a box muttering random strings of words that only sometimes cohere into meaningful sentences.

By the eighteenth and nineteenth centuries, European automata had moved beyond practical or ritualistic functions into the realm of entertainment. Jacques de Vaucanson's Digesting Duck (1739) and Wolfgang von Kempelen's chess-playing Turk (1770) enthralled audiences with their performances. The duck flapped its wings, swallowed bits of proffered grain, and even "digested" food (the duck's counterfeit droppings evidently elicited some surprise). Fortunately, the waste-producing mechanism was an illusion: the nibbled grain remained undigested in the automaton's artificial stomach. The Turk, similarly, was not an autonomous thinker but a cleverly concealed hoax: a diminutive operator (the device was quite small) guided its moves from a compartment hidden beneath its chessboard.

The impulse to invent self-moving machines did not end with these Enlightenment-era deceptions, of course. While Special Collections mainly curates the prehistory of robotics, it also holds more recent works, including a first edition of Ivan Sutherland's *A Walking Robot* (1983).[3] The book features several illustrations of Sutherland's "Trojan Cockroach," a machine with an insectile frame and articulated legs designed to traverse uneven terrain. Despite its steampunk and lumbering appearance (figure 9 appears to show the machine leaking large quantities of fluid), the robot's hulking form maintained a startling equilibrium. Threaded through with joints, cables, and servomechanisms, the Cockroach's legs distributed the device's weight to maintain balance with remarkable efficiency. And though it was developed as part of cutting-edge locomotion research — it was an important precursor to the development of robotic bipedal motion — the machine's segmented anatomy and creeping gait evoke something more ancient. Resembling a mechanical beast that might have appeared in Baldi's *On Automata*, among Hosokawa's *karakuri*, or in one of Leonardo da Vinci's notebooks, Sutherland's robot encapsulates humankind's enduring fascination with "machines that move themselves."

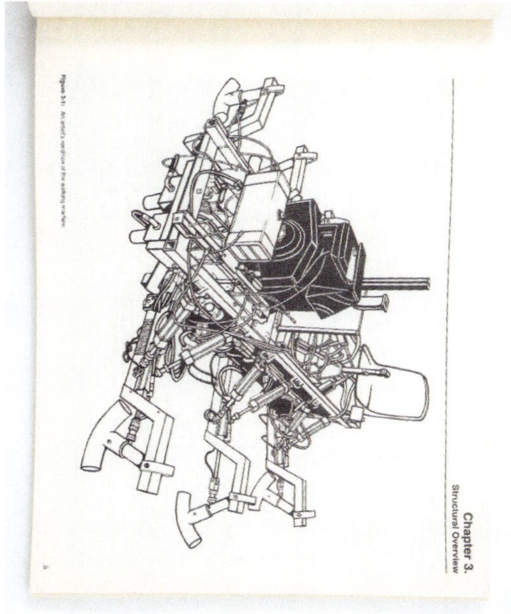

Figure 7 ↑ Ivan Sutherland's "Trojan Cockroach." From *A Walking Robot* (Pittsburgh: 1983). By permission of Carnegie Mellon University Libraries, Special Collections (Call number TJ211 .S9873 1983)

Figure 8 ↓ The “Trojan Cockroach” during a field test. By permission of the Carnegie Mellon University Archives.

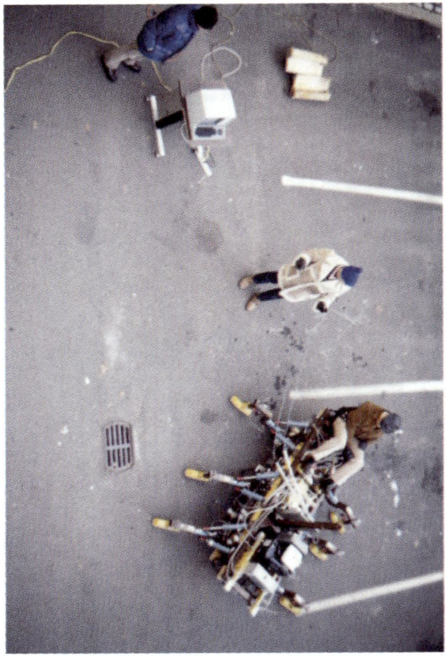

Figure 9 ↓ The “Trojan Cockroach” during a field test. By permission of the Carnegie Mellon University Archives.

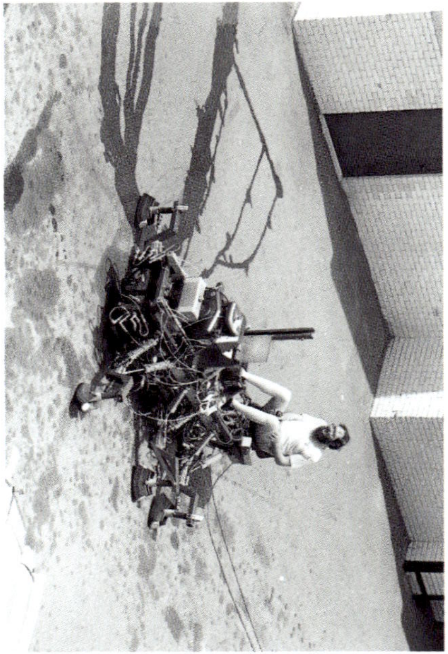

1 Borelli, Giovanni Alfonso. *De motu animalium*. Rome: Ex typographia Angeli Bernabo, 1680.

2 The work's title in Arabic is *Kitab fi ma'arifat al-hiyal al-handisaya* (كتاب في معرفة الحيل الهندسية).

3 See Carnegie Mellon University Archives' Robotics Project, which seeks to "preserve CMU's robotics legacy and contribute to a broader understanding of the people and technology that created the field." https://library.cmu.edu/university-archives/initiatives/robotics-project

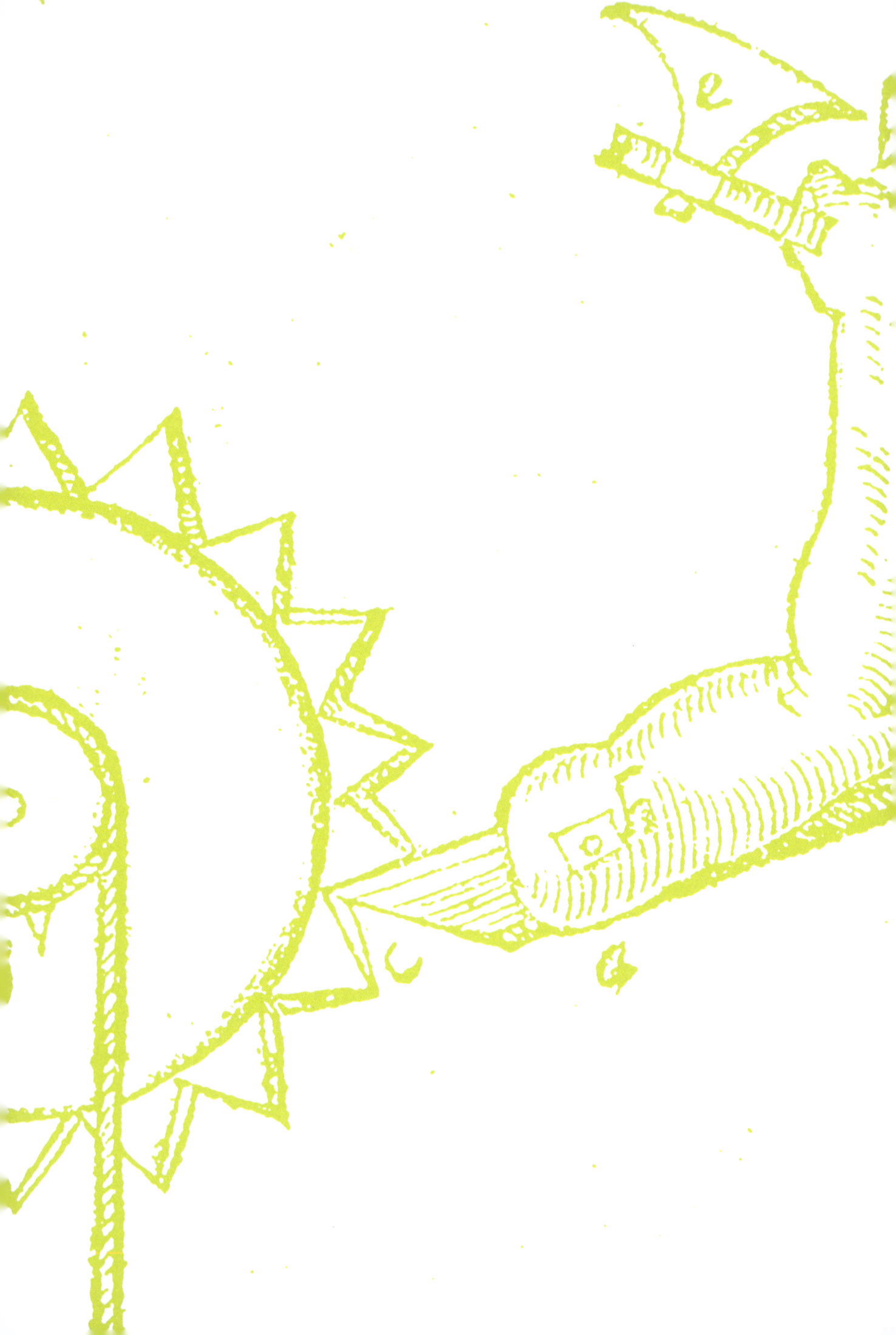
e
c
a

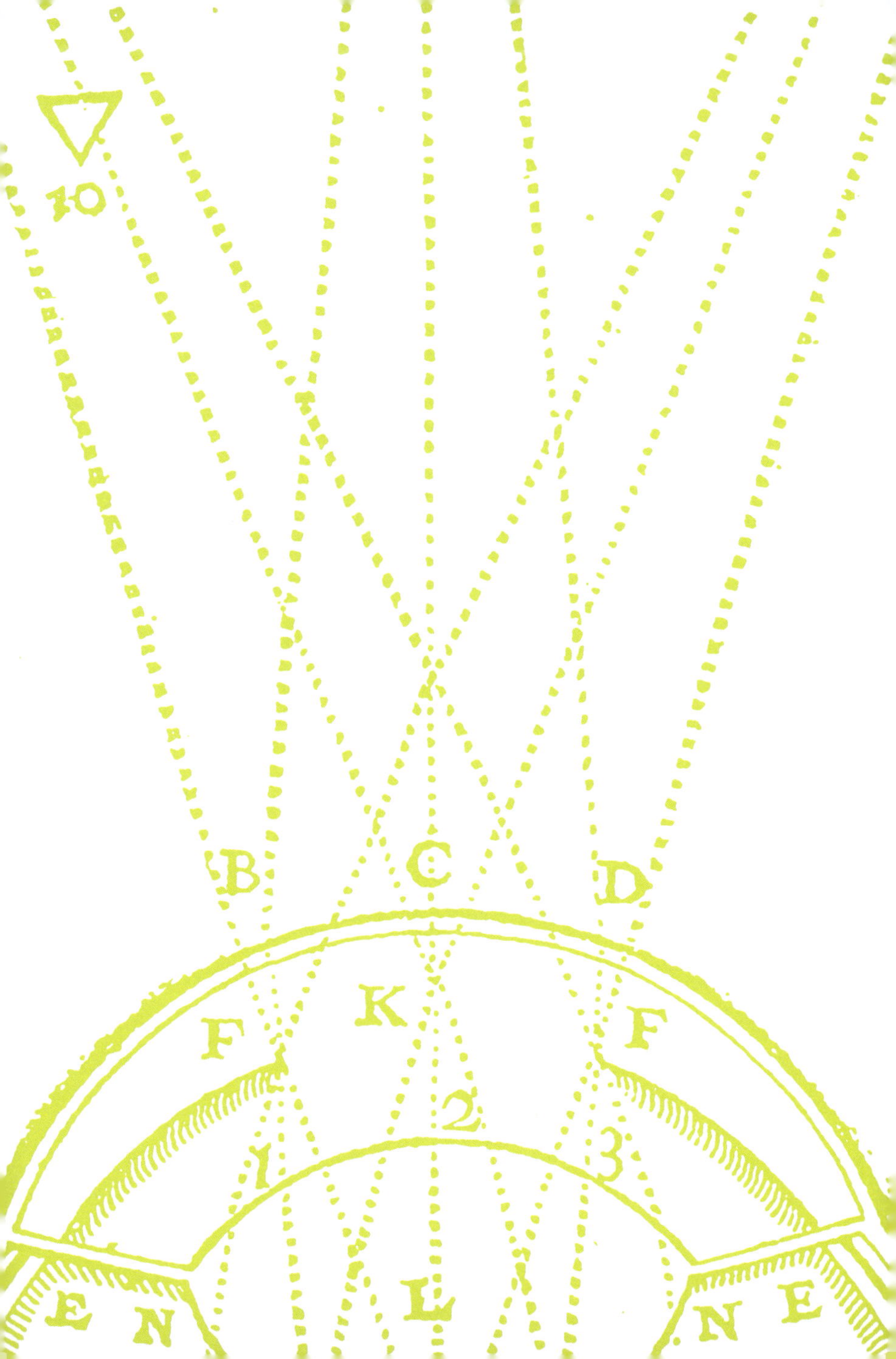

10
B
C
D
K
F
F
2
1
3
E N
L
N E

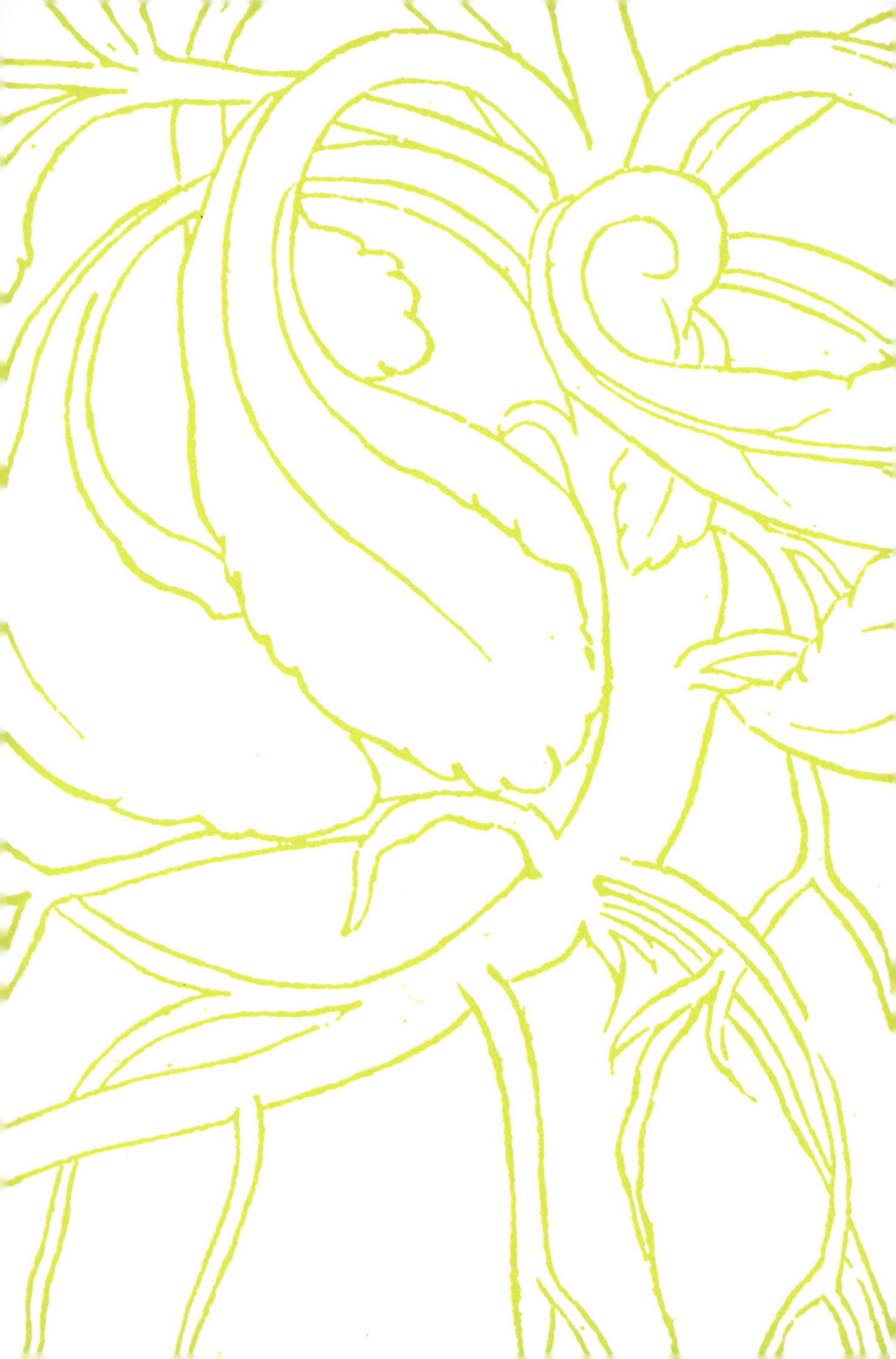

位法

用算盤已畢

手指指穴位

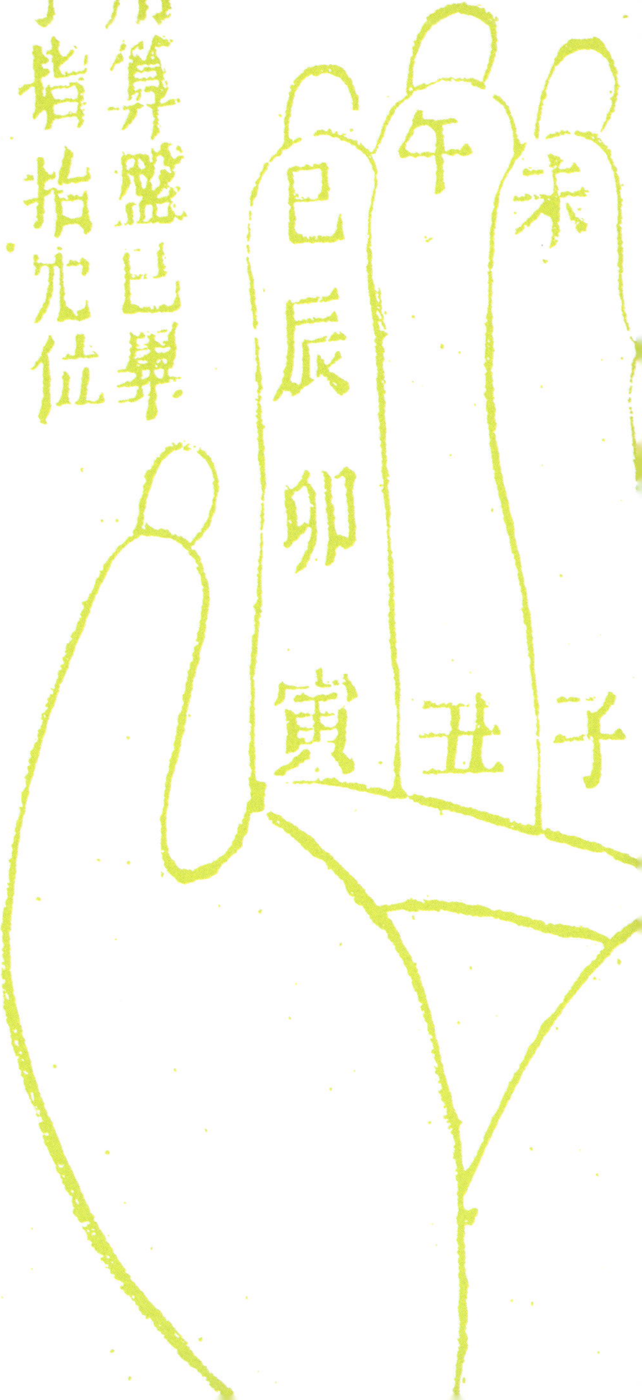

PART IV

Artificial Intelligence & Writing Machines

The dream of artificial or machine intelligence spans cultures and centuries. From a three-hundred-year-old fictional account of automatic text generation to postwar studies of computational heuristics, Special Collections holds a number of landmark works from the early history of AI.

SWIFT, Jonathan (1667–1745).
[Gulliver's] *Travels into Several Remote Nations of the World.* London, 1726.

Text generation before ChatGPT.

Jonathan Swift's *Gulliver's Travels* imagined a text-generating machine. An engraved illustration, shown here, depicts a device that randomly generates words to produce new knowledge. Swift's fictional account is a parody of humankind's misplaced trust in technology — a satirical portrait that is just as cutting today, three hundred years after its first publication.

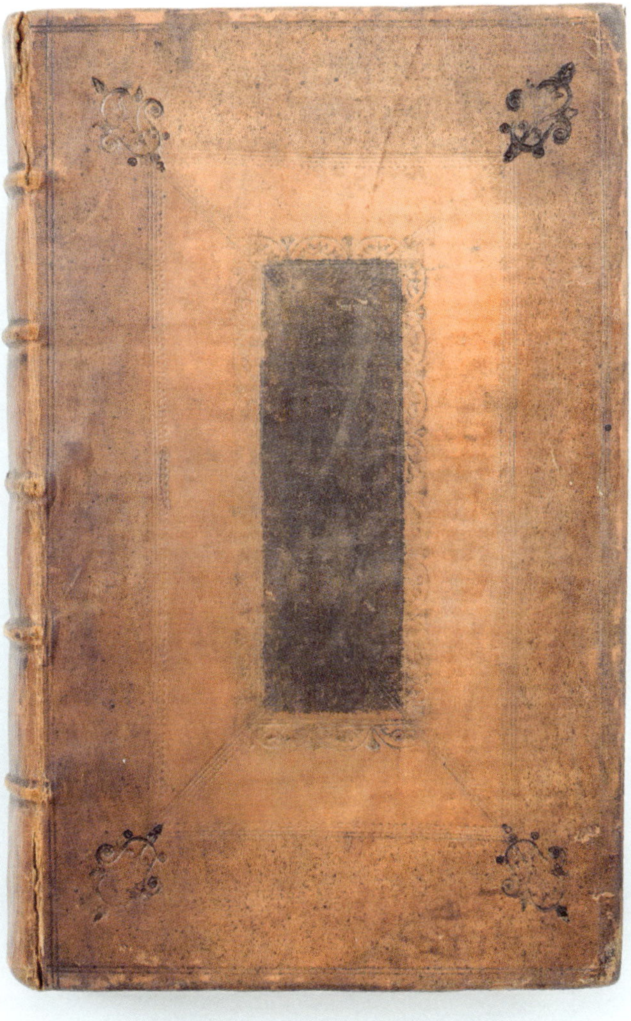

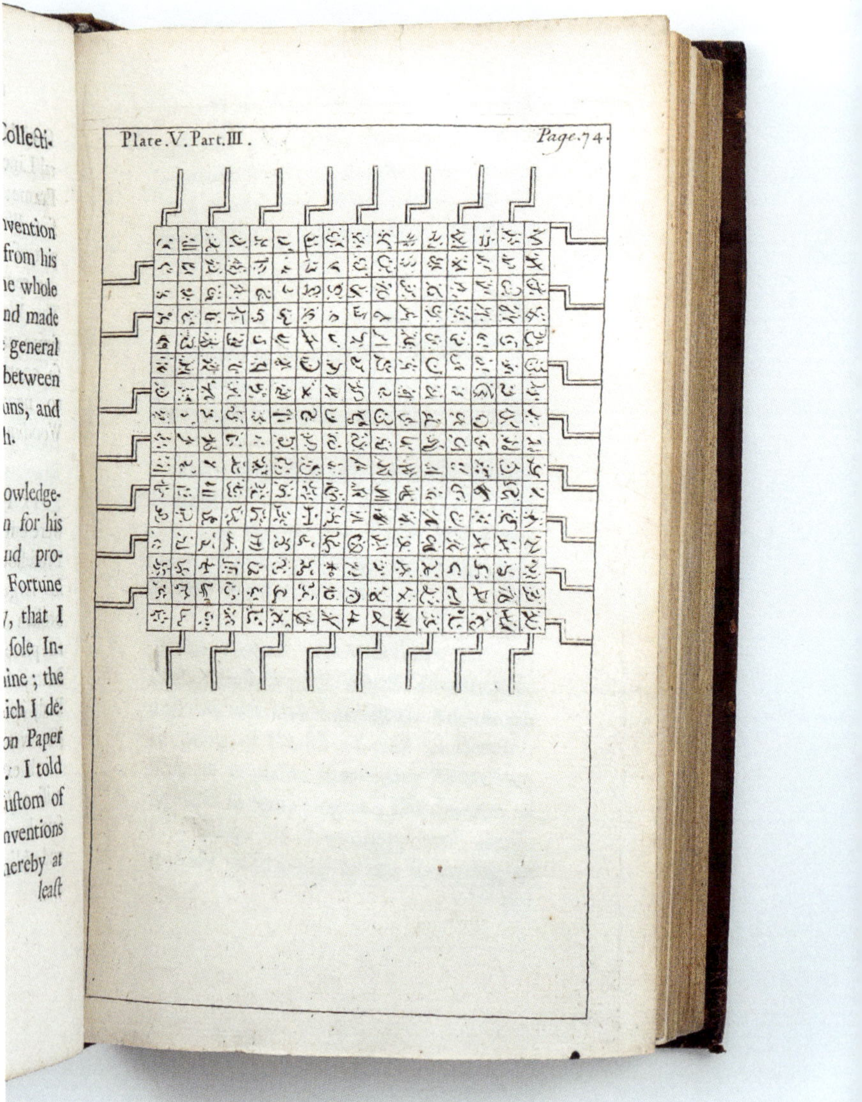
Plate.V.Part.III.
Page.74.

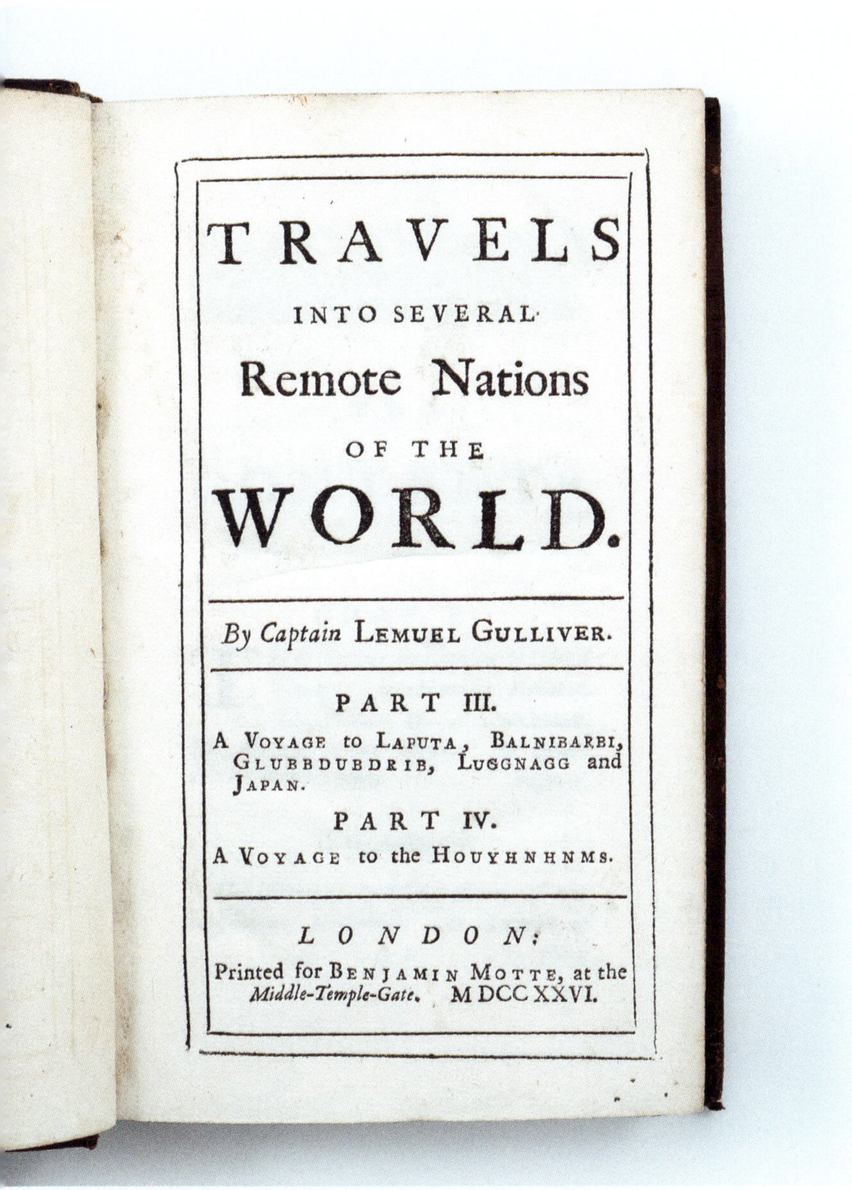

TRAVELS
INTO SEVERAL
Remote Nations
OF THE
WORLD.

By Captain LEMUEL GULLIVER.

PART III.
A VOYAGE to LAPUTA, BALNIBARBI, GLUBBDUBDRIB, LUGGNAGG and JAPAN.

PART IV.
A VOYAGE to the HOUYHNHNMS.

LONDON:
Printed for BENJAMIN MOTTE, at the *Middle-Temple-Gate.* MDCCXXVI.

SHELLEY, Mary (1797–1851).
Frankenstein; or, the Modern Prometheus.
London, 1818.

The first work of science fiction & a pathbreaking exploration of the ethics of AI.

Mary Shelley's *Frankenstein* remains one of the most influential works of English literature. It takes on added relevance in Special Collections, where it is shelved alongside works by Alan Turing and other pioneers of modern artificial intelligence. Shelley's "monster" originated the idea that intelligence can be made, though with unforeseeable and unnerving results. The book is far more than a chilling tale of reanimated flesh, then: it examines anxieties about artificial life and sentience that still linger, two centuries after its publication. Donated by Charles J. Rosenbloom, CMU's copy is one of the finest examples in existence and is preserved in its original paperboard binding.

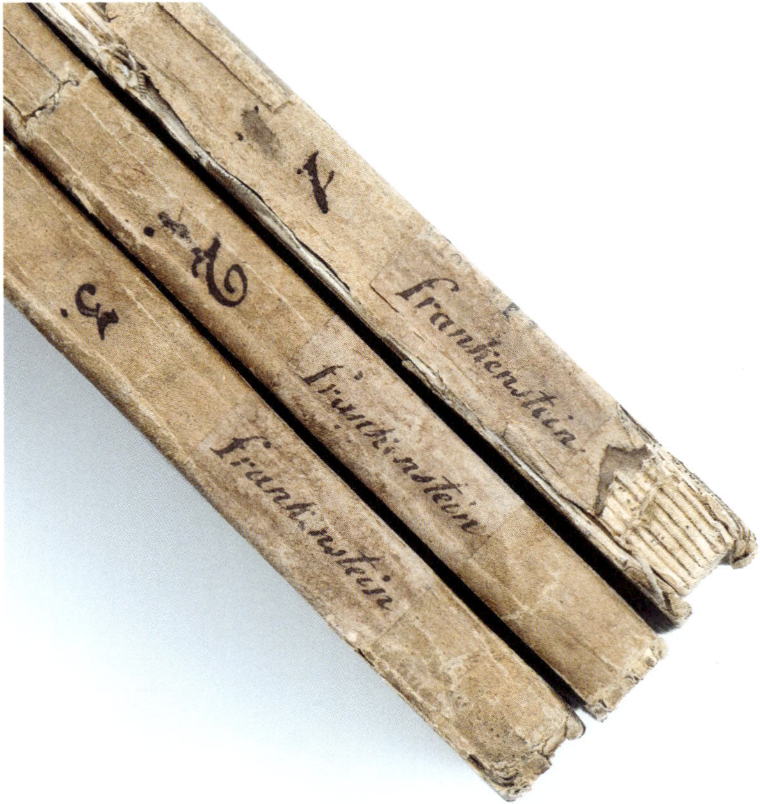

frankenstein
1
frankenstein
2
frankenstein
3

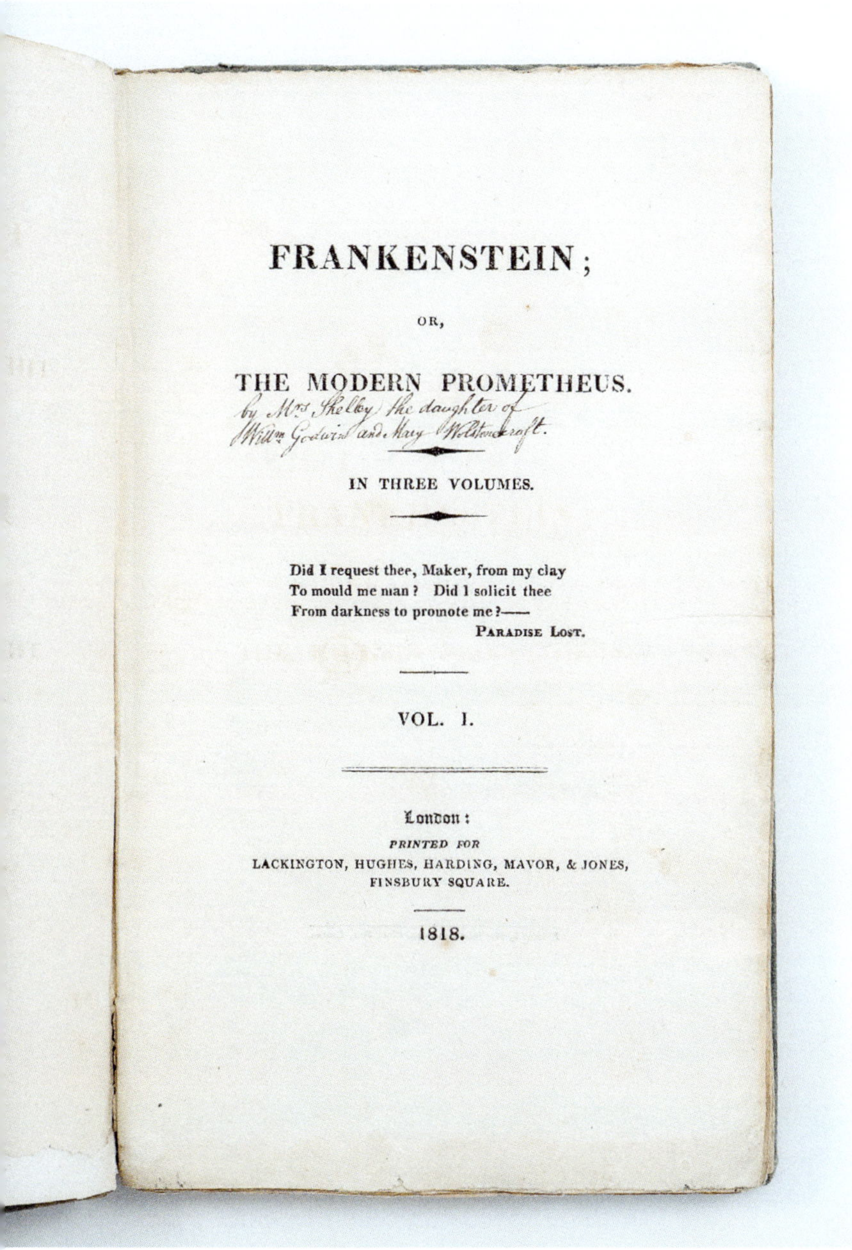

FRANKENSTEIN;

OR,

THE MODERN PROMETHEUS.

by Mrs Shelley the daughter of Willm Godwin and Mary Wolstonecraft.

IN THREE VOLUMES.

Did I request thee, Maker, from my clay
To mould me man? Did I solicit thee
From darkness to promote me?——
PARADISE LOST.

VOL. I.

London:
PRINTED FOR
LACKINGTON, HUGHES, HARDING, MAVOR, & JONES,
FINSBURY SQUARE.

1818.

SCHOTT, Gaspar (1608–1666).
Schola steganographica. Nuremberg, 1665.

The origin of machine translation.

Schott's *School of Secret Writing* describes a number of early cryptographic devices and linguistic machines. Shown here is the *Arca Glossotactica*, which translated common phrases into several languages. Designed by Jesuit scientist and scholar Athanasius Kircher (1602–1680), the device resembles a wooden computer: words and grammatical markers for tense, case, and number were stored behind sliding slats and drawers, ready for retrieval.

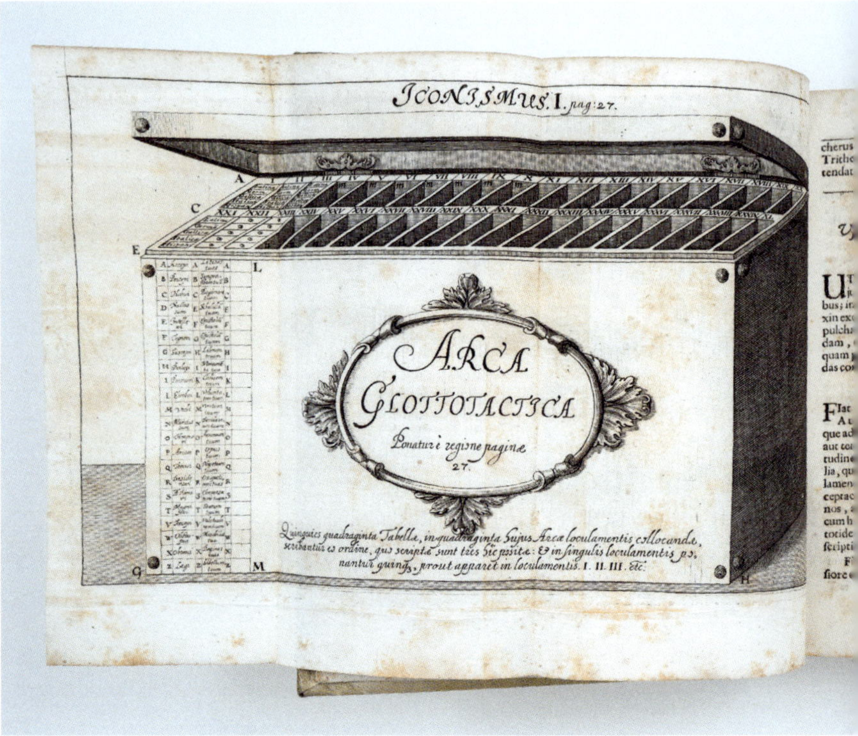
ICONISMUS. I. pag: 27.
ARCA GLOTTOTACTICA
Ponatur è regione paginæ 27.
Quinquies quadraginta Tabellæ, in quadraginta hujus Arcæ loculamentis collocandæ, scribantur eo ordine, quo scriptæ sunt tres hic positæ: & in singulis loculamentis ponantur quinque, prout apparet in loculamentis. I. II. III. etc.

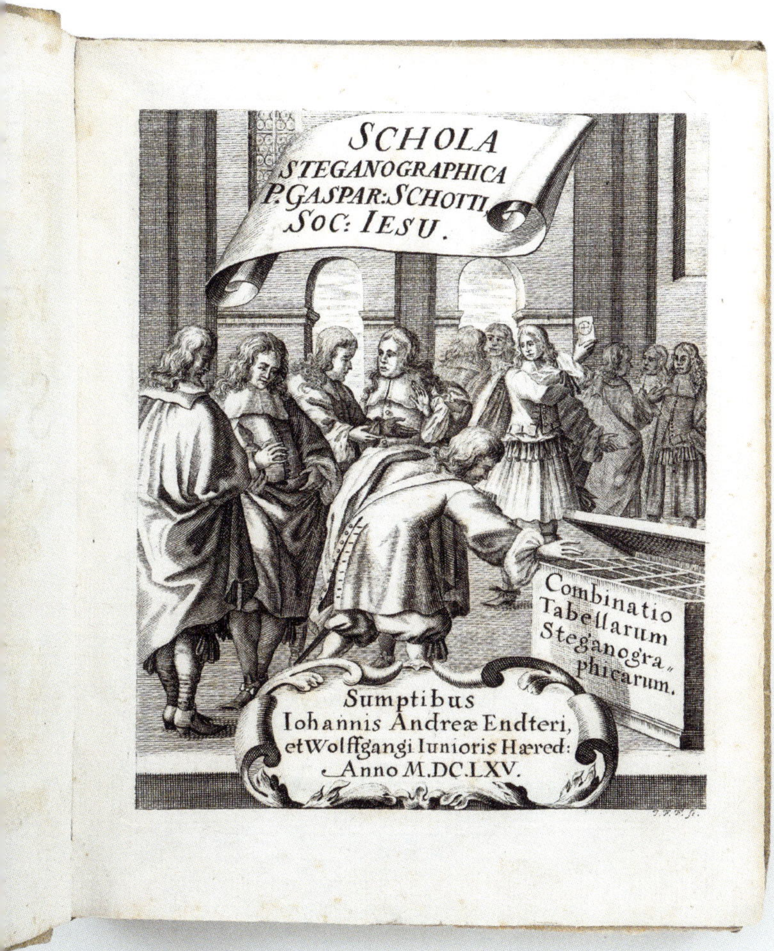
SCHOLA
STEGANOGRAPHICA
P. GASPAR: SCHOTTI,
SOC: IESU.
Combinatio
Tabellarum
Steganographicarum.
Sumptibus
Iohannis Andreæ Endteri,
et Wolffgangi Iunioris Hæred:
Anno M.DC.LXV.

BARDEEN, John (1908–1991) and
BRATTAIN, Walter Houser (1902–1987).
“Physical Principles Involved in Transistor Action.”
In *Bell System Technical Journal*, April 1949.

The invention of the solid-state transistor.

This paper outlined the theoretical basis of the solid-state transistor — a then-new device built using the semi-conductive metalloid, germanium. The invention marked a decisive moment in electronics and led to the replacement of failure-prone vacuum tubes with smaller, more reliable, and energy-efficient components. Published by Bell Labs in its *Technical Journal*, the paper helped launch the transistor era and ushered in the miniaturization of computing devices — a process that continues uninterrupted today.

large part of *holes*; that is of carriers of opposite sign to those normally in excess in the body of the semi-conductor.

The collector is biased in the reverse, or negative direction. Current flowing in the germanium toward the collector point provides an electric

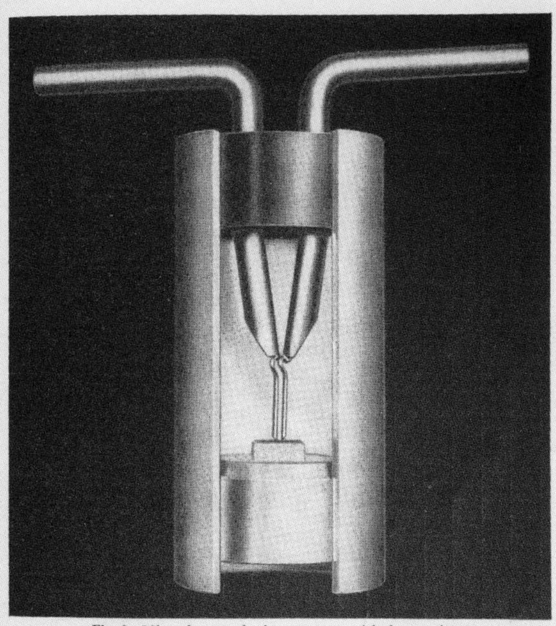

Fig. 2—Microphotograph of a cutaway model of a transistor

field which is in such a direction as to attract the holes flowing from the emitter. When the emitter and collector are placed in close proximity, a large part of the hole current from the emitter will flow to the collector and into the collector circuit. The nature of the collector contact is such as to provide a high resistance barrier to the flow of electrons from the metal to the semi-conductor, but there is little impediment to the flow of holes into

TURING, Alan (1912–1954).
"Computing Machinery and Intelligence,"
in *Mind*, October 1950.

The foundation of modern artificial intelligence and computational heuristics.

Turing's groundbreaking article "Computing Machinery and Intelligence" introduced the concept of heuristics, or "learning machines," to early AI. This was the idea that machines could acquire information by way of childlike observation. The article also introduced a thought experiment known today as the Turing Test, but which Turing called The Imitation Game. The Game was simple: a participant exchanged typed interactions with two respondents, a computer and a human being. After a set period of time, if the interrogator failed to distinguish one from the other, the computer would win. Such a machine could be said to think.

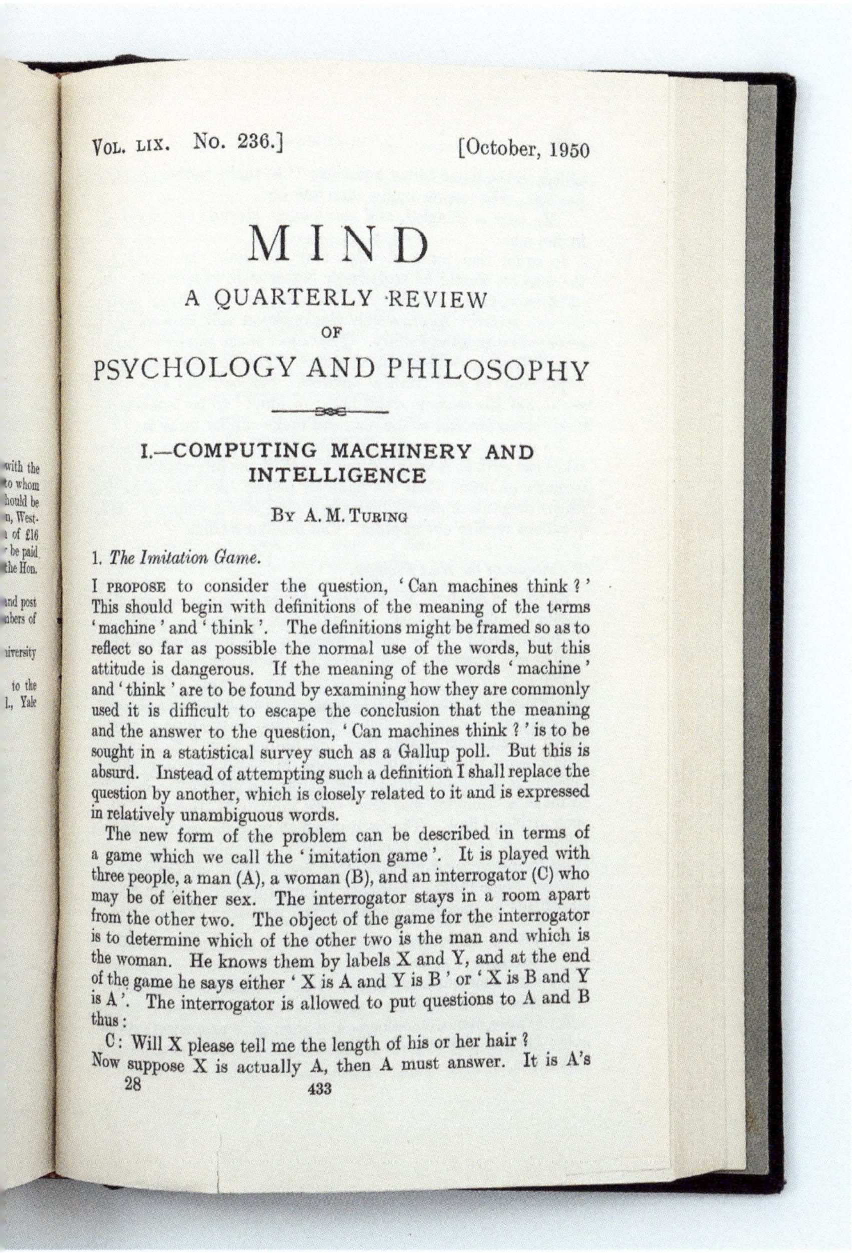

VOL. LIX. No. 236.] [October, 1950

MIND

A QUARTERLY REVIEW
OF
PSYCHOLOGY AND PHILOSOPHY

I.—COMPUTING MACHINERY AND INTELLIGENCE

BY A. M. TURING

1. *The Imitation Game.*

I PROPOSE to consider the question, 'Can machines think?' This should begin with definitions of the meaning of the terms 'machine' and 'think'. The definitions might be framed so as to reflect so far as possible the normal use of the words, but this attitude is dangerous. If the meaning of the words 'machine' and 'think' are to be found by examining how they are commonly used it is difficult to escape the conclusion that the meaning and the answer to the question, 'Can machines think?' is to be sought in a statistical survey such as a Gallup poll. But this is absurd. Instead of attempting such a definition I shall replace the question by another, which is closely related to it and is expressed in relatively unambiguous words.

The new form of the problem can be described in terms of a game which we call the 'imitation game'. It is played with three people, a man (A), a woman (B), and an interrogator (C) who may be of either sex. The interrogator stays in a room apart from the other two. The object of the game for the interrogator is to determine which of the other two is the man and which is the woman. He knows them by labels X and Y, and at the end of the game he says either 'X is A and Y is B' or 'X is B and Y is A'. The interrogator is allowed to put questions to A and B thus:

C: Will X please tell me the length of his or her hair?

Now suppose X is actually A, then A must answer. It is A's

28 433

FERRANTI.
Faster Than Thought:
The Ferranti Nimrod Digital Computer.
Manchester, 1951.

The first computer purpose-built for gaming and the first manual for a computer game.

Introduced at the 1951 Festival of Britain, the Nimrod was a computer programmed to play the mathematical strategy game, *Nim*. Designed by John Makepeace Bennett and built by Ferranti Ltd. in Manchester, UK, the computer relied on early artificial intelligence. Alan Turing famously defeated the Nimrod at the Festival and in the same year wrote an article on "Digital Computers Applied to Games."

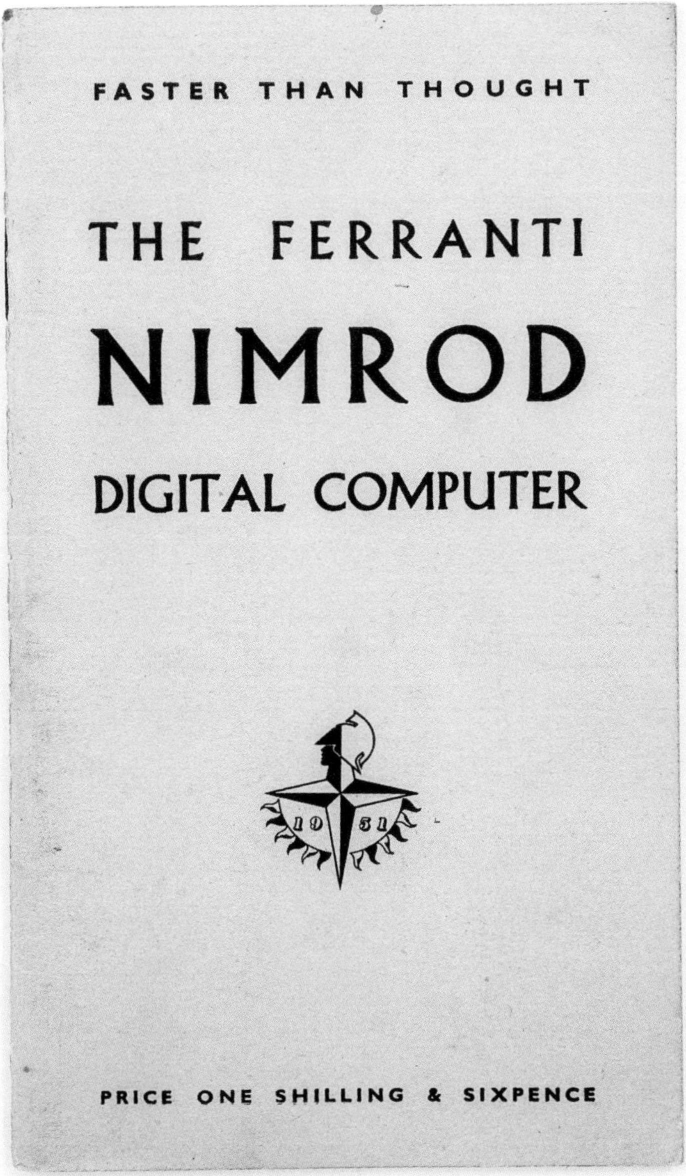

FASTER THAN THOUGHT

THE FERRANTI

NIMROD

DIGITAL COMPUTER

PRICE ONE SHILLING & SIXPENCE

Artificial Intelligence, Lockean Minds, & Cartesian Machines: The Enlightenment Roots of Machine Learning

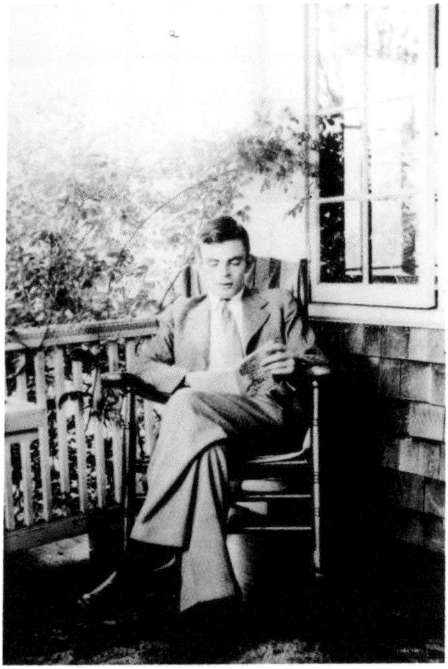

Figure 1 ↑ Alan Turing with a book, approximately 1936. By permission of the Institute for Advanced Study, Princeton University. Copyright holder unidentified.

In 1950 the British mathematician Alan Turing posed a question for research: “Can machines think?” The question — disquieting, paradoxical, a little tongue-in-cheek — opens Turing’s paper, “Computing Machinery & Intelligence,” published in the October 1950 issue of the scientific journal *Mind* (see “TURING,” page 215). Now a landmark text in the history of computing, the paper offered a sort of manifesto on artificial intelligence, six years before the term was coined.[1]

When “Computing Machinery & Intelligence” first appeared in print, Turing was Deputy Director at the University of Manchester’s Computing Laboratory,

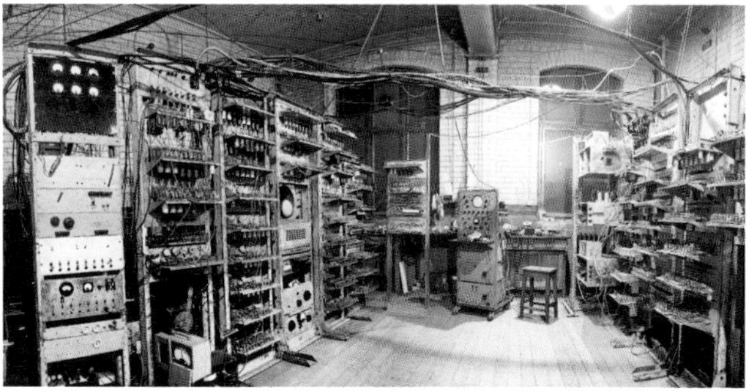

Figure 2 ↘ The Manchester "Mark 1," 1948.
By permission of the John Rylands Research Institute and Library, University of Manchester.

where he worked on programming the Manchester Mark 1, an early stored-program computer that occupied a towering row of steel-framed racks on the ground floor of the University's Electrotechnics building (figure 2). Turing's Manchester period was marked by writing of breathtaking inventiveness and a prevailing interest in machine intelligence. In 1948 he wrote a report on "Intelligent Machinery," which was followed by a string of publications and projects on the subject: in 1951, after witnessing the Nimrod computer at the Festival of Britain (see "FERRANTI," page 219), Turing wrote an article on "Digital Computers Applied to Games," created a chess-playing computer program, participated in a BBC program on the question "Can Digital Computers Think?," and delivered a paper titled "Intelligent machinery, a heretical theory" to Manchester's 51 Society.

Turing's most innovative and far-seeing essay from this period, "Computing Machinery & Intelligence," set out a thought experiment now known as the Turing Test (Turing called it "The Imitation Game"). The game's premise was simple: a participant would engage in a series of typed interactions with two hidden respondents — one human, the other circuit-bound. If, after a set period of time, the interrogator failed

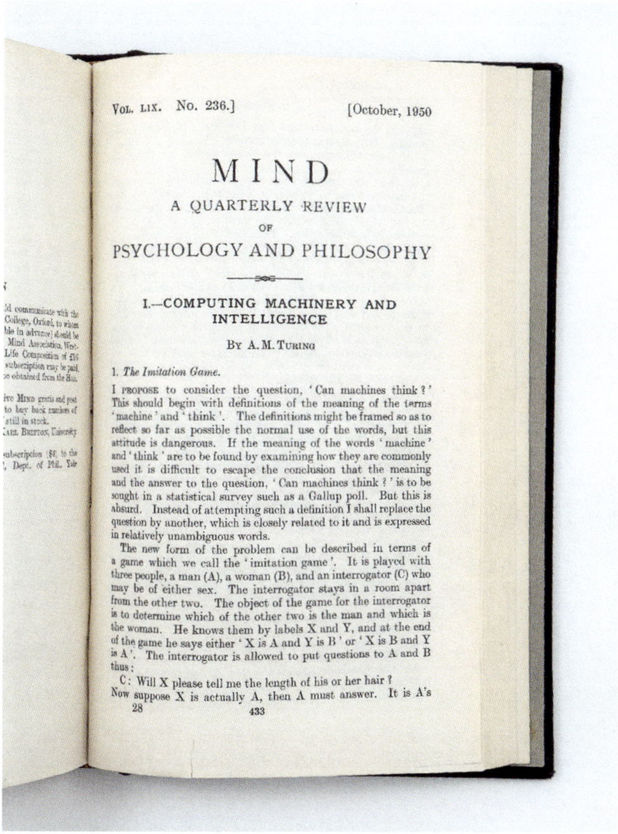
Vol. LIX. No. 236.] [October, 1950

MIND

A QUARTERLY REVIEW

OF

PSYCHOLOGY AND PHILOSOPHY

I.—COMPUTING MACHINERY AND INTELLIGENCE

By A. M. Turing

1. *The Imitation Game.*

I propose to consider the question, 'Can machines think ?' This should begin with definitions of the meaning of the terms 'machine' and 'think'. The definitions might be framed so as to reflect so far as possible the normal use of the words, but this attitude is dangerous. If the meaning of the words 'machine' and 'think' are to be found by examining how they are commonly used it is difficult to escape the conclusion that the meaning and the answer to the question, 'Can machines think ?' is to be sought in a statistical survey such as a Gallup poll. But this is absurd. Instead of attempting such a definition I shall replace the question by another, which is closely related to it and is expressed in relatively unambiguous words.

The new form of the problem can be described in terms of a game which we call the 'imitation game'. It is played with three people, a man (A), a woman (B), and an interrogator (C) who may be of either sex. The interrogator stays in a room apart from the other two. The object of the game for the interrogator is to determine which of the other two is the man and which is the woman. He knows them by labels X and Y, and at the end of the game he says either 'X is A and Y is B' or 'X is B and Y is A'. The interrogator is allowed to put questions to A and B thus:

C: Will X please tell me the length of his or her hair ?

Now suppose X is actually A, then A must answer. It is A's

28 433

Figure 3 ↑ Title page, "Computing Machinery & Intelligence." By permission of Carnegie Mellon University Libraries, Special Collections (Call number B1 .M65).

to distinguish between the two, the computer would, in effect, win. Having performed a convincing imitation of human cognition and written speech, such a machine could be said to think.

Turing's paper prefigured the now routine interactions we have with our own intelligent machines. Even if most of our devices have not yet attained a human standard of intelligence, we imbue them with personalities that give the impression of mindful behavior: Siri's chirpy replies veil code of deep complexity, her

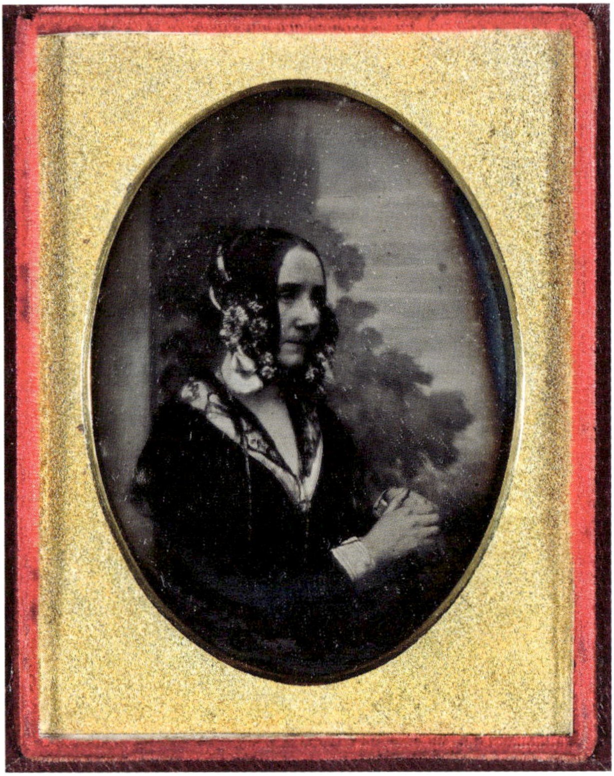

Figure 4 ↓ Augusta Ada King, Countess of Lovelace. Daguerreotype by Antoine Claudet, approximately 1843. Public domain.

phrasing designed to mimic surprise, humor, and empathy; the chatbot Claude (the AI startup Anthropic's entrant in the global race for increasingly powerful large language models) is praised for its humanlike qualities, including — unnervingly — its "willingness to express opinions" (Roose 2024).

Besides advancing an elusive benchmark by which to judge the purported intelligence of machines, Turing's paper offered a new theory of algorithmic learning. In one particularly compelling section, "Learning Machines," Turing hinted at how computers might be programmed to acquire and assimilate information rather than merely retrieve what they are instructed to know. This was the introduction of heuristics to the field of artificial intelligence — the idea that automated decision making could be based on training rather than exhaustive programming. In adopting this approach, Turing broke from older, more abstract notions of artificial thought: if machine intelligence is a problem to be solved, he suggested, the solution lies in the methodical education of digital minds.

Here Turing was writing against a narrow view of machine aptitude that in 1950 already had a long history: Ada Lovelace (1815–1852), writing in 1843, argued that a machine performs only what "we know how to order it to perform. It can follow analysis; but it has no power of anticipating any analytical relations or truths. Its province is to assist us in making available what we are already acquainted with" (Menabrea and Lovelace 1843, 722).

Lovelace's machine is little more than an automaton, capable only of parroting its maker's programming — a formidable device, to be sure, but one that performs only by rote. Lovelace, who worked on programming Charles Babbage's analytical engine, is credited with authoring the earliest machine algorithm: coded instructions for calculating Bernoulli Numbers, a sequence of rational numbers important in number theory and combinatorics (Menabrea and Lovelace 1843, 722–731). Intended for execution by a machine, Lovelace's algorithm was essentially the first computer program. Drawing on this experience, Lovelace objected to the idea that calculating machines might deliver anything resembling human thought. They are predictable by design, incapable of spontaneity.

The roots of Lovelace's objection are older still. Writing in 1637, the French scientist and mathematician René Descartes (see "DESCARTES," page 313) theorized that a machine might be programmed to react to a variety of prompts, but that such a machine would never attain to or even reliably imitate human reasoning or speech:

> For we can well conceive of a machine being so made that it utters words, and even that it utters words appropriate to the bodily actions that will cause some change in its organs (such as, if one touches it in certain places, it asks what one wants to say to it, or, if in another place, it cries out that one is hurting it, and the like). But it could not arrange its words differently so as to respond to the sense of all that will be said in its presence, as even the dullest men can do (Descartes 1999, 32).

Descartes' description of a machine that responds to touch is an astonishingly early account of the Turing Test, even if the French mathematician found the idea of a fully responsive and articulate mechanical mind absurd. Also remarkable is the fact that Descartes' "dullest men" could be replaced by Turing's more generous category, children. But the same applies: three centuries before Turing, Descartes viewed the idea of a machine capable of answering basic questions, let alone complex dialog, as a kind of technological folly; all machines, Descartes concluded, are doomed to lose Turing's game.

This wariness of artificial life seems to work against Descartes' own view that all bodily functions could be reduced to purely mechanical (or hydraulic, or pneumatic) processes — each organ driven by the equivalent of "counterweights and wheels" (Riskin 2016, 46). Human anatomy was, according to Descartes, an intricate, composite machine made up of even smaller machines. One story, almost certainly apocryphal, has it that Descartes himself fashioned an automaton — "a mechanical replica of his illegitimate daughter, Francine, who had died [...] at the age of five" (Maisano 2007, 63). But for Descartes the boundary between mechanical body and rational soul was inviolable: while an automaton might perform the material and mechanical functions of the body, speech, emotion, and cognition all required an incorporeal, vital soul.

In "Computing Machinery & Intelligence," Turing reformulated Lovelace's and Descartes' objection in a playful mood, noting that their argument amounted to a basic premise: that "a machine can never 'take us by surprise'" (450). Citing his own experience,Turing asserted the opposite view, claiming that "machines take me by surprise with great frequency" (450). Already in 1950, Turing saw a future in which digital brains might prove Lovelace and Descartes wrong, exhibiting a sort of intelligence that was not only quick-witted but unpredictable and even charismatic.

But the capacity to surprise is not everything. And Turing arguably overlooked the importance of consciousness, the object that Descartes and Lovelace were at pains to address. Can a machine lay claim to the human faculty of intelligence without being self-aware (without, that is, "the baggage of consciousness and subjective experience," as Blaise Agüera y Arcas recently puts it [2025, "Interlude: The Prehistory of Computing"])? This should not suggest that Turing viewed consciousness as trifling — "I do not wish to give the impression that I think there is no mystery about consciousness," he wrote — but he ultimately evaded the question by eliminating it from the scope of his work: "I do not think these mysteries necessarily need to be solved before we can answer the question with which we are concerned in this paper" (447).

Enraptured by visions of mischievous machine minds, Turing confronted a paradox of artificial intelligence: if a machine models human thought to a degree that makes it indistinguishable from the real item, what necessary proof remains? The way out of the paradox, Turing realized, required a rejection of how Descartes, Lovelace, and others had defined a thinking machine. Following the line of argument set out by Lovelace and quoted above, Turing allowed that a machine might theoretically be built with sufficient memory to hold a universe of knowledge — Lovelace's machine assistant that "[makes] available what we are already acquainted with." But even if this theorized machine were to succeed in executing its omniscient programming, Turing agreed that it would inevitably fall into Lovelace's and Descartes' trap: forever echoing what it already knows, hemmed in by its coded expertise. Turing instead embraced the disorder of experience:

> Instead of trying to produce a programme to simulate the adult mind, why not rather try to produce one which simulates the child's? If this were then subjected to an appropriate course of education one would obtain the adult brain. Presumably the child brain is something like a notebook as one buys it from the stationer's. Rather little mechanism, and lots of blank sheets [...]. Our hope is that there is so little mechanism in the child brain that something like it can be easily programmed." (456)

Turing's machine was as much a product of fairy tale as science — a "child machine" made real by way of algorithmic enchantment. Such a machine, Turing noted, will doubtless "make mistakes at times" but also charm us with "new and very interesting

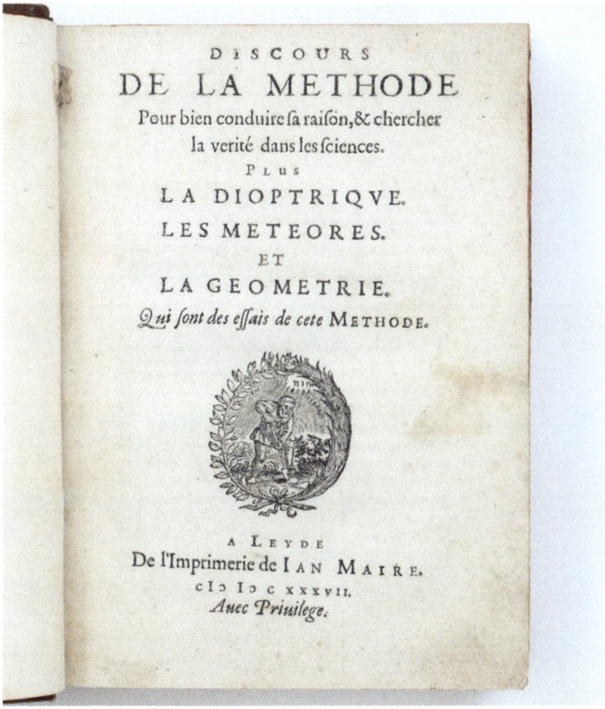

DISCOURS
DE LA METHODE
Pour bien conduire ſa raiſon, & chercher
la verité dans les ſciences.
PLUS
LA DIOPTRIQVE.
LES METEORES.
ET
LA GEOMETRIE.
Qui ſont des eſſais de cete METHODE.

A LEYDE
De l'Imprimerie de IAN MAIRE.
CIↃ IↃ C XXXVII.
Auec Priuilege.

Figure 5 ↑ René Descartes, *Discours de la méthode* (1637). By permission of Posner Memorial Collection, Carnegie Mellon University Libraries, Special Collections (Call number Q155 .D43).

statements" (Turing 1951, 2 [97]). Impressionable, spontaneous, adaptive, and imperfect, such a machine could surprise us and even model our own impulse to form conceptual frameworks with which to investigate and describe the world.

Turing's ideas owe much to another seventeenth-century theorist of mind. In the 1690 *Essay Concerning Humane Understanding* (see "LOCKE," page 521), English philosopher John Locke likened a child's brain to a ream of unmarked paper:

> Let us then suppose the Mind to be, as we say, white Paper, void of all Characters, without any Ideas; How comes it to be

> furnished? Whence comes it by that vast store, which the busy and boundless fancy of Man has painted on it, with an almost endless variety? Whence has it all the materials of Reason and Knowledge? To this I answer, in one word, From Experience: In that, all our Knowledge is founded; and from that it ultimately derives itself. Our Observation employ'd either about external, sensible Objects; or about the internal Operations of our Minds, perceived and reflected on by our selves, is that, which supplies our Understandings with all the materials of thinking. These two are the Fountains of Knowledge, from whence all the Ideas we have, or can naturally have, do spring (Locke 1690, 37).

Turing's "notebook [...] of blank sheets" resorts to Locke's papery metaphor: far from a preformed (or preprogrammed) and coherent whole, human and machine minds require a process of induction into the infinite variety of *experience* (the first edition of Locke's *Essay* emphasizes the word in italics). Only by way of this piecemeal, experiential learning is the mind furnished with "all the materials of thinking" — a Lockean turn of phrase that seems to anticipate Turing's writing on the subject.

Seventy-five years after their first publication, Turing's ideas seem doubly quaint: our machines have surpassed Turing's wildest imaginings, while his innocent account of a knowledge-hungry artificial brain troubles as much as it inspires. While we might want our digital assistants — ChatGPT and its kind — to surprise us with their brilliance and alacrity, our appetite for the unexpected ends there. If pressed, most of us would admit to a preference

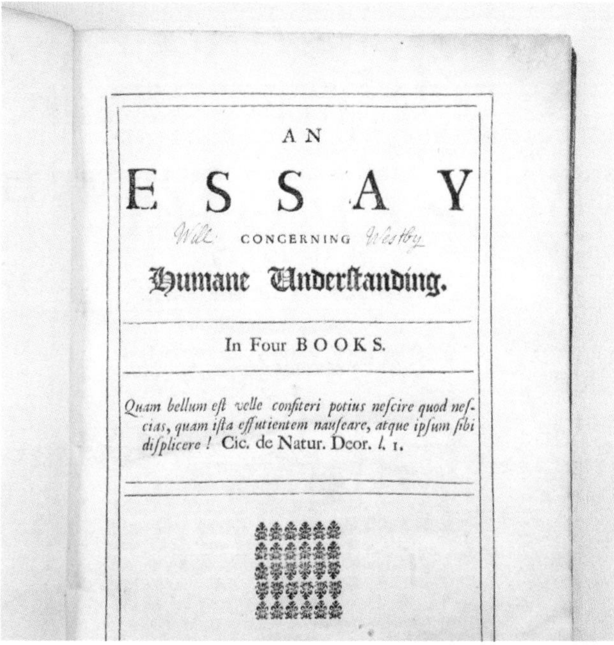

AN
ESSAY
CONCERNING
Humane Understanding.

In Four BOOKS.

Quam bellum est velle confiteri potius nescire quod nescias, quam ista effutientem nauseare, atque ipsum sibi displicere! Cic. de Natur. Deor. *l.* 1.

Figure 6 ↑ John Locke, *An Essay Concerning Humane Understanding* (1690). By permission of Carnegie Mellon University Libraries, Special Collections. (Call number B1290 1690).

for Lovelace's code-bound assistant rather than Turing's inquisitive child machine. In the intervening decades, too, we have come to rely on our machines to a degree that even Turing could not have predicted. They are entangled in the fabric of our lives, ensuring our society's dependable humming. Entire economies are now networked, raising the possibility of machine-driven societal collapse. We are, according to dire predictions that appear with some regularity in the periodical press, one solar storm away from a new dark age. But we also have a more mature appreciation for the potential for harm that comes with introducing error-prone, unfettered, or biased algorithms into public discourse.

Yet Turing's account of a childlike and expansive AI still offers an alluring, if untaken path. It was also

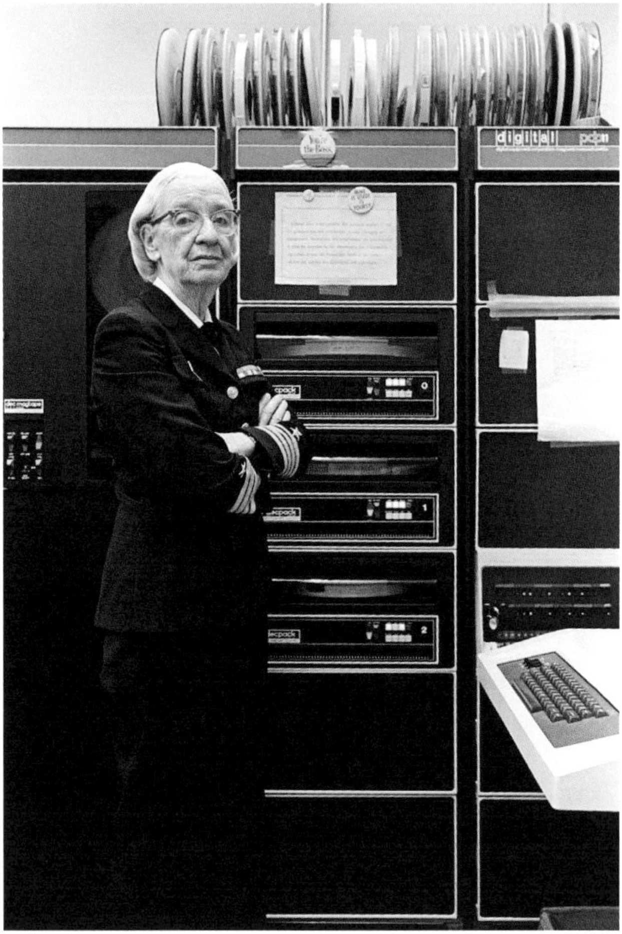

Figure 7 → Grace Murray Hopper in her office in Washington DC, 1978. Photograph by Lynn Gilbert. By permission of Lynn Gilbert.

influential: Grace Murray Hopper (1906–1992), American mathematician and computer scientist, wrote a 1952 paper titled "The Education of a Computer," in which she described the mathematical "education" of the UNIVAC — the hulking general purpose digital computer built by the Eckert-Mauchly Computer Corporation in 1951 — as "fully equivalent to that of a college sophomore" (Hopper 2021 (1952),

178). Turing was not alone in anthropomorphizing learning machines, then. Meanwhile, heuristics now forms the basis of current research in machine learning — an outcome Turing all but predicted.

The legacy of Turing's writing on machine intelligence is clouded by his death by apparent suicide in 1954. This followed his forced sterilization after he was convicted for homosexuality under England's regressive sexual morality laws (Hodges 1983, 456–475). What if Turing had lived to continue his line of inquiry? While this is a tragically unresolvable question, Turing's ideas remain ripe for revival and reassessment.

⋆ ⋆ ⋆

Special Collections holds a first edition of Turing's "Computing Machinery and Intelligence." Until 2022 Special Collections' copy was in circulation, available for withdrawal from the Libraries' off-site shelving facility. Rare in its own right, the issue of the journal that contains Turing's paper was transferred to Special Collections, where it is shelved under secure and environmentally controlled conditions and made available for research and instruction.

The rediscovery of CMU's copy of Turing's paper was the result of an ongoing project that seeks to identify and withdraw from circulation copies of rare and important works in the history of computing. So far the project has flagged more than one hundred titles for transfer to Special Collections, including works by Grace Murray Hopper, Claude Shannon, and Alonso Church. Now housed in the Posner Center, Turing's paper joins the Traub-McCorduck Collection — an important group of books and calculating devices donated in memory of Joseph Traub (1932–2015), a pioneer in information-based complexity and a former head of CMU's School of Computer Science (see "TRAUB-McCORDUCK COLLECTION," page 293). Pamela McCorduck (1940–2021), who gifted the collection to CMU Libraries, echoed Turing in the title of her book about the history of AI, *Machines Who Think* (1979). Like Turing, McCorduck viewed machine intelligence as a form of charismatic, if artificial life — more like a being *who* thinks than a machine *that* calculates. The Traub-McCorduck collection aptly holds two examples of the Enigma cipher machine, which Turing helped decode at Bletchley Park (see "ENIGMA," page 129).

1 John McCarthy (1927–2011) is credited with inventing the term "artificial intelligence" in advance of the 1956 Dartmouth Summer Research Project on Artificial Intelligence.

Uor t
les esse
nores
gere d
partē e
contin
non se
ter dif
sas du
¶ Pl
ppend
nē circuli cōtinēs corda
icupat. ¶ Angulus aūt p
Supra arcū angulus con

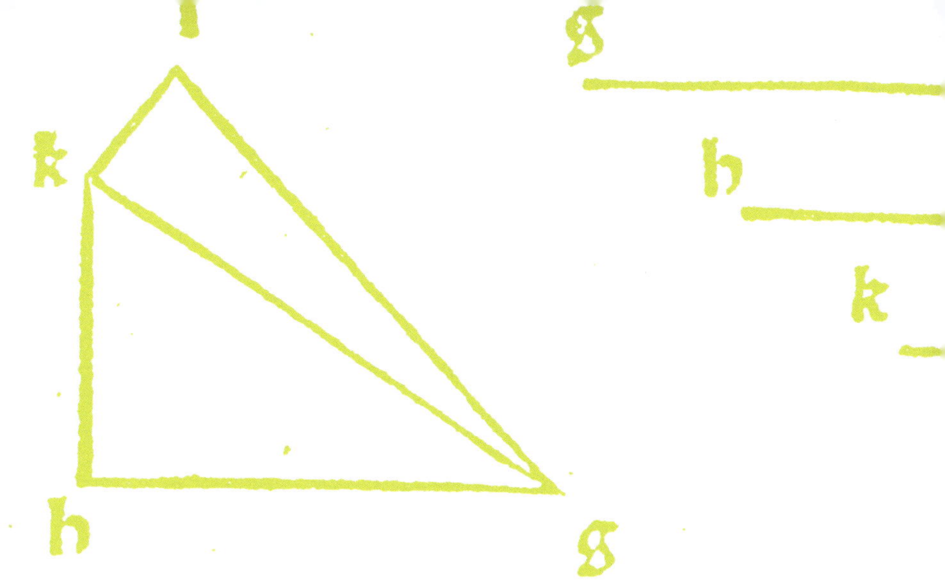
l
g
k
h
k
h
g

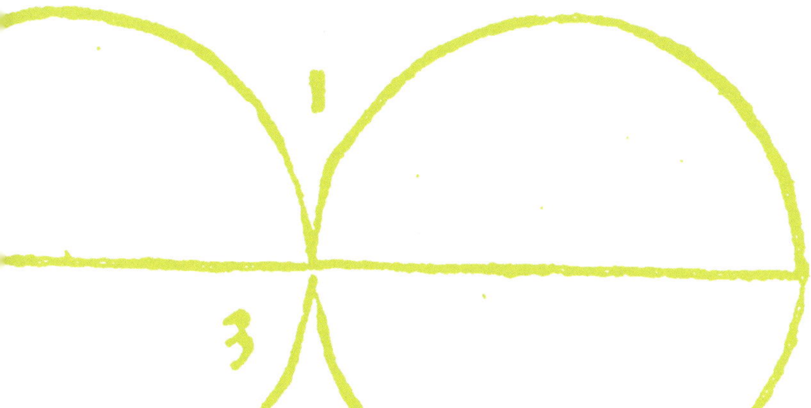
1
3
2

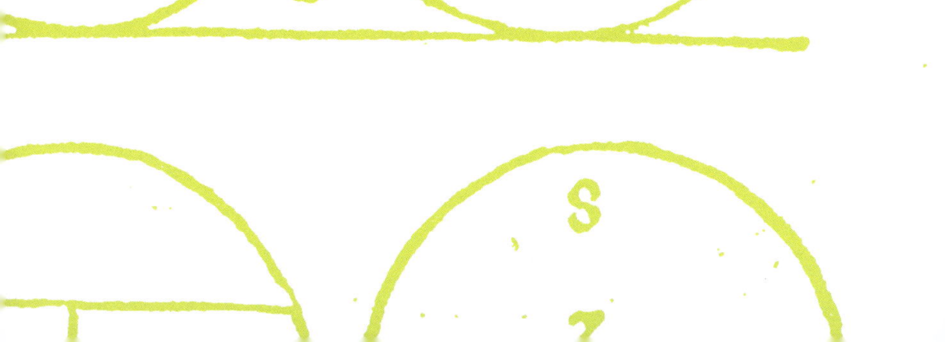
S

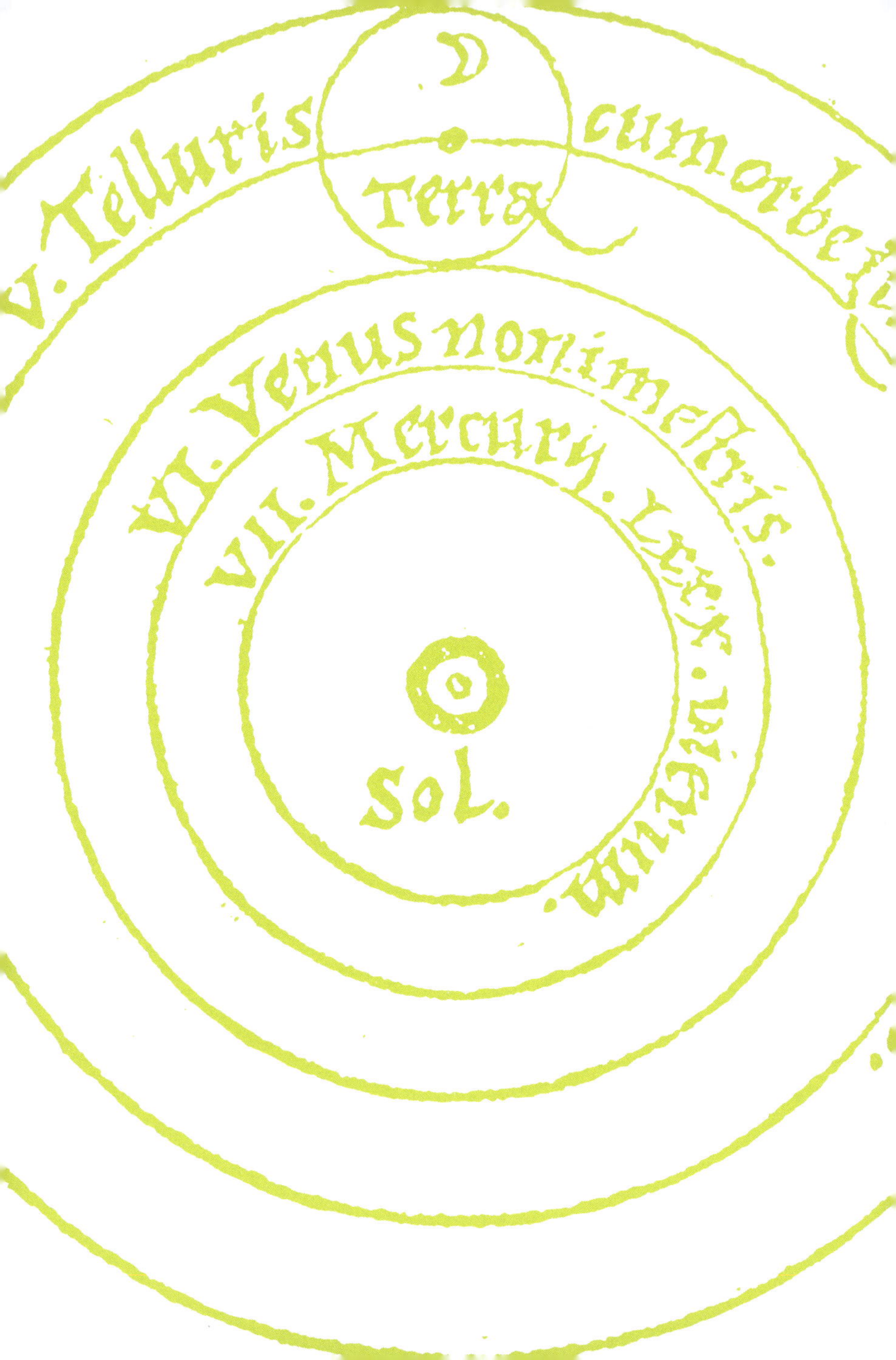

V. Telluris cum orbe lu
Terra
VI. Venus nonimestris.
VII. Mercurij. LXXX. dierum.
Sol.

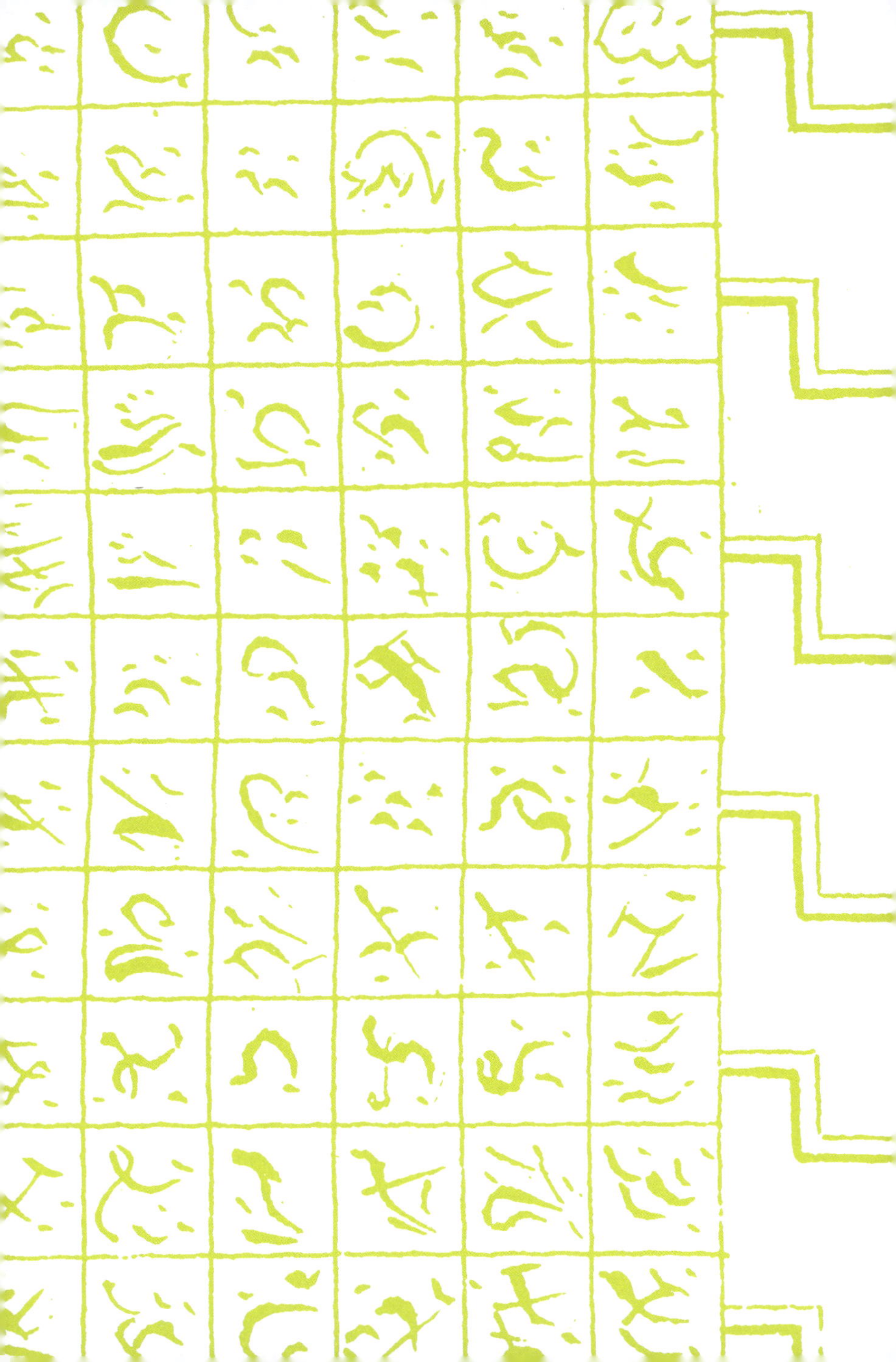

PART V

Mechanical Computing

Early instruments for computation looked more like clockwork than transistor computers. But the mathematics behind these cogwheel calculators and adding machines evolved to become modern computer science. The story of this centuries-long evolution is rife with oddities and fascinating prototypes, a number of which are held in Special Collections.

NAPIER, John (1550–1617).
Napier's rods calculating device.
Cambridge, England, circa 1690.

The earliest portable calculator (besides the abacus).

This is the oldest calculating device in Special Collections. Its inventor, John Napier, who also invented logarithms, coined the term *rabdology* to describe a method for calculating with wooden rods. These rods were carved or stamped with numbers that reduced complex mathematical operations to rudimentary arithmetic. Sets of Napier's rods were immediately popular and were manufactured into the nineteenth century. Made of pear wood, the rectangular case for this set is incised with lines made to resemble the binding of a small book. It is likely that the device was kept upright on a student's bookshelf in seventeenth-century Cambridge.

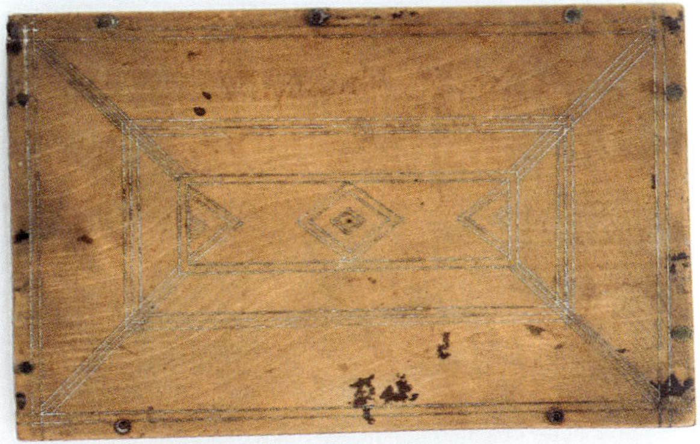

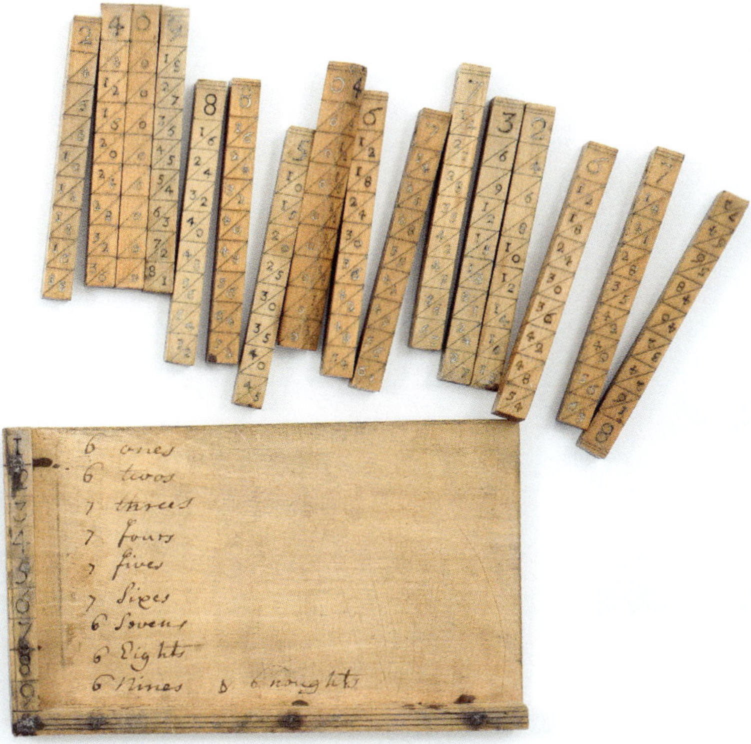
6 ones
6 twos
7 threes
7 fours
7 fives
7 sixes
6 sevens
6 eights
6 nines & 6 noughts

NAPIER, John (1550–1617).
Raddologia, overo arimmetica virgolare in due libri divisa. Verona, 1623.

This is an early Italian edition of Napier's work describing the theory and operation of his calculating rods, an example of which is described in the previous entry. The first edition of the work was printed in Edinburgh, Scotland in 1617; this is considered to be the first book on mechanical computation.

Napier's book on rod-computing also contains an illustrated description of another calculating device, called the Promptuary. Forming a complex apparatus of labeled rods and shuttered panels inside a box-like frame, the Promptuary could be programmed to convert multi-digit mathematical operations into basic lattice multiplication. Napier's depiction of the Promptuary bears an uncanny resemblance to early mainframe computers.

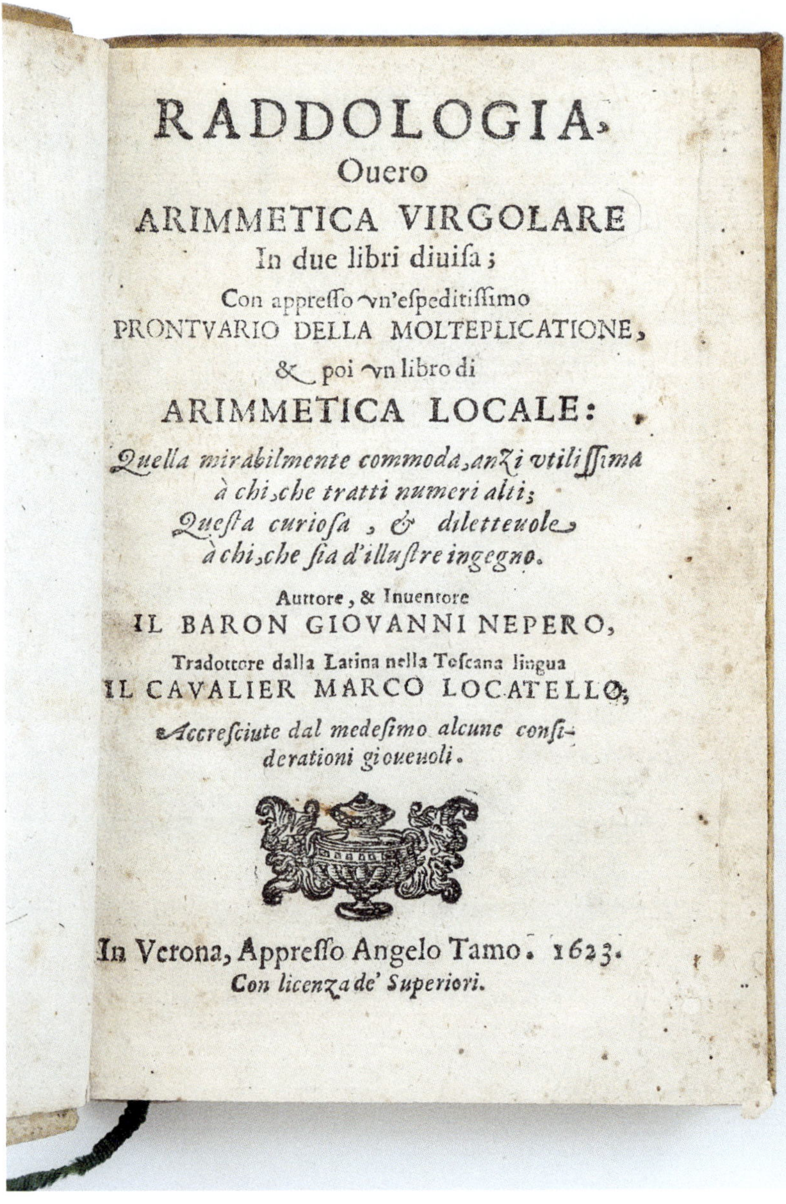

RADDOLOGIA,
Ouero
ARIMMETICA VIRGOLARE
In due libri diuisa;
Con appresso vn'espeditissimo
PRONTVARIO DELLA MOLTEPLICATIONE,
& poi vn libro di
ARIMMETICA LOCALE:
Quella mirabilmente commoda, anzi vtilissima
à chi, che tratti numeri alti;
Questa curiosa, & dilettevole
à chi, che sia d'illustre ingegno.
Auttore, & Inuentore
IL BARON GIOVANNI NEPERO,
Tradottore dalla Latina nella Toscana lingua
IL CAVALIER MARCO LOCATELLO;
Accresciute dal medesimo alcune considerationi giouenoli.

In Verona, Appresso Angelo Tamo. 1623.
Con licenza de' Superiori.

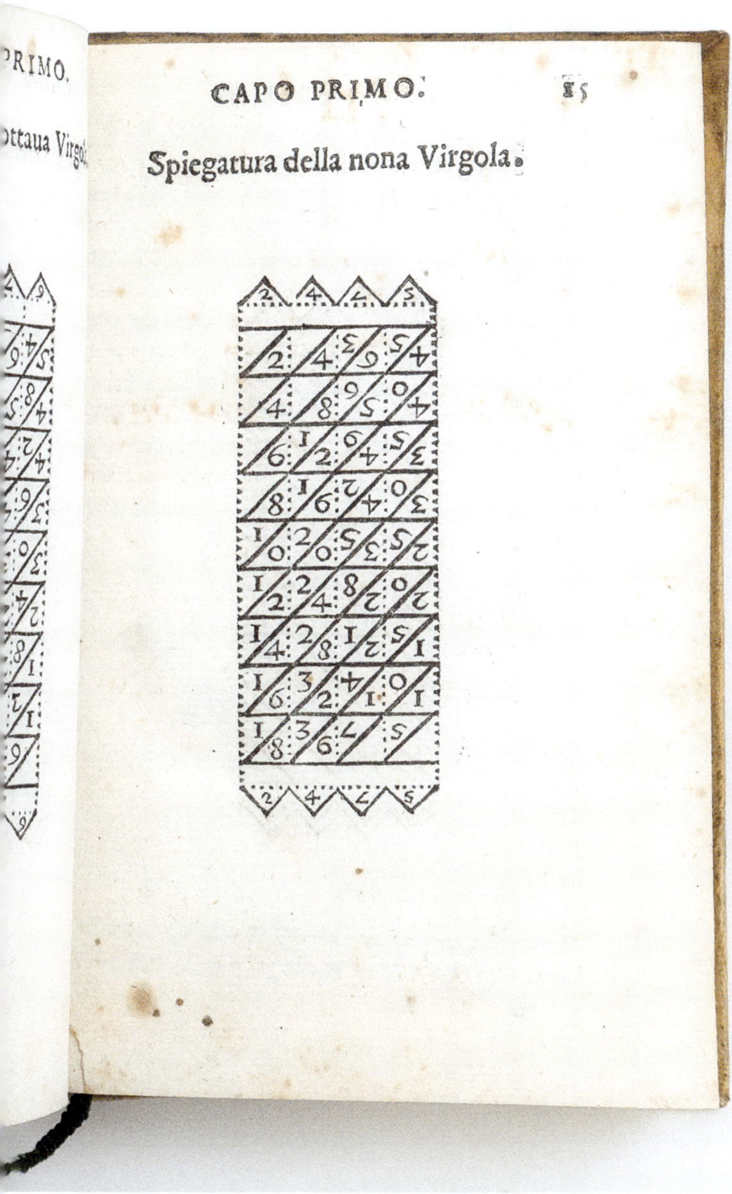
CAPO PRIMO. 15
Spiegatura della nona Virgola.

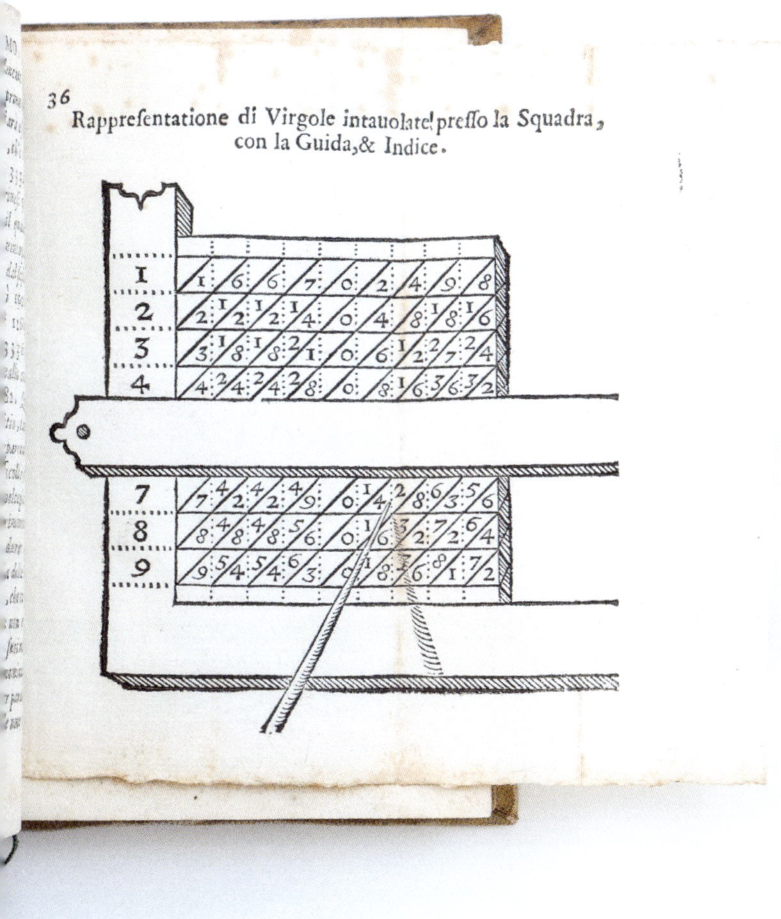
36
Rappreſentatione di Virgole intauolate preſſo la Squadra, con la Guida, & Indice.

ISOMURA Yoshinori (–1710).
増補筭法闕疑抄 [*Zōho sanpō ketsugishō*].
Kyoto, 1684.

The earliest known Japanese book on mechanical calculation and early Sudoku.

This book includes several illustrations of the Japanese abacus. It also describes mathematical games, diagrams, and combinatorics, including "magic squares" that form the early basis of the number game, Sudoku. Only two other copies of this edition are held in the United States.

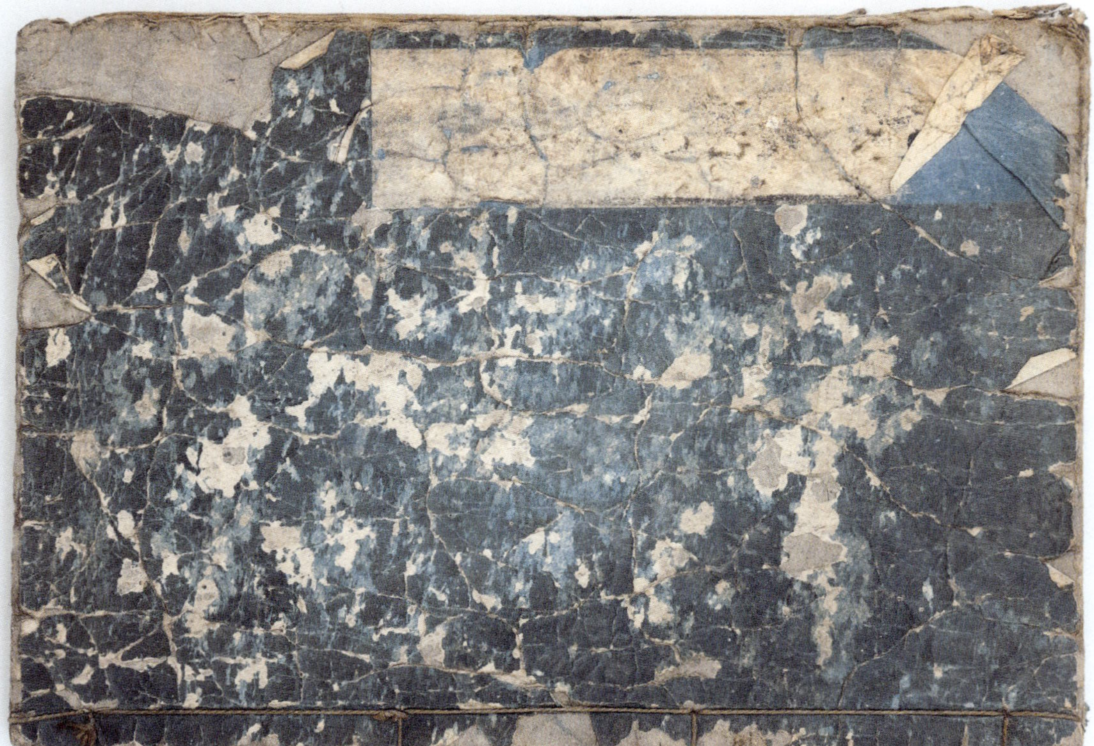

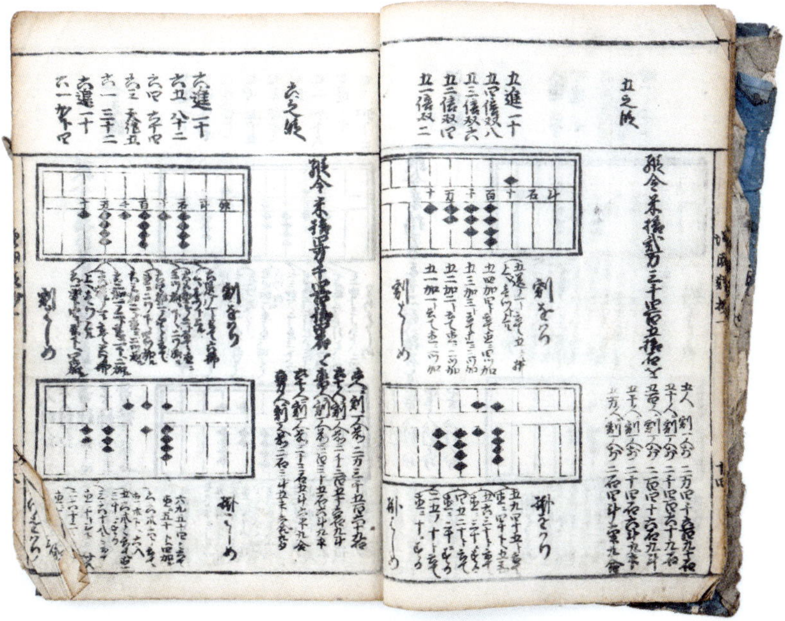

CHENG Dawei (1533–1606).
新增算法統宗大全
[*Xin zeng suan fa tong zong da quan*].
[China]: Sao ye shan fang, 1883.

A late edition of the first book to include an illustration of the Chinese abacus.

This book points to a long and inventive tradition of writing on computational instruments in early modern Asia. Cheng Dawei was a mathematician who collected calculating machines and books on computational techniques in sixteenth-century China. This 1883 printing of Cheng's book on the Zhusuan (Chinese: 珠算 — literally, "bead calculation"), bound in six volumes, describes the complex strategies and manual skill required to operate the suanpan, or Chinese abacus. First published in 1592, Cheng's book remained popular for centuries and is now considered a classic in the history of Chinese mathematics.

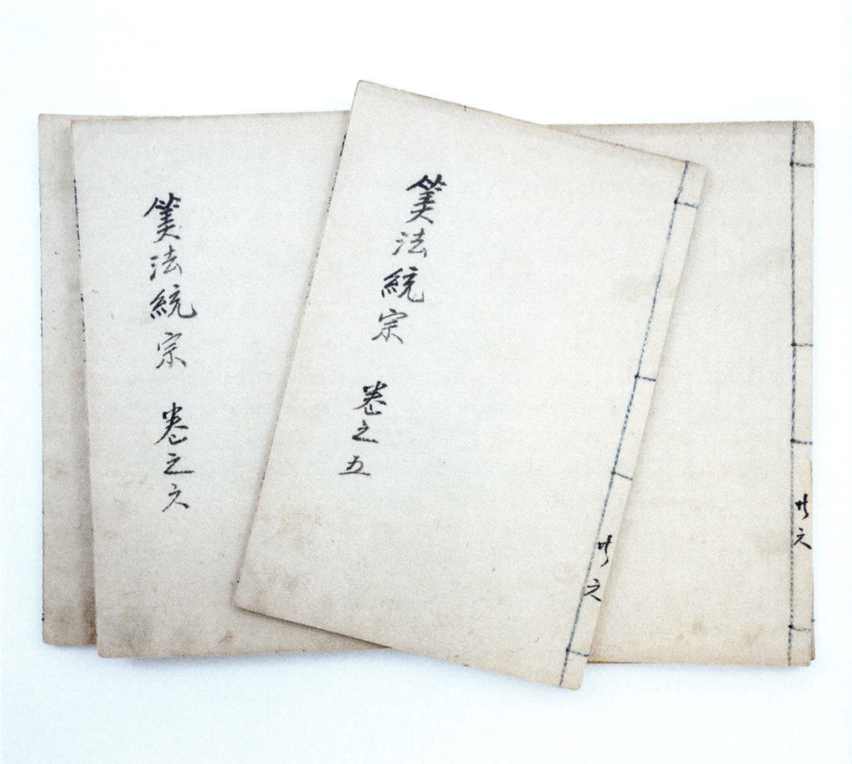
筭法統宗
卷之六
筭法統宗
卷之五

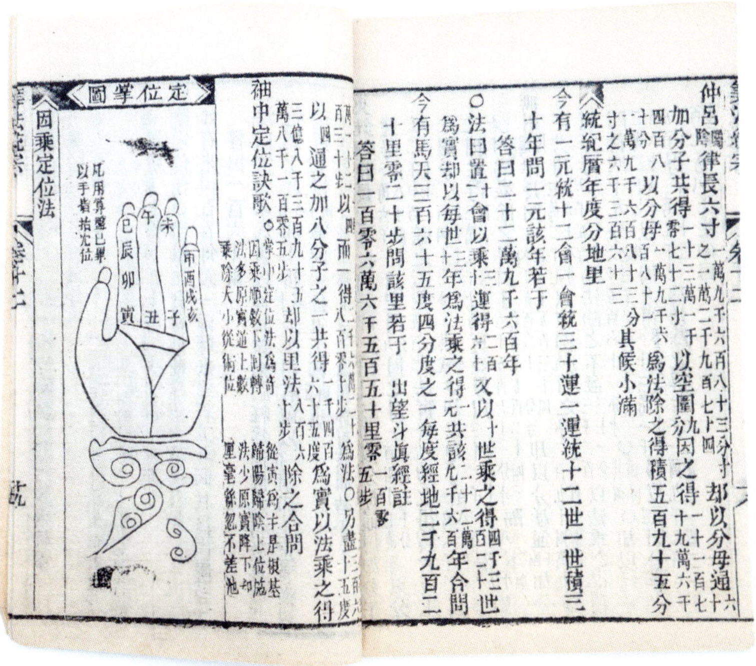

統紀曆年度分地里

今有一元統十二會一會統三十運一運統十二世一世積三十年問一元該年若干

答曰一十二萬九千六百年

○法曰置十二會以乘三十運得三百六十又以十二世乘之得四千三百二十世為實却以每世三十年為法乘之得共該一十二萬九千六百年合問

今有周天三百六十五度四分度之一每度經地二千九百三十二里零二十步問該里若干 出望斗真經註

因乘定位法

定位掌圖

原本直指算法統宗卷之二

新安　賓渠程大位汝思甫　編

分別法實左右圖

法　實

初學盤式

右　左

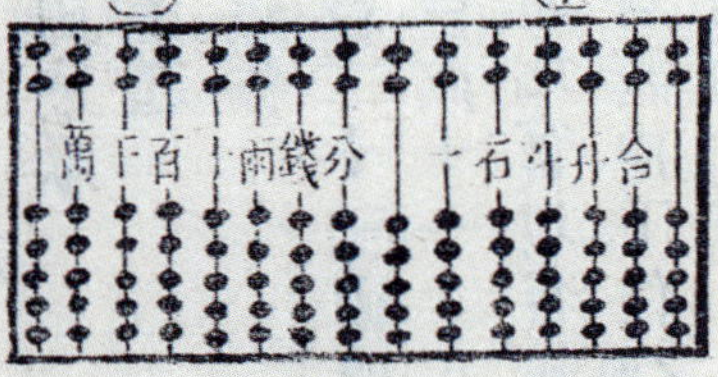

法之末位

法之首位

實之末位

實之首位

法爲母　靜

爲次位下位

實爲子　動

爲前位上位

〔九因〕

○凡二至九乘位者用此置物爲實以價爲法呼九九合數言
十就身言如隔位從末位算起用九歸還原

LEIBNIZ, Gottfried Wilhelm von (1646–1716). "Brevis descriptio machinae arithmeticae, cum figura," in *Miscellanea Berolinensia ad incrementum scientiarum.* Berlin, 1710.

LEIBNIZ, Gottfried Wilhelm von (1646–1716). Stepped reckoner mechanical calculator (replica).

An illustrated description of the first general purpose mechanical calculator and a twentieth-century replica.

Known for inventing calculus independently of Isaac Newton and for his groundbreaking work on binary code, Leibniz was also a pioneer of computer design. His "Brief Description of an Arithmetic Machine," published in 1710, announced the invention of the stepped reckoner, a complex mechanical calculator capable of all four arithmetic operations — the first of its kind. The advantage of mechanical computers, or "arithmetic machines," Leibniz wrote, was their ability to make calculations "*leicht*, *geschwind*, *gewiß*" — easy, fast, and reliable. Leibniz's paper includes an engraved illustration showing the device with its hinged lid opened.

Special Collections is unique in holding both Leibniz's printed description of his calculating device and a twentieth-century replica. Only one of Leibniz's original machines survives today.

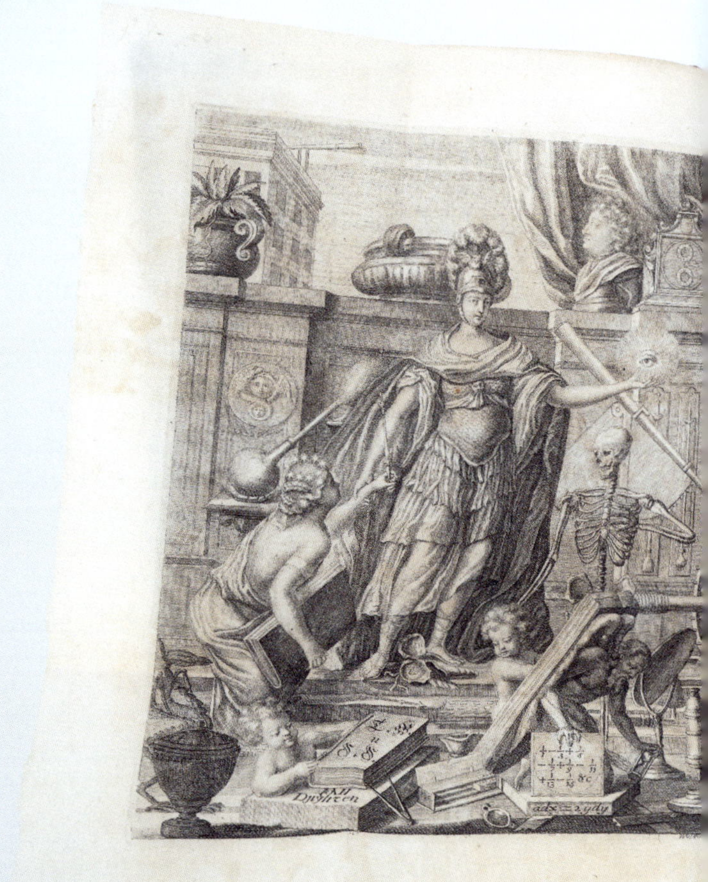
adx = 2ydy

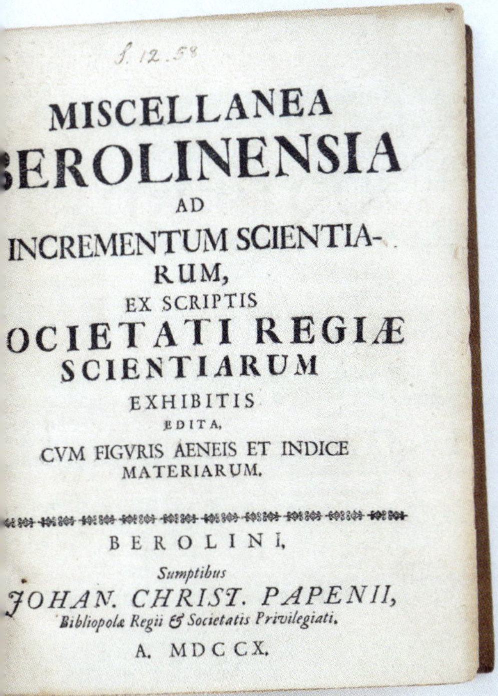

MISCELLANEA
BEROLINENSIA
AD
INCREMENTUM SCIENTIA-
RUM,
EX SCRIPTIS
SOCIETATI REGIÆ
SCIENTIARUM
EXHIBITIS
EDITA,
CUM FIGURIS AENEIS ET INDICE
MATERIARUM.

BEROLINI,
Sumptibus
JOHAN. CHRIST. PAPENII,
Bibliopolæ Regii & Societatis Privilegiati.
A. MDCCX.

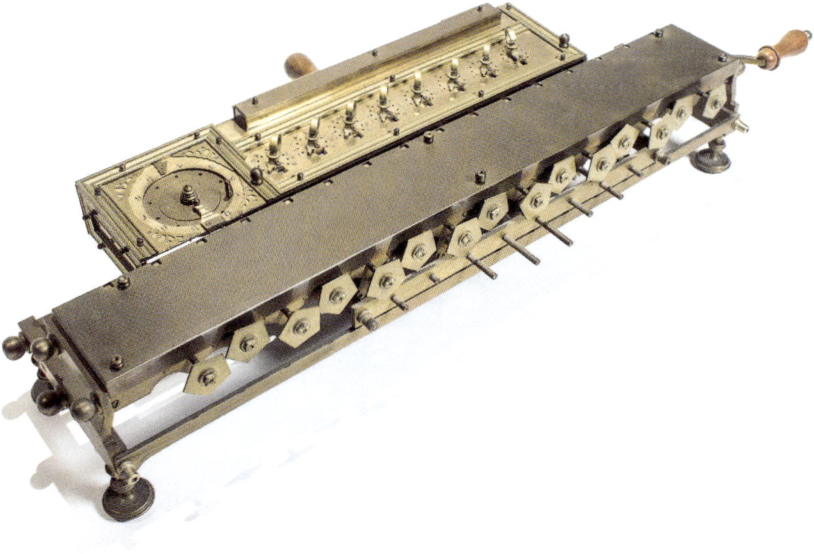

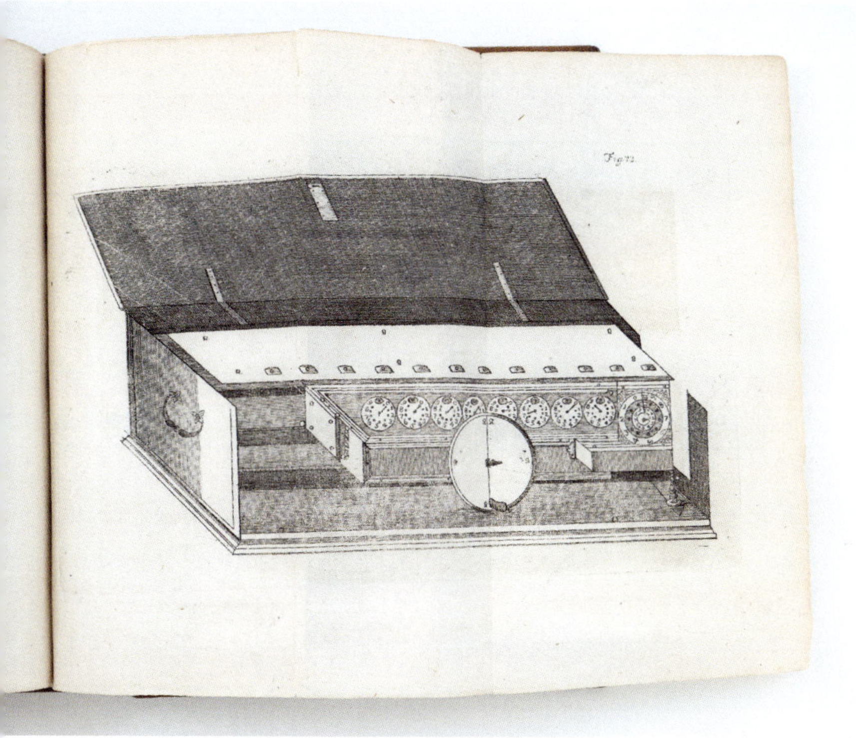

PASCAL, Blaise (1623–1662).
Pascaline mechanical calculator (replica).

Pascal's "thinking machine."

Special Collections holds a replica of a calculating machine built by Blaise Pascal in the 1630s. Unlike Leibniz's later stepped reckoner, the Pascaline was only capable of addition and subtraction. Described by Pascal as a "*machine à penser*" (thinking machine), one variant of the device was designed to speed financial accounting and returned quantities in the French currency of *sous, deniers,* and *livres*.

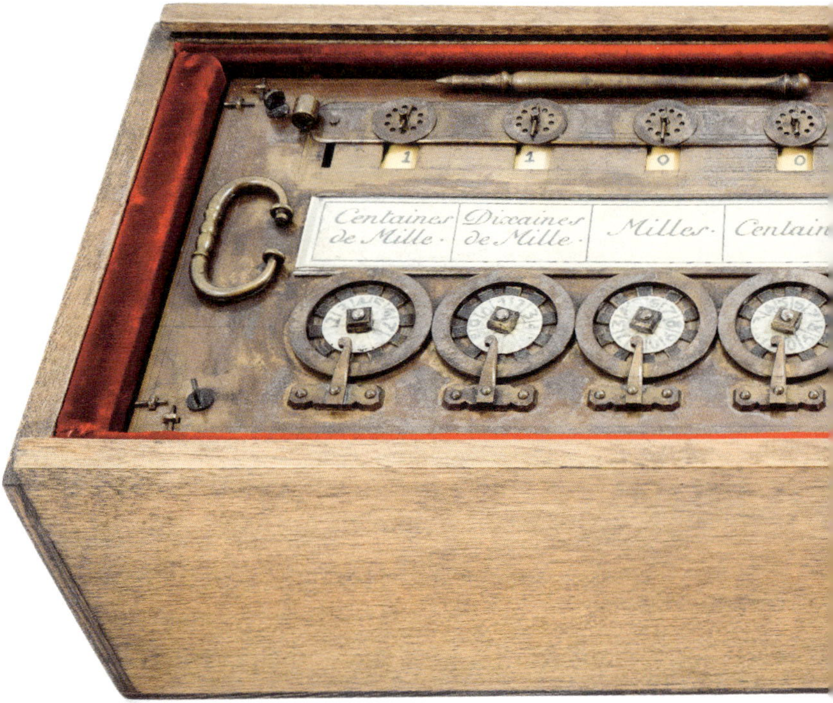
1
1
0
0
Centaines de Mille.
Dixaines de Mille.
Milles.

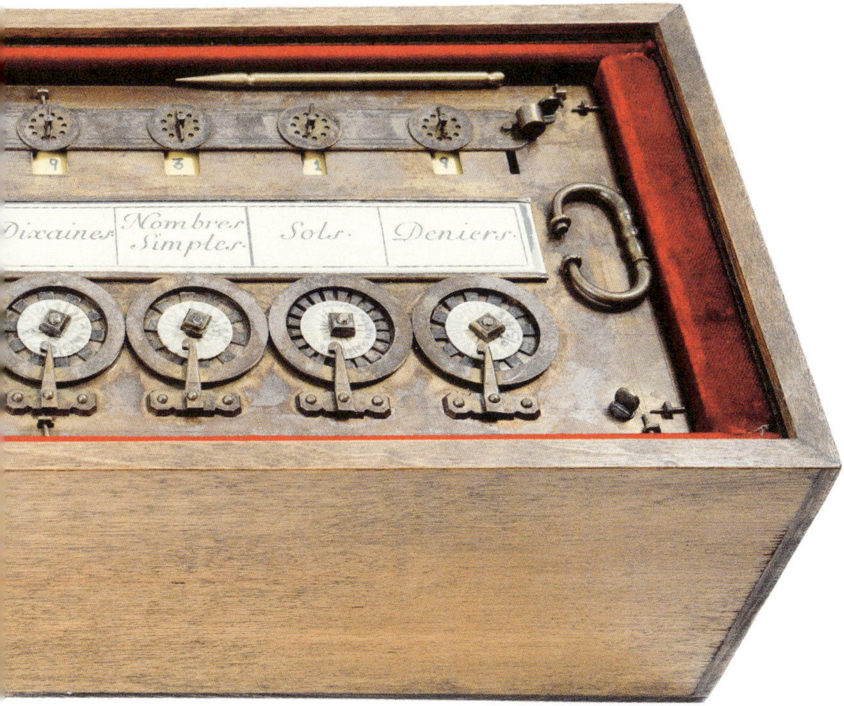
Dixaines
Nombres Simples
Sols.
Deniers.

PASCAL, Blaise (1623–1662).
Oeuvres. The Hague, 1779.

Printed posthumously in five volumes, this is the first collected edition of Pascal's works. Shown is an illustration with notes on Pascal's calculating machine (see previous entry). This is the first printed account of the device's design and operation. Special Collections' copy survives in its original paperboard binding.

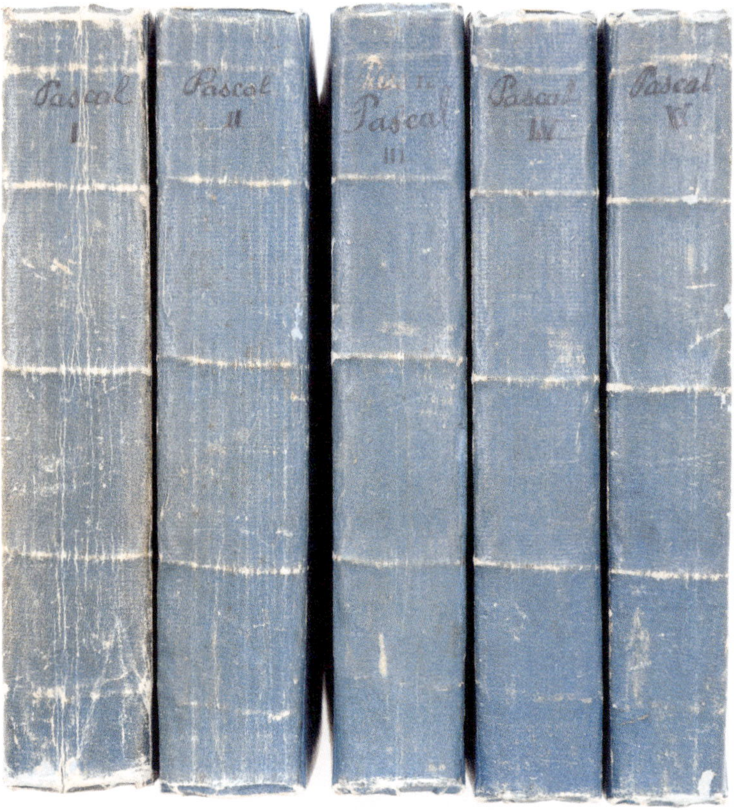
Pascal
I
Pascal
II
Pascal
III
Pascal
IV
Pascal
V

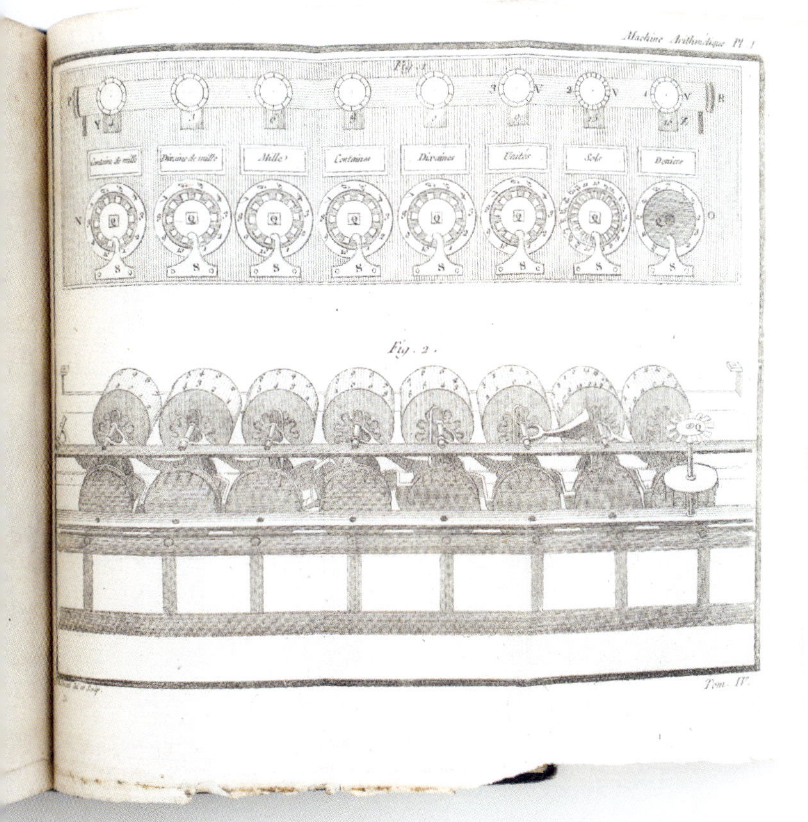
Machine Arithmétique Pl. I
Fig. 1
Centaine de mille
Dizaine de mille
Mille
Centaines
Dizaines
Unités
Sols
Deniers
Fig. 2
Tom. IV.

GENAILLE, Henri (1856–1903)
and LUCAS, Edouard (1842–1891).
Les réglettes multiplicatrices: appareils à calculs exacts et instantanés pour simplifier la multiplication et la division. Paris, 1885.

A nineteenth-century pocket calculator — if you have big pockets.

This device was designed to simplify multiplication. Offering a refinement on Napier's earlier rods (see "NAPIER," page 247), Genaille innovated a system of numbered rods ("réglettes") that gave numerical results without the need to carry digits. Brilliant in design and operation, the device's monochromatic appearance is also strikingly modern. Though initially popular, the Genaille rods were soon obsolete as mechanical calculators were made more widely available.

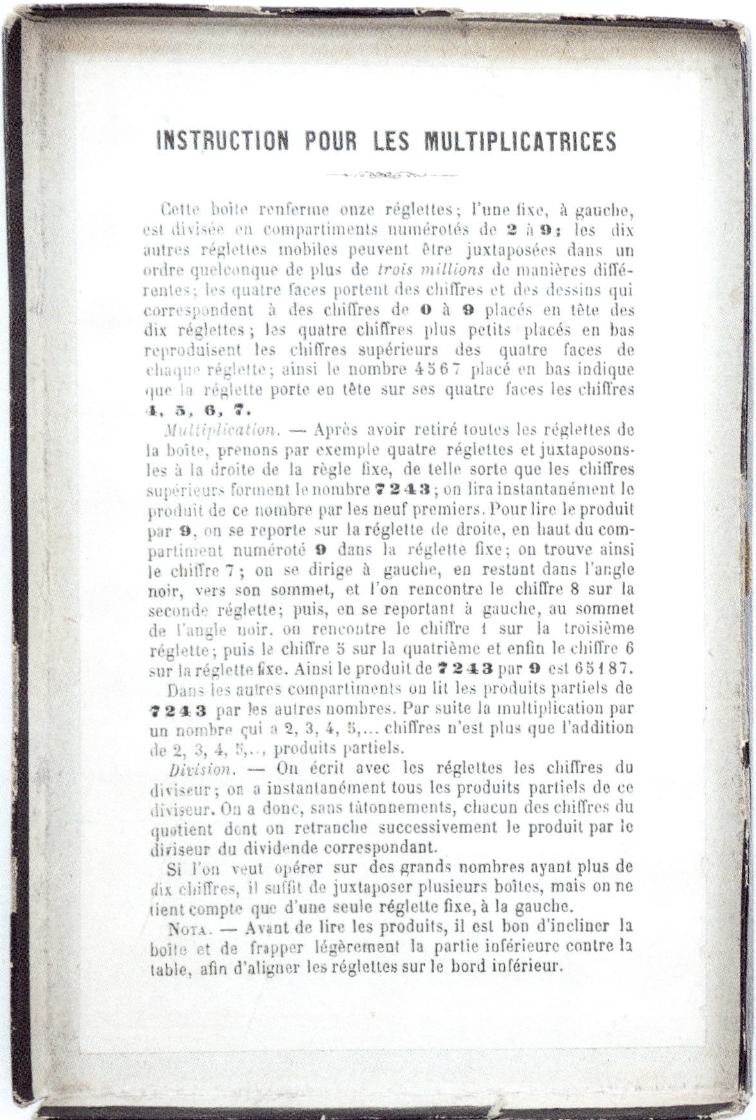

INSTRUCTION POUR LES MULTIPLICATRICES

Cette boîte renferme onze réglettes; l'une fixe, à gauche, est divisée en compartiments numérotés de **2** à **9**; les dix autres réglettes mobiles peuvent être juxtaposées dans un ordre quelconque de plus de *trois millions* de manières différentes; les quatre faces portent des chiffres et des dessins qui correspondent à des chiffres de **0** à **9** placés en tête des dix réglettes; les quatre chiffres plus petits placés en bas reproduisent les chiffres supérieurs des quatre faces de chaque réglette; ainsi le nombre 4567 placé en bas indique que la réglette porte en tête sur ses quatre faces les chiffres **4**, **5**, **6**, **7**.

Multiplication. — Après avoir retiré toutes les réglettes de la boîte, prenons par exemple quatre réglettes et juxtaposons-les à la droite de la règle fixe, de telle sorte que les chiffres supérieurs forment le nombre **7243**; on lira instantanément le produit de ce nombre par les neuf premiers. Pour lire le produit par **9**, on se reporte sur la réglette de droite, en haut du compartiment numéroté **9** dans la réglette fixe; on trouve ainsi le chiffre 7; on se dirige à gauche, en restant dans l'angle noir, vers son sommet, et l'on rencontre le chiffre 8 sur la seconde réglette; puis, en se reportant à gauche, au sommet de l'angle noir, on rencontre le chiffre 1 sur la troisième réglette; puis le chiffre 5 sur la quatrième et enfin le chiffre 6 sur la réglette fixe. Ainsi le produit de **7243** par **9** est 65187.

Dans les autres compartiments on lit les produits partiels de **7243** par les autres nombres. Par suite la multiplication par un nombre qui a 2, 3, 4, 5,... chiffres n'est plus que l'addition de 2, 3, 4, 5,.., produits partiels.

Division. — On écrit avec les réglettes les chiffres du diviseur; on a instantanément tous les produits partiels de ce diviseur. On a donc, sans tâtonnements, chacun des chiffres du quotient dont on retranche successivement le produit par le diviseur du dividende correspondant.

Si l'on veut opérer sur des grands nombres ayant plus de dix chiffres, il suffit de juxtaposer plusieurs boîtes, mais on ne tient compte que d'une seule réglette fixe, à la gauche.

Nota. — Avant de lire les produits, il est bon d'incliner la boîte et de frapper légèrement la partie inférieure contre la table, afin d'aligner les réglettes sur le bord inférieur.

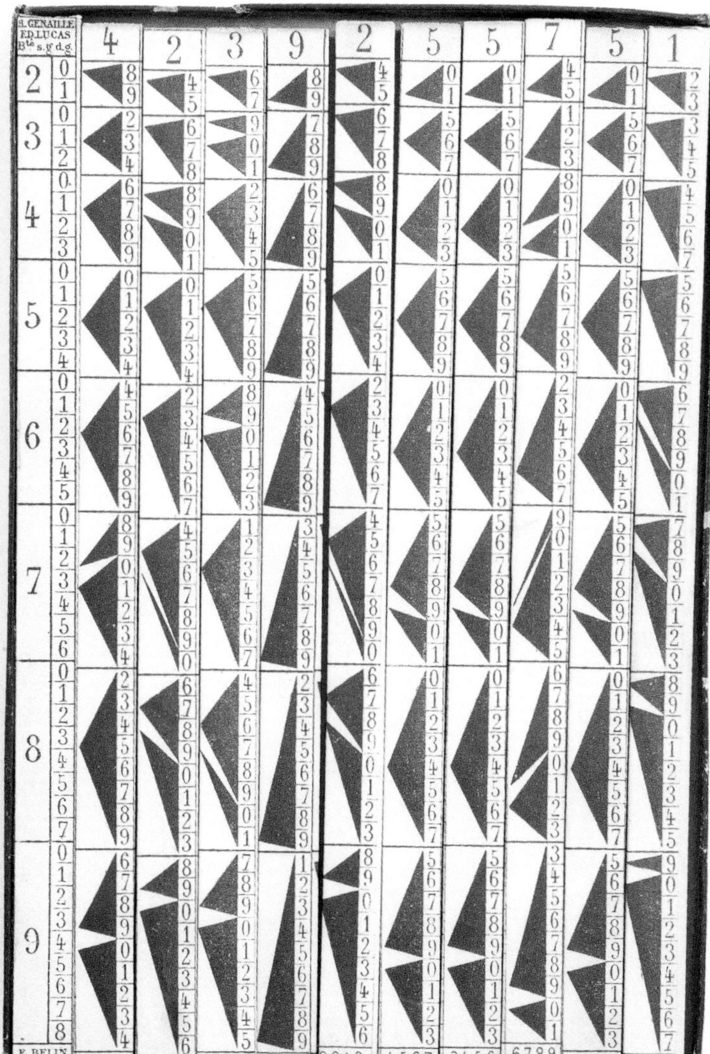
H. GENAILLE
ED. LUCAS
Bte s.g d.g.
E. BELIN

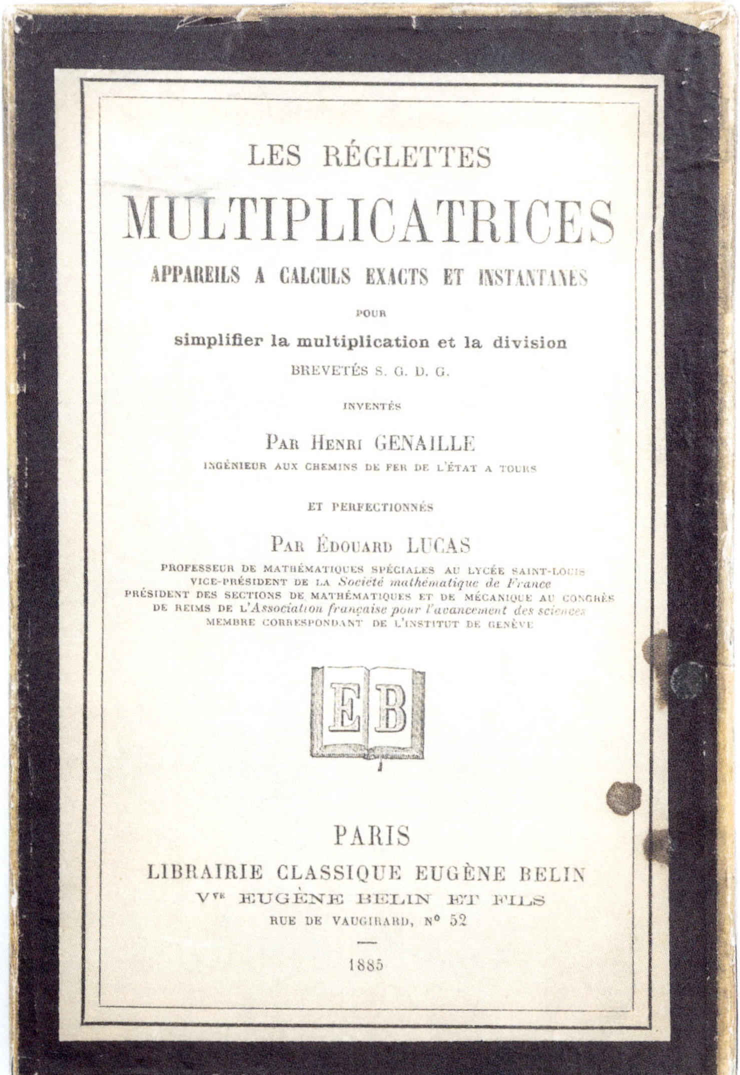
LES RÉGLETTES
MULTIPLICATRICES
APPAREILS A CALCULS EXACTS ET INSTANTANÉS
POUR
simplifier la multiplication et la division
BREVETÉS S. G. D. G.
INVENTÉS
PAR HENRI GENAILLE
INGÉNIEUR AUX CHEMINS DE FER DE L'ÉTAT A TOURS
ET PERFECTIONNÉS
PAR ÉDOUARD LUCAS
PROFESSEUR DE MATHÉMATIQUES SPÉCIALES AU LYCÉE SAINT-LOUIS
VICE-PRÉSIDENT DE LA Société mathématique de France
PRÉSIDENT DES SECTIONS DE MATHÉMATIQUES ET DE MÉCANIQUE AU CONGRÈS
DE REIMS DE L'Association française pour l'avancement des sciences
MEMBRE CORRESPONDANT DE L'INSTITUT DE GENÈVE
EB
PARIS
LIBRAIRIE CLASSIQUE EUGÈNE BELIN
VVE EUGÈNE BELIN ET FILS
RUE DE VAUGIRARD, N° 52
1885

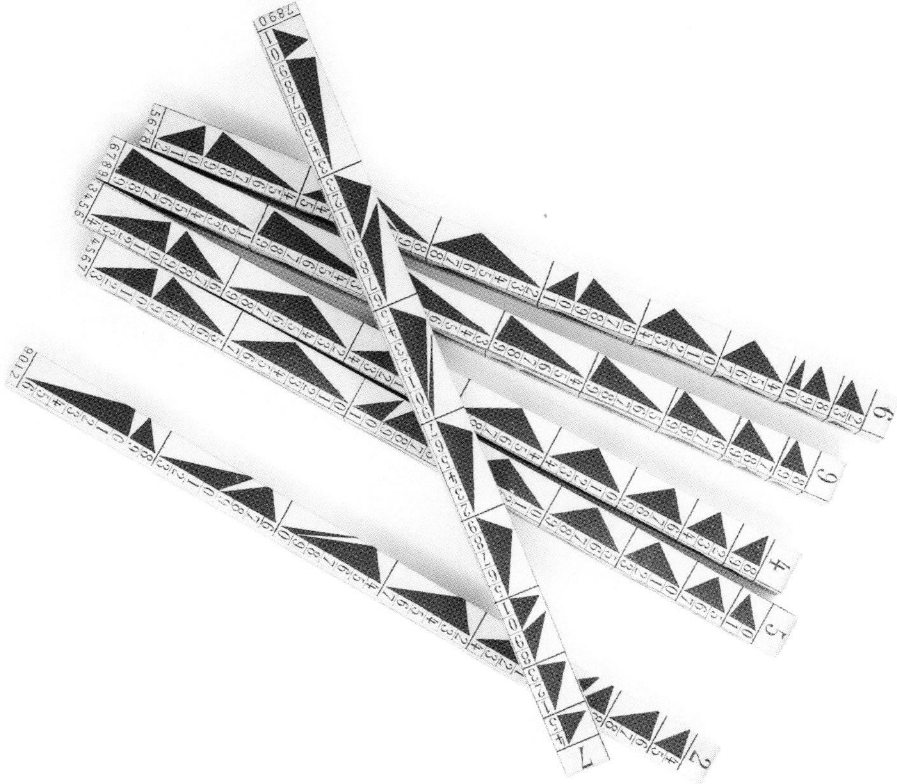

COLMAR, Charles Xavier Thomas de
(1785–1870).
Arithmomètre, Model T1858A.
Paris, circa 1860.

Introduced in 1820, Colmar's arithmometer was the first commercially successful mechanical calculator. It was capable of all four arithmetic operations: addition, subtraction, multiplication, and division. Special Collections holds six examples of the device in the Traub-McCorduck collection. The example shown here, stamped with the serial number 11, dates to approximately 1860.

↓ The TRAUB-McCORDUCK COLLECTION of the HISTORY OF COMPUTING

CMU's T1858A Arithmomètre came to Special Collections in 2018 with the Traub-McCorduck Collection of the History of Computing. Assembled by Joseph F. Traub (1932–2015; head of CMU School of Computer Science, 1971–79), the collection illustrates the evolution of machine calculation and encryption from the Renaissance to the twentieth century. Pamela McCorduck (1940–2021), an authority on the early history of AI and Traub's wife, donated the collection to CMU Libraries (see "TRAUB-McCORDUCK COLLECTION," page 293).

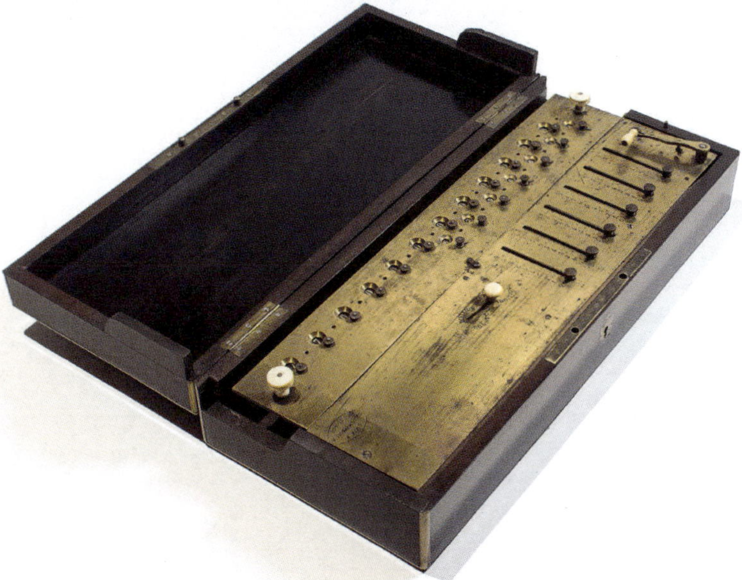

06	159	212	265	318	371	424	477
08	162	216	270	324	378	432	486
10	165	220	275	330	385	440	495
12	168	224	280	336	392	448	504
14	171	228	285	342	399	456	513
16	174	232	290	348	406	464	522
18	177	236	295	354	413	472	531
20	180	240	300	360	420	480	540
22	183	244	305	366	427	488	549
24	186	248	310	372	434	496	558
26	189	252	315	378	441	504	567
28	192	256	320	384	448	512	576
130	195	260	325	390	455	520	585
132	198	264	330	396	462	528	594
134	201	268	335	402	469	536	603
136	204	272	340	408	476	544	612
138	207	276	345	414	483	552	621
140	210	280	350	420	490	560	630
142	213	284	355	426	497	568	639
144	216	288	360	432	504	576	648
146	219	292	365	438	511	584	657
148	222	296	370	444	518	592	666
150	225	300	375	450	525	600	675
152	228	304	380	456	532	608	684

u o t i e n t

2	3	4	5	6	7	8	9
160	240	320	400	480	560	640	720

i v i d e n d

168	252	336	420	504	588	672	756
170	255	340	425	510	595	680	765
172	258	344	430	516	602	688	774
174	261	348	435	522	609	696	783
176	264	352	440	528	616	704	792
178	267	356	445	534	623	712	801
180	270	360	450	540	630	720	810
182	273	364	455	546	637	728	819
184	276	368	460	552	644	736	828
186	279	372	465	558	651	744	837
188	282	376	470	564	658	752	846
190	285	380	475	570	665	760	855
192	288	384	480	576	672	768	864
194	291	388	485	582	679	776	873
196	294	392	490	588	686	784	882

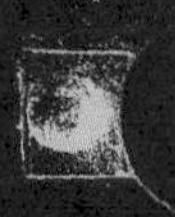

Safety

If the mac
shipped these
be fastened
marked
the carriage
extreme ri

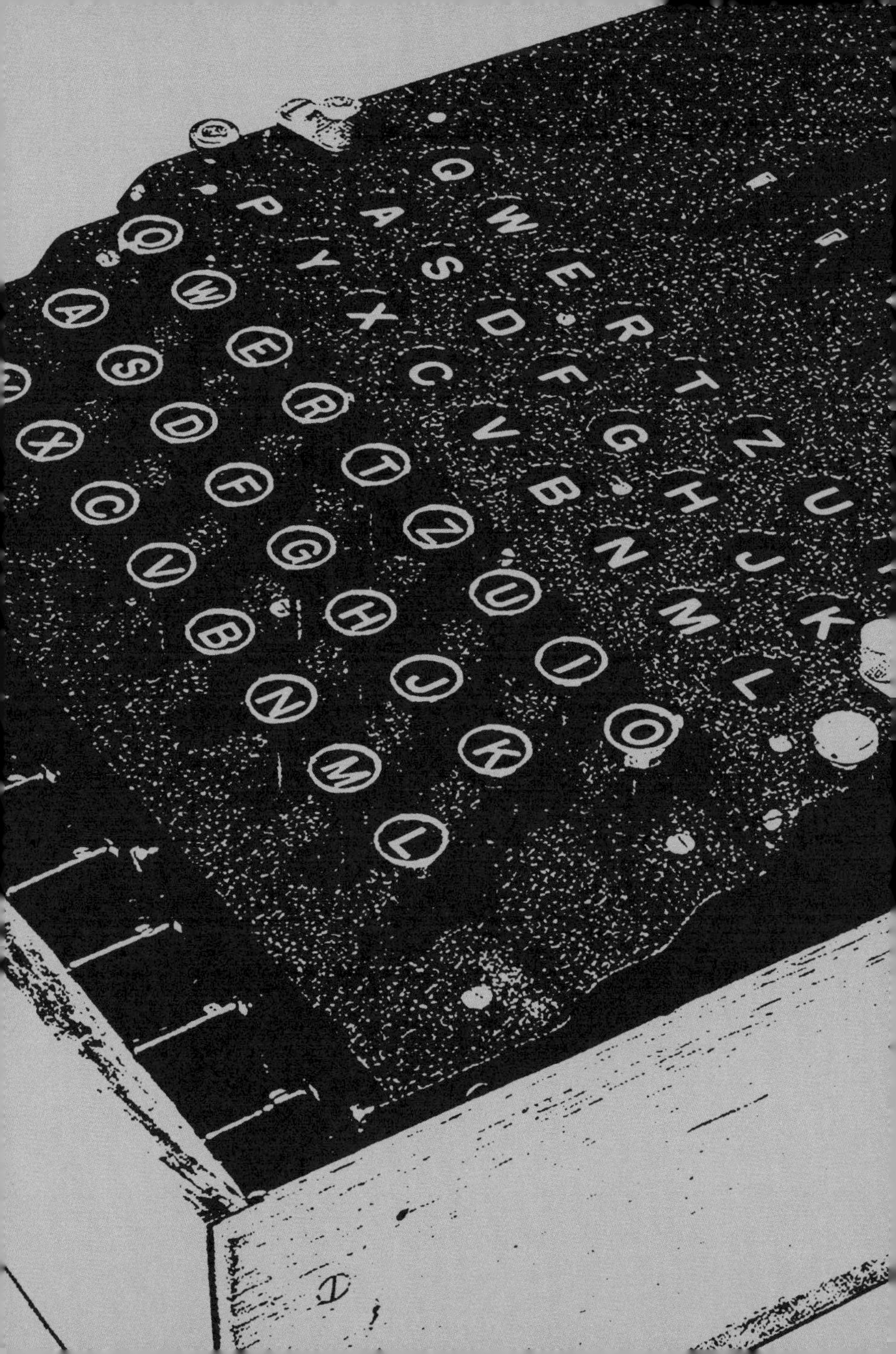
Q W E R T Z U I O
A S D F G H J K
P Y X C V B N M L

Ancestral Machines: The Traub-McCorduck Collection of the History of Computing

The Traub-McCorduck Collection gathers books and artifacts that illustrate the history of machine calculation and encryption from the Renaissance to the twentieth century. Assembled by Joseph F. Traub (1932–2015; head of CMU School of Computer Science, 1971–1979), the collection was donated to CMU Libraries in 2018 by Pamela McCorduck (1940–2021), an authority on the early history of AI and Traub's wife. The collection includes two World War II-era Enigma cipher machines, several Arithmometer-type "stepped-drum" calculators, instruments for precision measurement, and a number of rare books on cryptography, number theory, and mathematics.

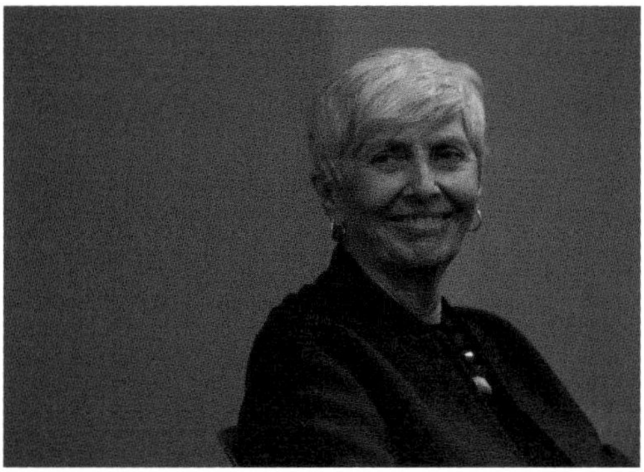

↑ Pamela McCorduck, 1940–2021

Entries are individually labeled with model names or titles and dates of manufacture or publication. For the machines, dates are often approximate: manufactured in multiples, few of these devices were date-stamped or otherwise serialized. More detailed descriptions and metadata can be found on the Carnegie Mellon University Libraries' online catalog: https://library.cmu.edu.

1 Peerless Calculator
Type: Mechanical stepped-drum calculator
Keuffel & Esser Company (USA)
ca. 1907

An American stepped-drum calculator used for commercial and engineering applications.

2 Brunswiga 13Z
Type: Mechanical pinwheel calculator
Brunsviga-Maschinenwerke, Grimme, Natalis & Company A.-G. (Germany)
ca. 1935

A German pinwheel calculator designed for rapid and reliable business arithmetic.

3 NAPIER, John (1550–1617).
Mirifici Logarithmorum Canonis Descriptio.
Edinburgh, 1620.

The first published description of logarithms — a breakthrough that transformed calculation by reducing multiplication and division to simple addition and subtraction. The term mirifici *("wondrous, miraculous") in the title reflects Napier's own sense of the method's transformative, almost otherworldly power.*

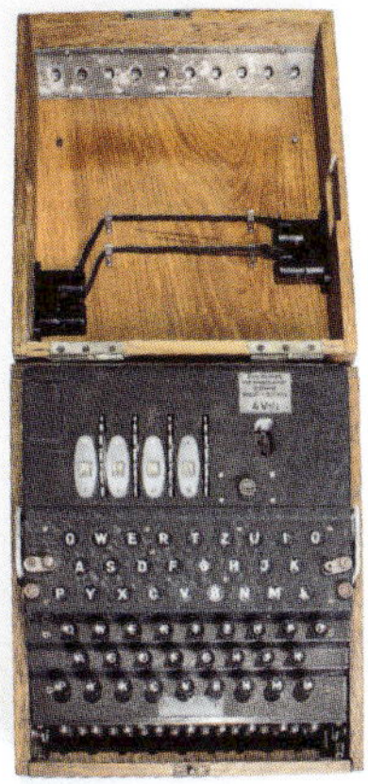

4 Enigma, Model M4
Type: Electro-mechanical cipher machine
Ertel-Werk für Feinmechanik (Germany)
ca. 1942

A naval cipher machine used by the German Kriegsmarine *(see "ENIGMA," page 129).*

5 Original-Odhner Model 127 / 602 No. 5
Type: Mechanical pinwheel calculator
AB Original-Odhner (Sweden)
ca. 1937

A Swedish refinement of the pinwheel calculator design, widely used in Europe.

6 Trinks-Brunsviga Model A
Type: Mechanical pinwheel calculator
Grimme, Natalis & Company (Germany)
ca. 1900

One of the earliest Brunsviga pinwheel calculators, used in scientific and financial settings.

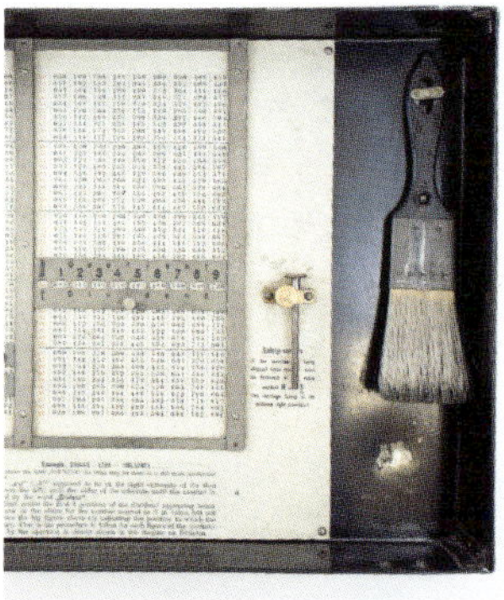

7 Millionaire Calculator (Serial No. 2380)
Type: Mechanical direct-multiplication calculator
Hans W. Egli (Switzerland)
ca. 1900

A Swiss-engineered calculator capable of direct multiplication (computing a full product in a single step without repeated addition). The Millionaire was one of the first commercially successful machines to offer this ability. The example held in the Traub-McCorduck Collection includes a mounted table of figures and a cleaning brush stored in the lid.

8 Felix M Arithmometer
Type: Mechanical pinwheel calculator
National Calculating Machine Factory (Russia)
ca. 1960

A Soviet adaptation of the Odhner pinwheel calculator, widely used behind the Iron Curtain.

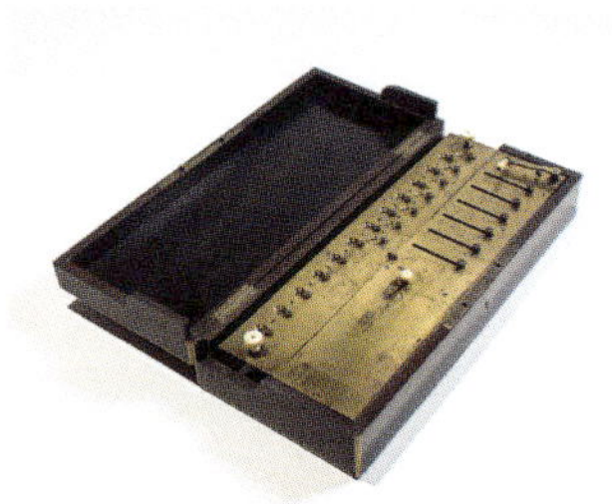

9 Arithmomètre, Model T1858 A (Serial No. 11)
Type: Mechanical stepped-drum calculator
Thomas de Colmar (France)
ca. 1858

The arithmometer is a mechanical calculator that performed basic arithmetic operations using geared wheels and a hand crank, streamlining calculations in the nineteenth century. This type operates on the basis of the stepped drum (German, Staffelwalze*) or "Leibniz gear," invented by Gottfried Wilhelm von Leibniz (1646–1716) in the seventeenth century (see "LEIBNIZ," page 267).*

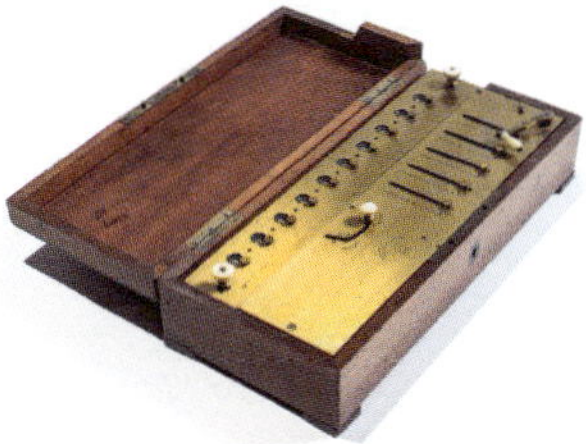

10 Arithmomètre (Serial No. 502)
Type: Mechanical stepped-drum calculator
Thomas de Colmar (France)
ca. 1863

An advanced nineteenth-century office tool that mechanized multi-digit calculations.

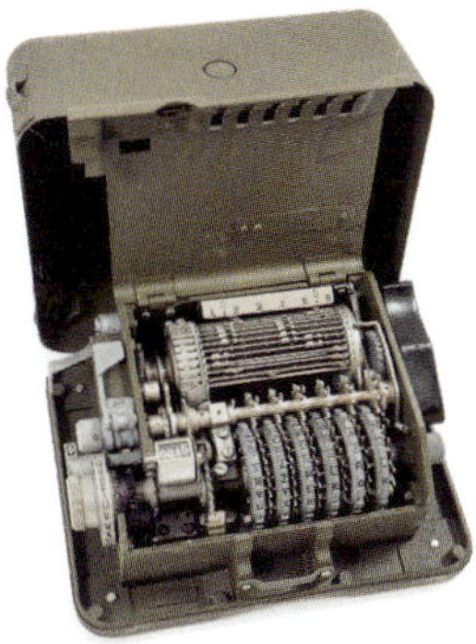

11 M-209-B Cipher Machine
Type: Pin-and-lug cipher machine
LC Smith & Corona Typewriters, Incorporated (USA)
ca. 1944

A portable World War II cipher device used by the United States military for field encryption.

12 Felt & Tarrant Comptometer
Type: Mechanical key-driven calculator
Felt & Tarrant Manufacturing Company (USA)
ca. 1895

One of the earliest successful key-driven calculators, enabling rapid commercial computation.

13 Felt & Tarrant Comptometer, Model J
Type: Mechanical key-driven calculator
Felt & Tarrant Manufacturing Company (USA)
ca. 1915

A refined version of the original Comptometer, optimized for speed and ergonomics.

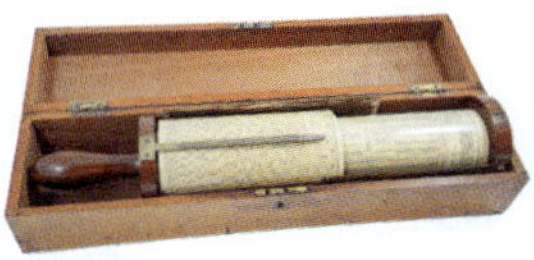

14 Calculating Slide Rule
Type: Cylindrical slide rule calculator
W.F. Stanley (England)
ca. 1880

A cylindrical slide rule offering precision engineering calculations before electronic tools.

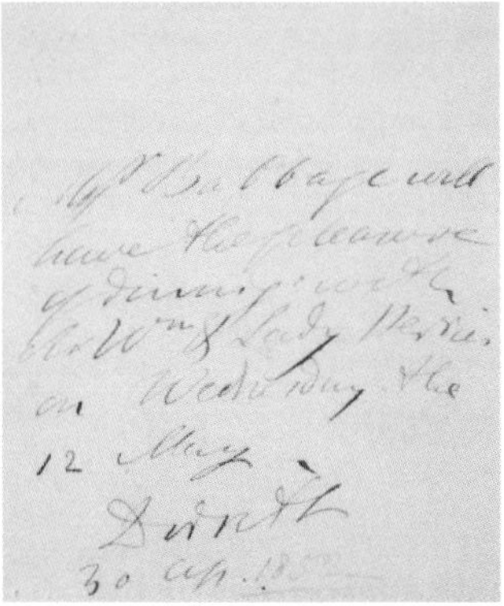

15 BABBAGE, Charles (1791–1871)
Autograph letter signed to Sir William and Lady Nevris?
London, England
1852

An invitation to dine with Babbage, whose mechanical designs anticipated modern computing.

Transcription: Mr Babbage will have the pleasure of dining with Rv. Wm/ & Lady Nevris [?] on Wednesday the 12 May [Dorset Street?] 30 Ap.[ril] 1852

16 BABBAGE, Charles (1791–1871).
The Ninth Bridgewater Treatise.
Philadelphia, 1841.

In The Ninth Bridgewater Treatise *Babbage used his own Difference Engine as a metaphor for a divinely created universe. Here was a machine that appeared to follow predetermined mathematical patterns but that could be programmed to alter its behavior. Babbage argued by analogy that irregularities in nature need not contradict natural law if those laws were designed to change at specified intervals. The universe, Babbage suggested, is less like a piece of clockwork than a programmable machine.*

The copy held in the Traub-McCorduck collection is from the first American edition. The first English edition appeared in 1837.

17 TIM ("Time is Money") Calculator
Type: Mechanical stepped-drum calculator
Ludwig Spitz & Company (Germany)
ca. 1910

A German stepped-drum calculator marketed to businesses for time-saving arithmetic.

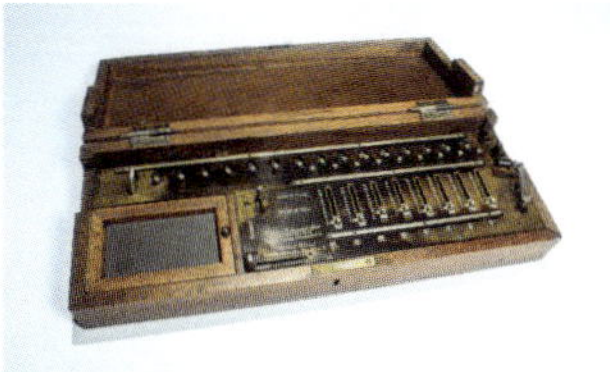

Archimedes Calculator
18 Type: Mechanical stepped-drum calculator
Reinhold Pöthig (Germany)
ca. 1910

Used for routine arithmetic, this German-made device employed a stepped-drum mechanism.

Rapid Calculator
19 Type: Mechanical pinwheel calculator
The Rapid Calculator Company (USA)
ca. 1920

Designed for speed, this American-made machine performed addition and subtraction via pinwheel mechanics.

Arithmomètre
20 Type: Mechanical stepped-drum calculator
Veuve Louis Payen (France)
ca. 1910

An advanced nineteenth-century office tool that mechanized multi-digit calculations.

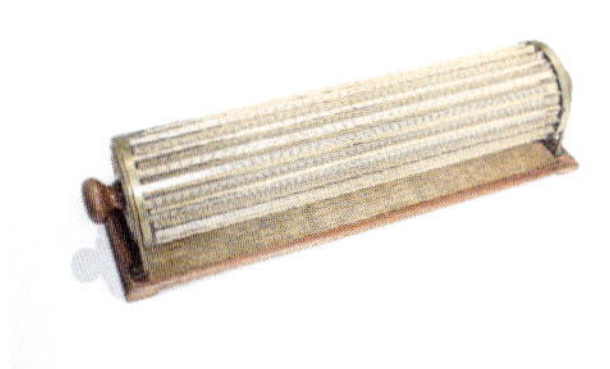

Thacher's Calculating
21 Instrument, Model 4012
Type: Cylindrical slide rule calculator
Keuffel & Esser Company (USA)
ca. 1910

A cylindrical slide rule designed for accuracy over wide numerical ranges.

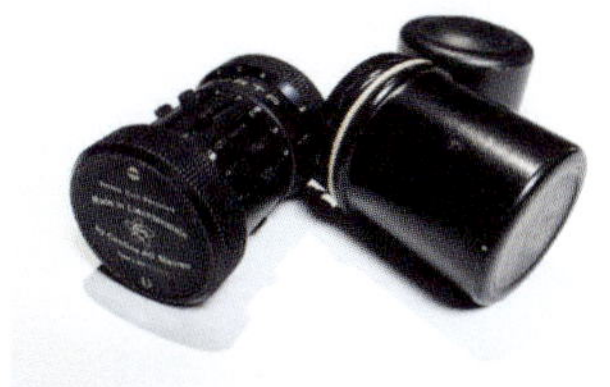

Curta Type I
22 Type: Miniature mechanical stepped-drum calculator
Contina AG Mauren (Liechtenstein)
ca. 1960

A compact, high-precision handheld calculator valued by engineers before the electronic era.

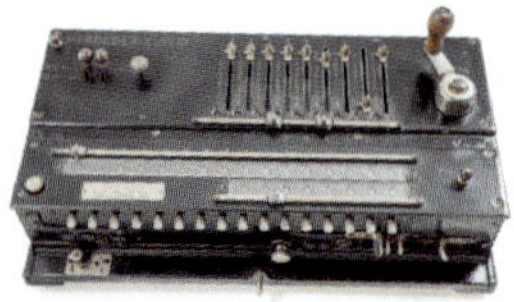

Mercedes Euklid M1
23 Type: Mechanical proportional-level calculator
Mercedes Bureau Maschinen-Werke (Germany)
ca. 1922

A proportional-lever machine capable of more complex arithmetic, bridging hand and office machines.

The Adder
24 Type: Mechanical rotary adding machine
C.H. Webb (USA)
ca. 1880

An early American rotary calculator designed for basic addition, notable for its compact form.

Ivory sector

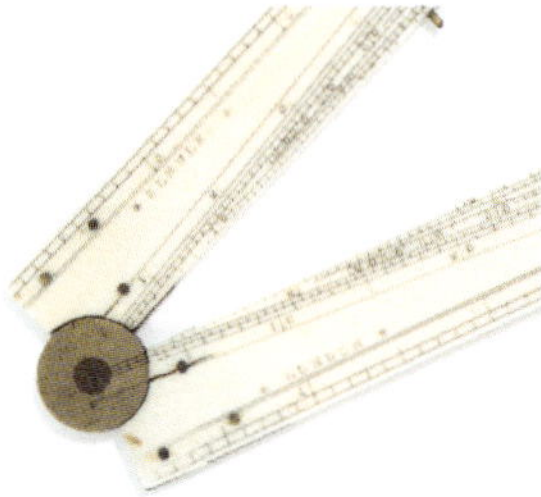

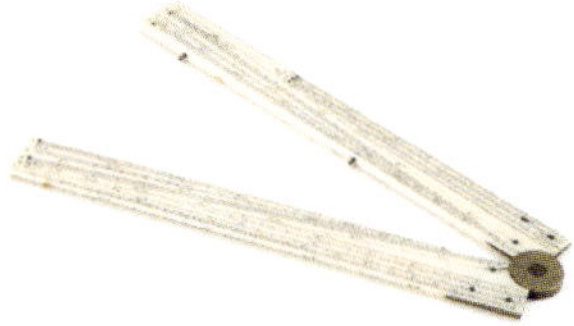

25 Type: Proportional compass or Sector
John Bleuler (England)
ca. 1800

A scientific instrument for geometric and proportional calculations in surveying and gunnery.

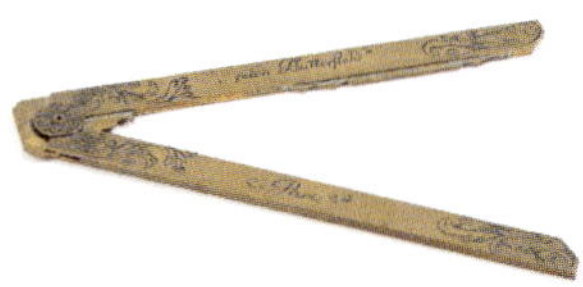

Brass ruler, "Pied de Roy"
26 Type: Ruler
Michael Butterfield (France)
ca. 1700

A pre-metric measuring instrument based on the French royal foot, used in technical trades.

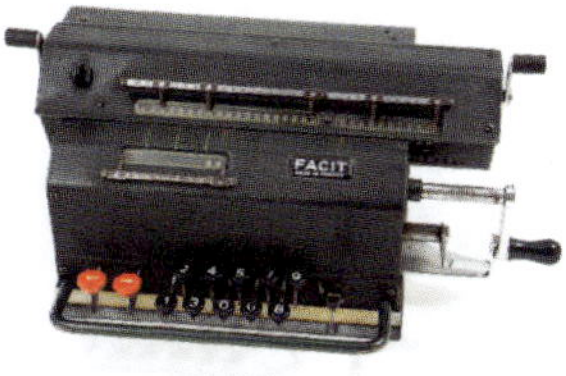

Facit, Model TK
27 Type: Mechanical key-driven calculator
Aktiebolaget Facit (Sweden)
ca. 1940

A Swedish key-driven calculator popular in European offices in the early twentieth century.

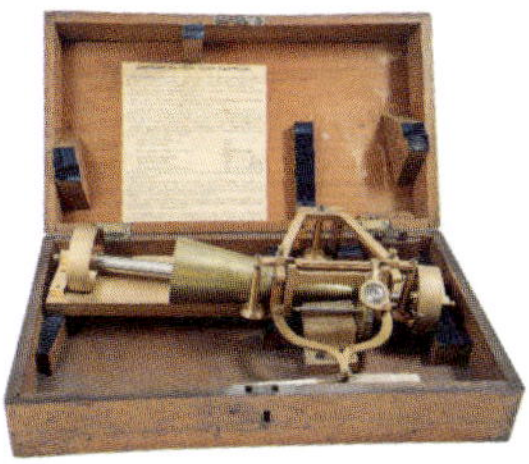

Sang's Platometer
28 Type: Planimeter
John Sang (Scotland)
ca. 1855

A planimeter is a precision instrument used to measure the area of irregular shapes on a flat surface, such as maps or drawings, by tracing their perimeter.

Adix Adding Machine

29 Type: Mechanical adding machine
Pallweber & Bordt (Germany)
ca. 1905

A German desktop adder that simplified bookkeeping before electronic calculators.

The Calcumeter

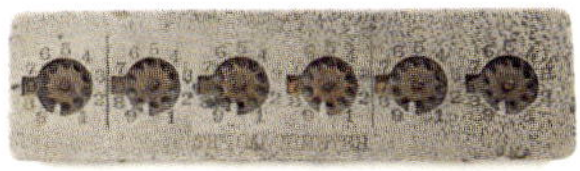

30 Type: Mechanical adding machine
H.N. Morse (USA)
ca. 1907

Pascaline (Replica)

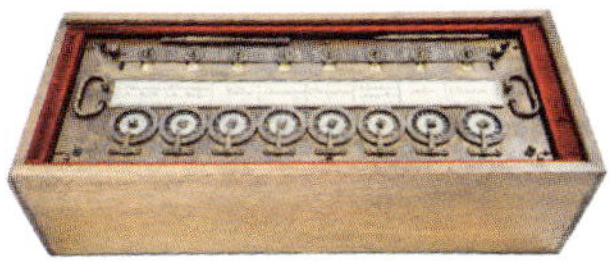

31 Inventor: Blaise Pascal (1623–1662)
Modelist: Roberto Ambrogio Guatelli (1909–1993)
ca. 1960 (original ca. 1640)

A reproduction of Pascal's original seventeenth-century adding machine, designed to automate tax collection and financial accounting *(see "PASCAL," page 273).*

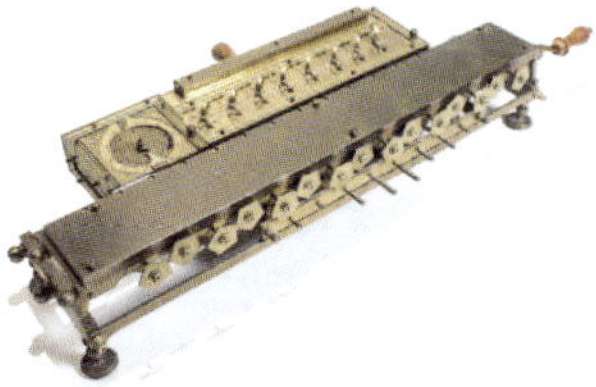

Leibniz Calculator, or Stepped-
32 Drum Calculator (Replica)
Inventor: Gottfried Wilhelm von Leibniz (1646–1716)
Modelist: Roberto Ambrogio Guatelli (1909–1993)
ca. 1960 (original ca. 1690)

A replica of the first general-purpose mechanical calculator, developed in the seventeenth century (see "LEIBNIZ," page 267).

Tate Arithmometer
33 Type: Mechanical stepped-drum calculator
Charles & Edward Layton (England)
ca. 1900

An English variant of the stepped-drum calculator, used in technical and business fields.

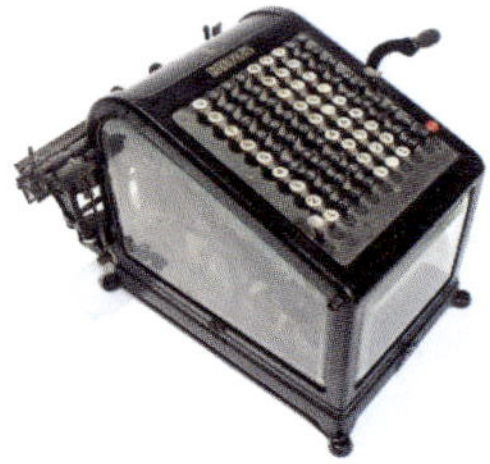

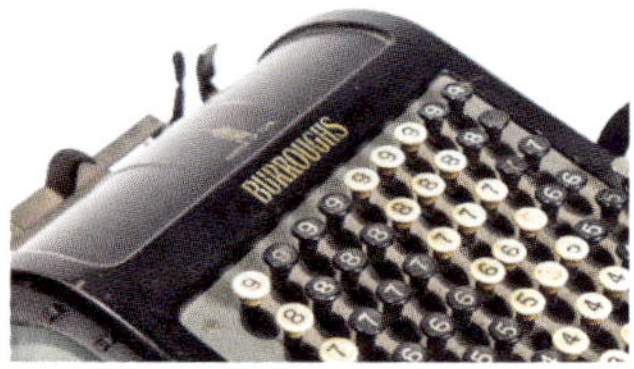

Burroughs Class 1, Model 3
34 Type: Mechanical adding machine
Burroughs Adding Machine Company
ca. 1905

One of the earliest adding machines marketed for American business use.

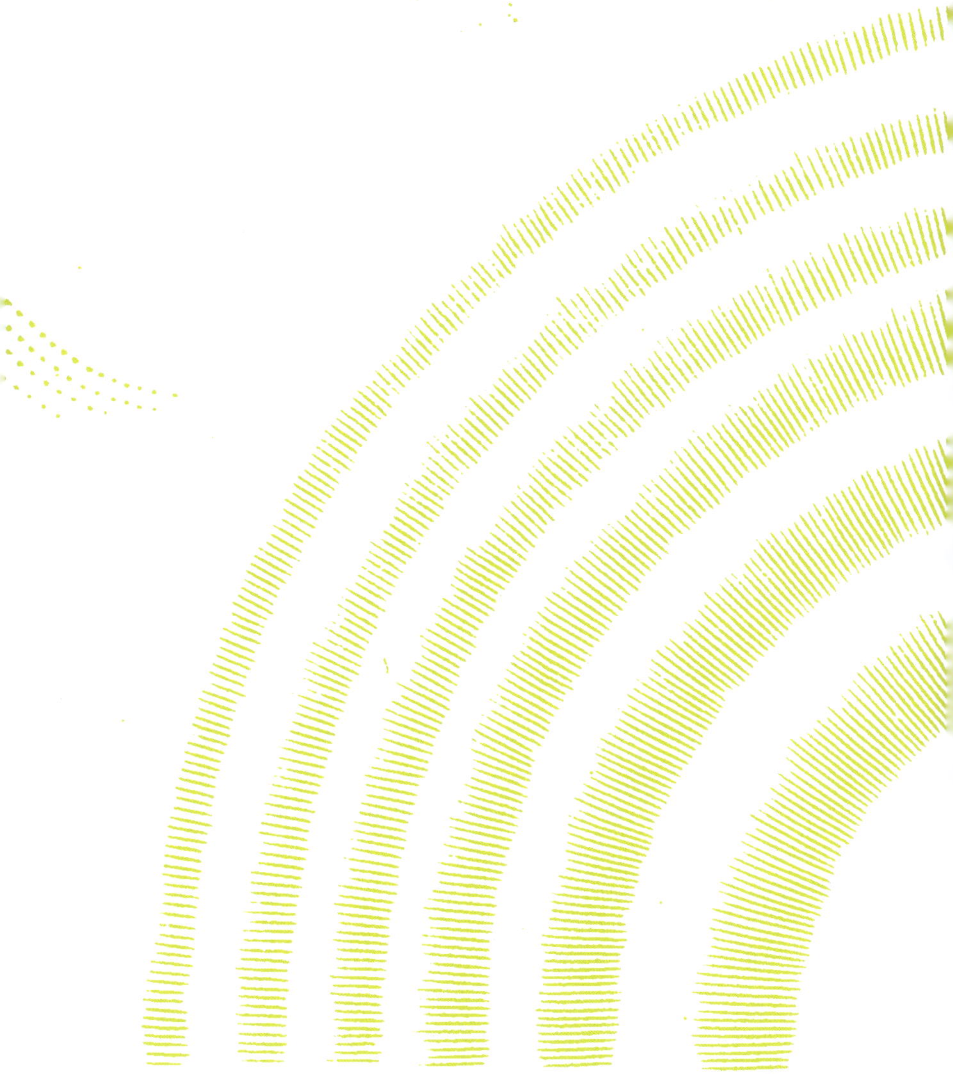
zyxvtsr qponmlkih gfed

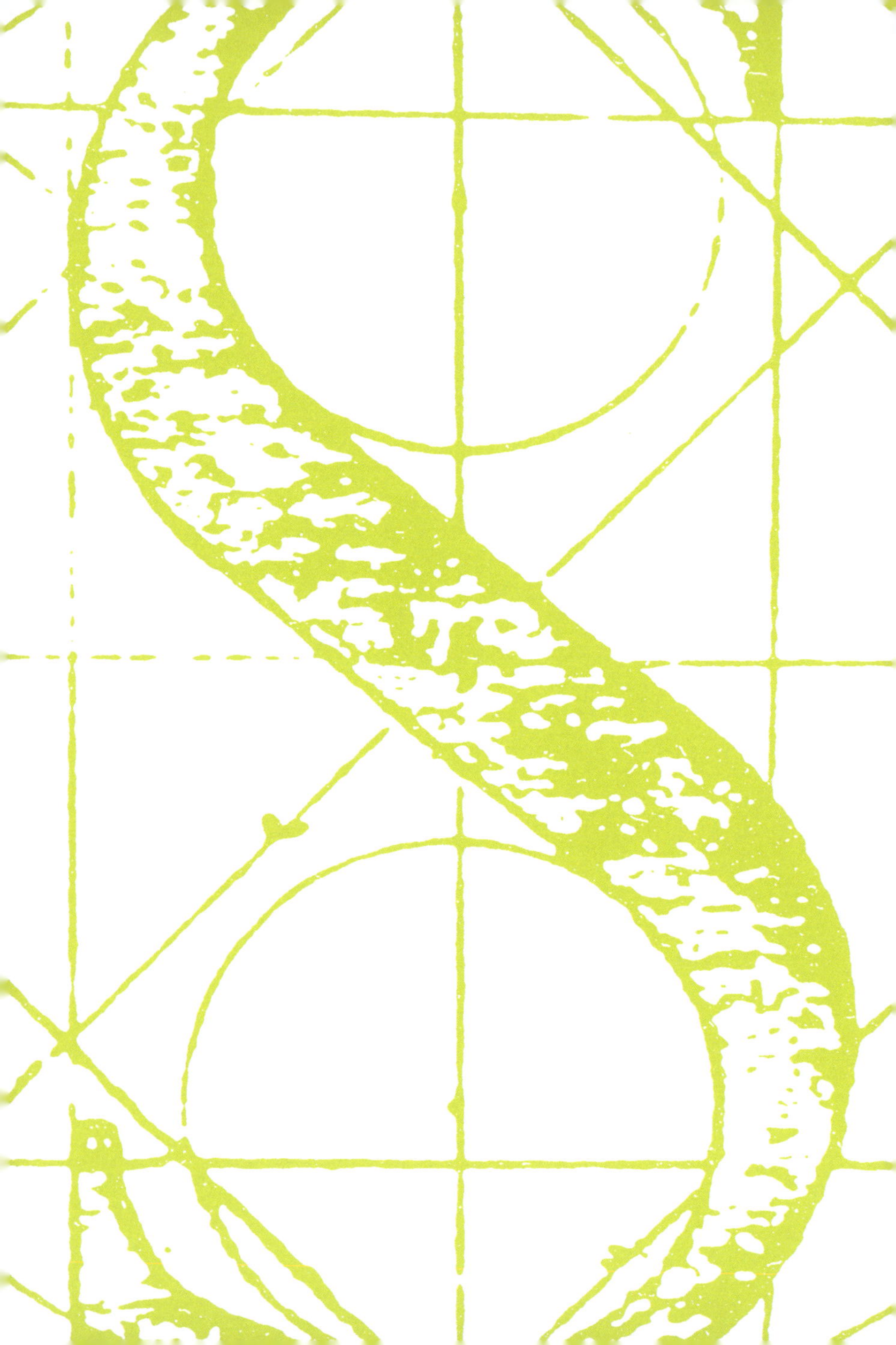

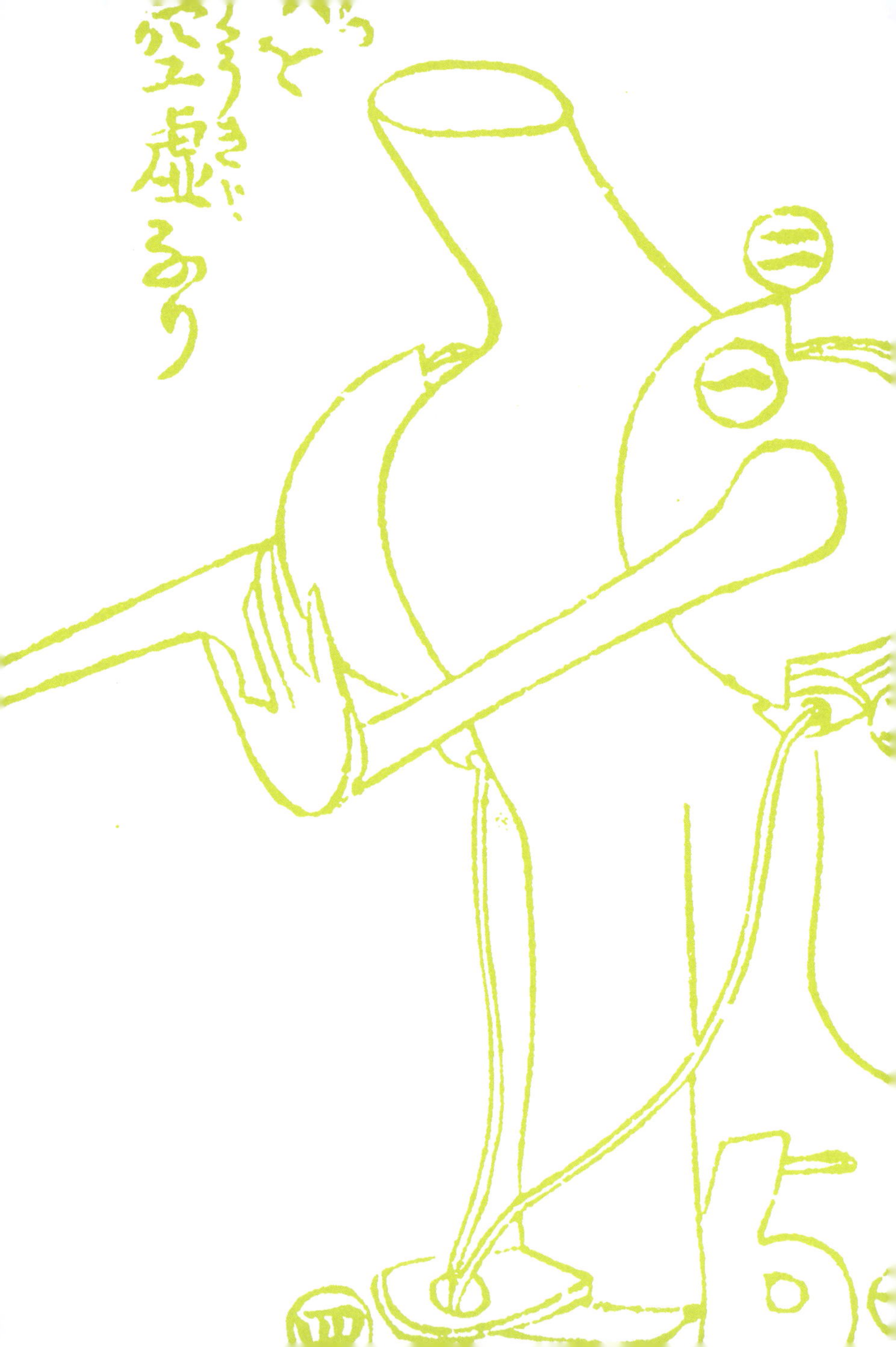
と
空虚あり

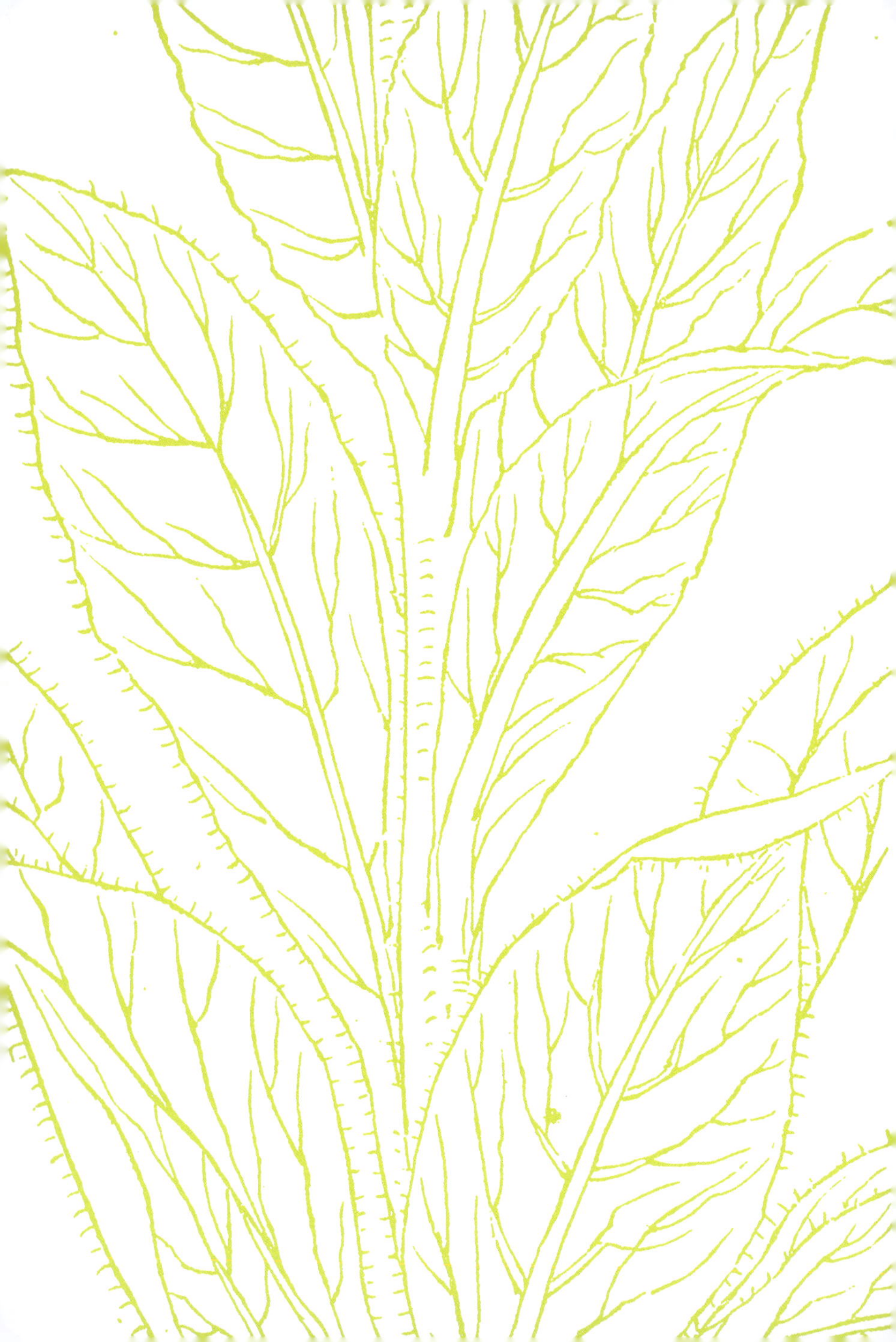

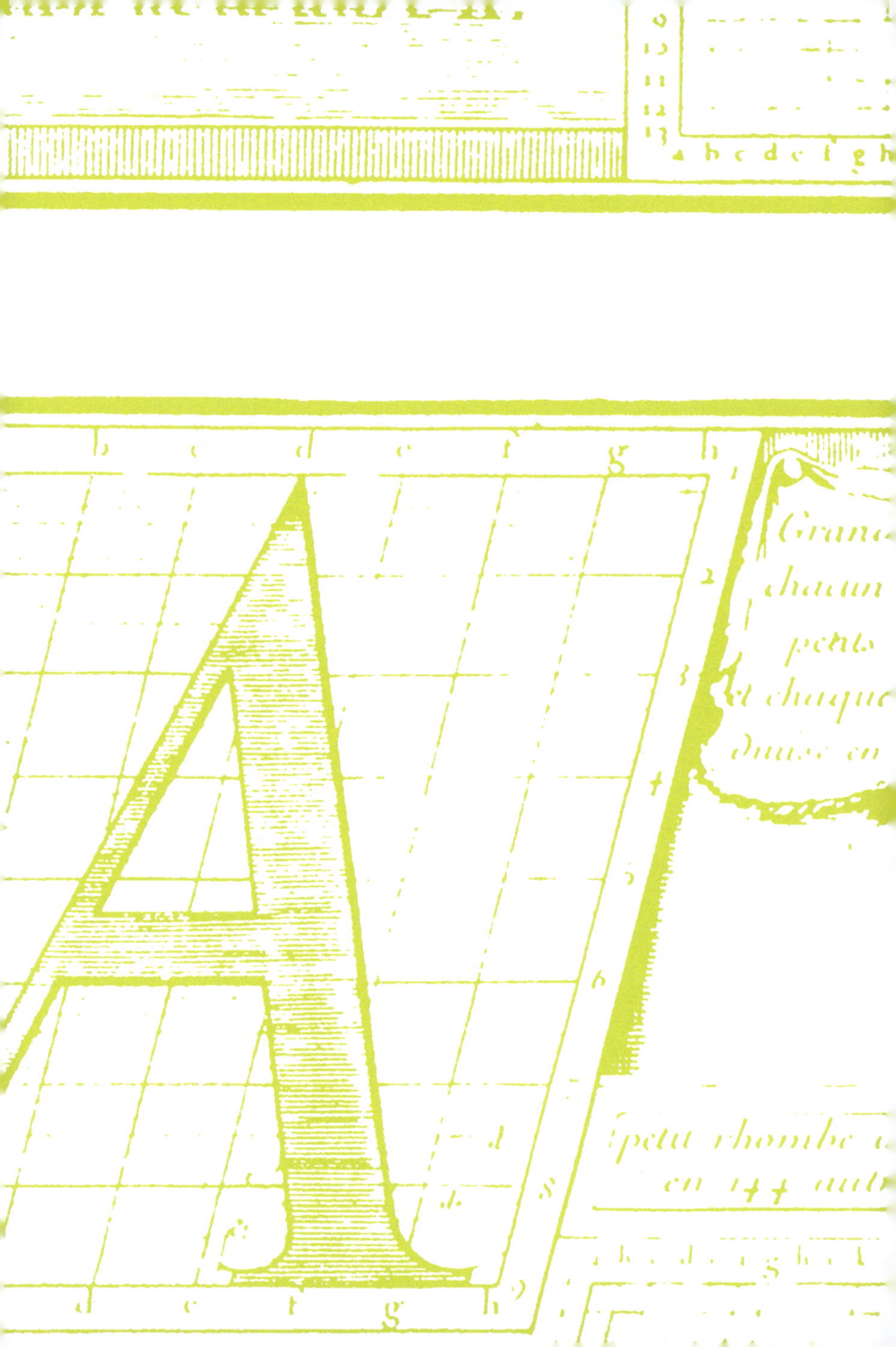

PART VI

Scientific Inquiry & Theory Building

The history of science is embodied in early printed books. Special Collections holds many of the most important works from this history, including first editions by Newton, Darwin, Copernicus, Curie, and Einstein. Some of these names are chiseled above the entrance to the Posner Center. Gathered here is a selection of major scientific works in Special Collections.

DESCARTES, René (1596–1650).
Discours de la méthode. Leiden, 1637.

The origin of the scientific method.

In *Discours de la méthode* Descartes laid out the principles of the scientific method. In the woodcut illustration shown, Descartes investigates the function of the human eye. Like other seventeenth-century scientists, Descartes likened the parts of human anatomy to components of a complex machine. Here, the eye is shown in cutaway, depicted as an optical instrument: lines trace three images that enter the eye at three points. The images refract through the eye's lens and vitreous humor and imprint on the retina at points T, S, and R.

What is most striking about this illustration is that it shows a gruesome experiment Descartes actually performed. The accompanying text conducts the reader through the experiment: take the eye of a freshly dead man and cut away the retinal wall at the back of the eye, leaving the lens and cornea intact. In a darkened room, set the eye in a hole cut into a tablet and face the tablet toward a brightly illuminated object. By placing a piece of paper or an eggshell behind the sectioned eye, an image of the object will appear as clearly as it would when imprinted onto the retina.

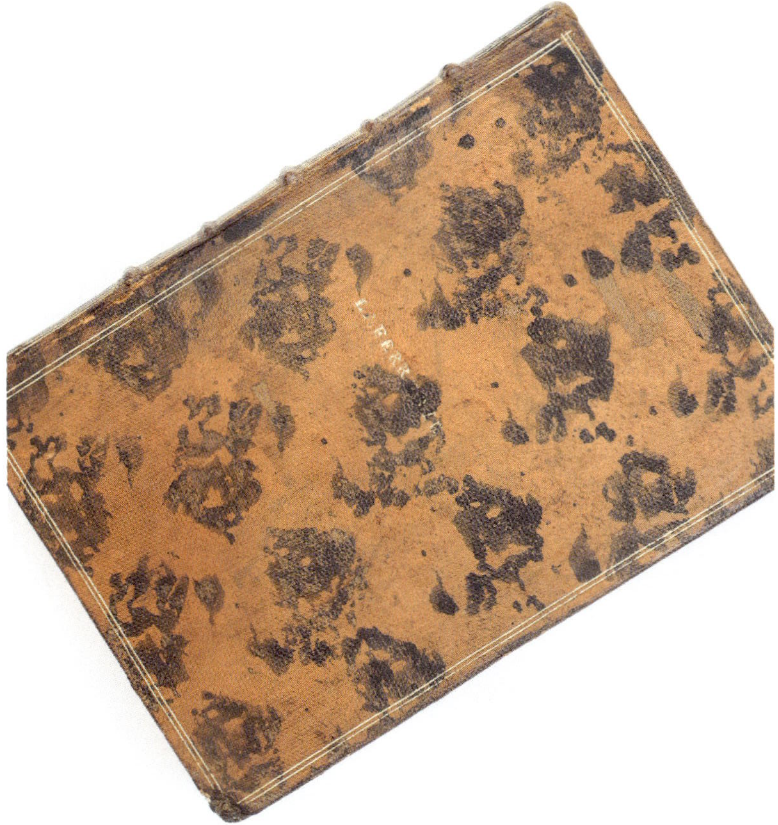

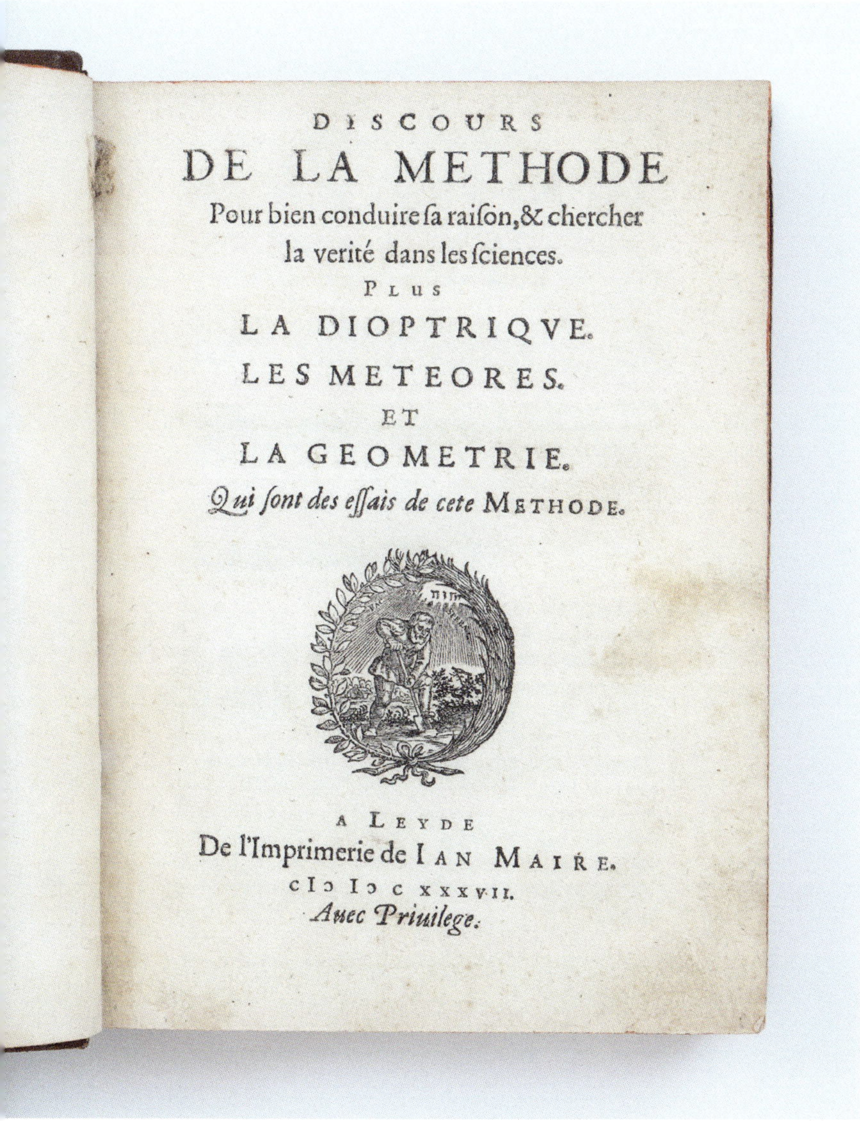
DISCOURS

DE LA METHODE

Pour bien conduire ſa raiſon, & chercher la verité dans les ſciences.

PLUS

LA DIOPTRIQVE.

LES METEORES.

ET

LA GEOMETRIE.

Qui ſont des eſſais de cete METHODE.

A LEYDE

De l'Imprimerie de IAN MAIRE.

CIↃ IↃ C XXXVII.

Auec Priuilege.

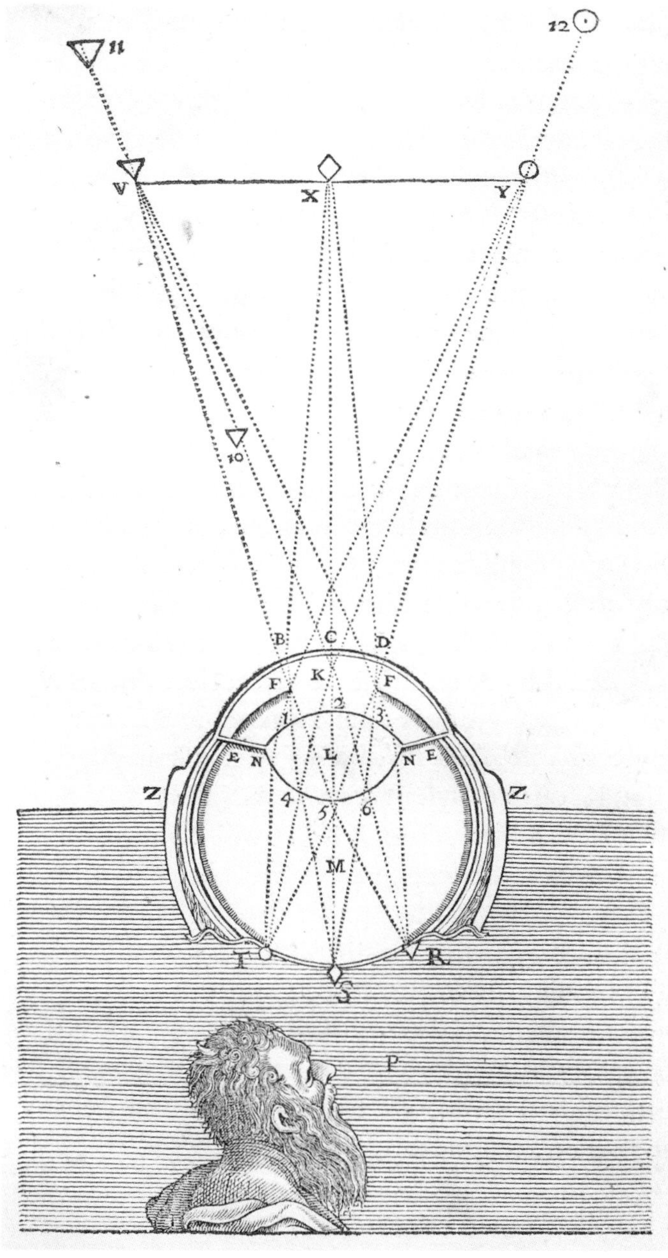
11
12
V
X
Y
10
B
C
D
K
F
F
1
2
3
E N
L
N E
Z
Z
4
5
6
M
T
S
R
P

PASCAL, Blaise (1623–1662).
Traitez de l’équilibre des liqueurs,
et de la pesanteur de la masse de l’air.
Paris, 1663.

The first printed account of Pascal's Law.

Pascal's posthumous treatise on the "weight and mass of air" offered the first correct account of pressure and atmospheric weight. Drawing on earlier work by Simon Stevin (1548–1620), Galileo, and Evangelista Torricelli (1608–1647), Pascal's work describes a series of experiments on air pressure and fluid equilibrium — most famously a 1648 experiment in which his brother-in-law ascended the Puy de Dôme in central France with a barometer to demonstrate that air has measurable weight. A landmark in the study of hydrostatics, this work helped establish empirical methods in physics and meteorology.

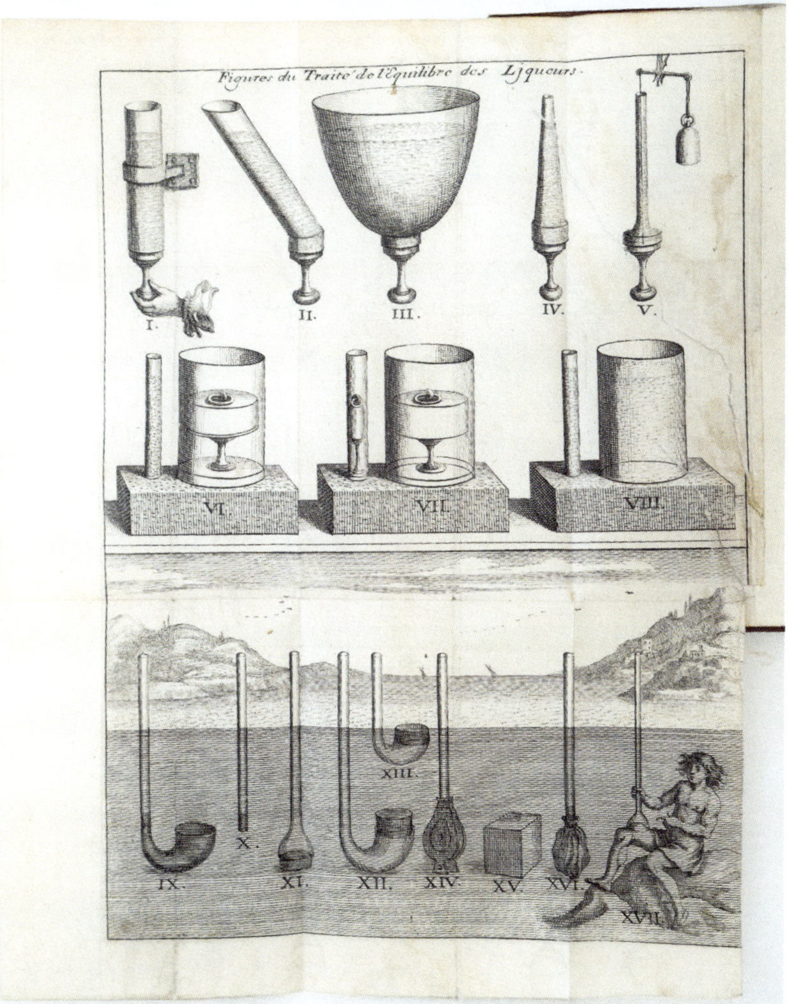
Figures du Traité de l'Equilibre des Liqueurs.
I.
II.
III.
IV.
V.
VI.
VII.
VIII.
IX.
X.
XI.
XII.
XIII.
XIV.
XV.
XVI.
XVII.

DARWIN, Charles (1809–1882).
On the Origin of Species by Means of Natural Selection. London, 1859.

The first comprehensive account of natural selection and the foundation of evolutionary biology.

Called "the most important single work in science," Darwin's *Origin* offered a new account of biological diversity, explaining how species diverge and evolve over vast stretches of time. Special Collections holds two copies of the *Origin*. One copy (shown) is inscribed in Darwin's handwriting to his close friend, John Maurice Herbert (1808–1882). A third copy of the *Origin* is held at CMU's Hunt Institute for Botanical Documentation (see "HUNT BOTANICAL LIBRARY," page 551).

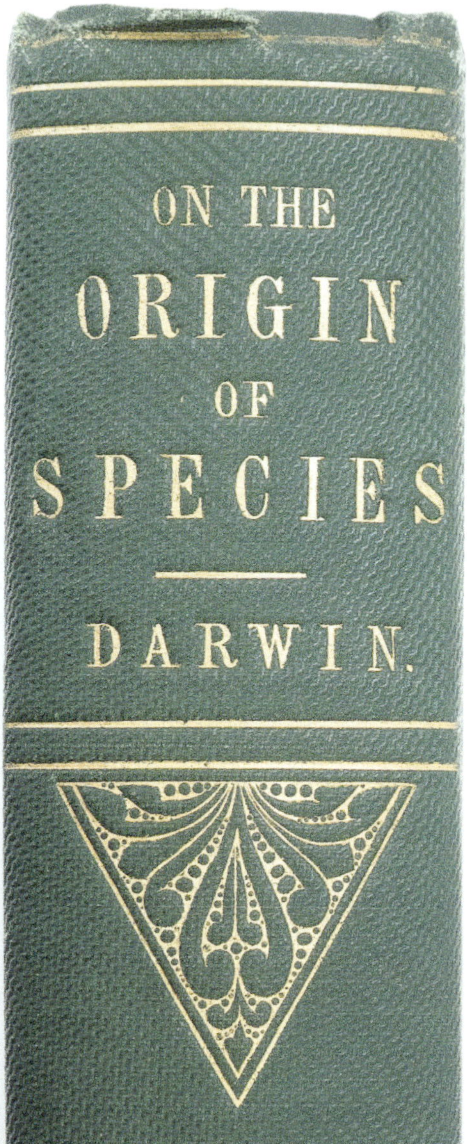
ON THE
ORIGIN
OF
SPECIES
DARWIN.

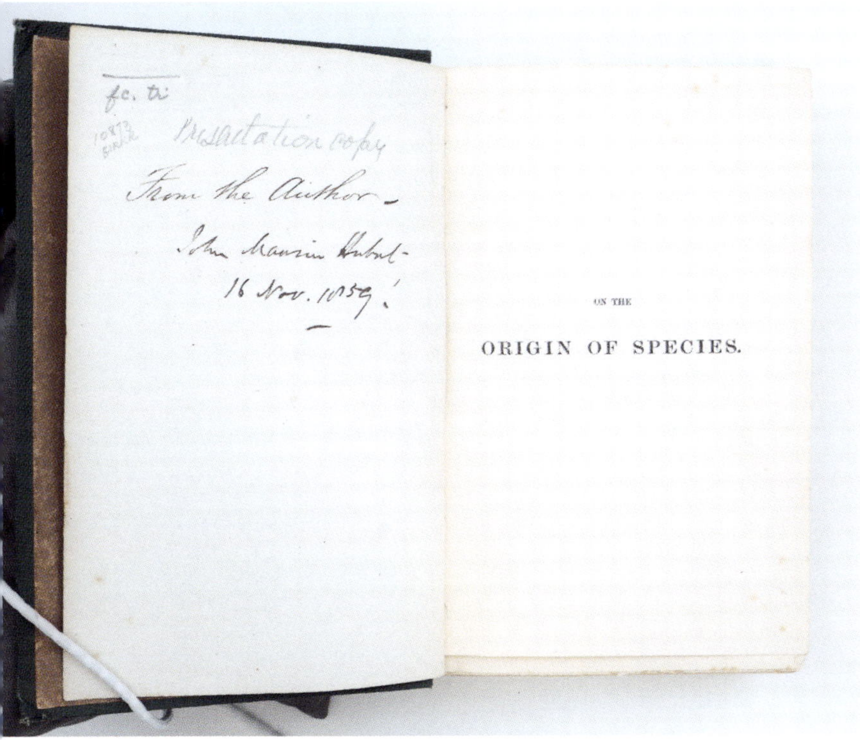
Presentation copy
From the Author
16 Nov. 1859
ON THE
ORIGIN OF SPECIES.

DARWIN, Charles (1809–1882).
Insectivorous Plants. London, 1875.

Darwin's groundbreaking study of botanical carnivory.

In *Insectivorous Plants*, Darwin documented his observations on the mechanisms by which carnivorous plants ensnare and digest insects. Published nearly two decades after *On the Origin of Species*, the work provided evidence of natural selection at work, illustrating how organisms evolve adaptive strategies to thrive in diverse environments.

While conducting research for the book, Darwin imported a living herbarium of carnivorous plants from around the world, keeping his specimens in what Darwin biographers Adrian Desmond and James Moore describe as a "rogues' gallery" in clay pots, "displaying every devious means of gluing, trapping and drowning their prey." Darwin, a competent draughtsman, drew Figures 7 and 8 in the book.

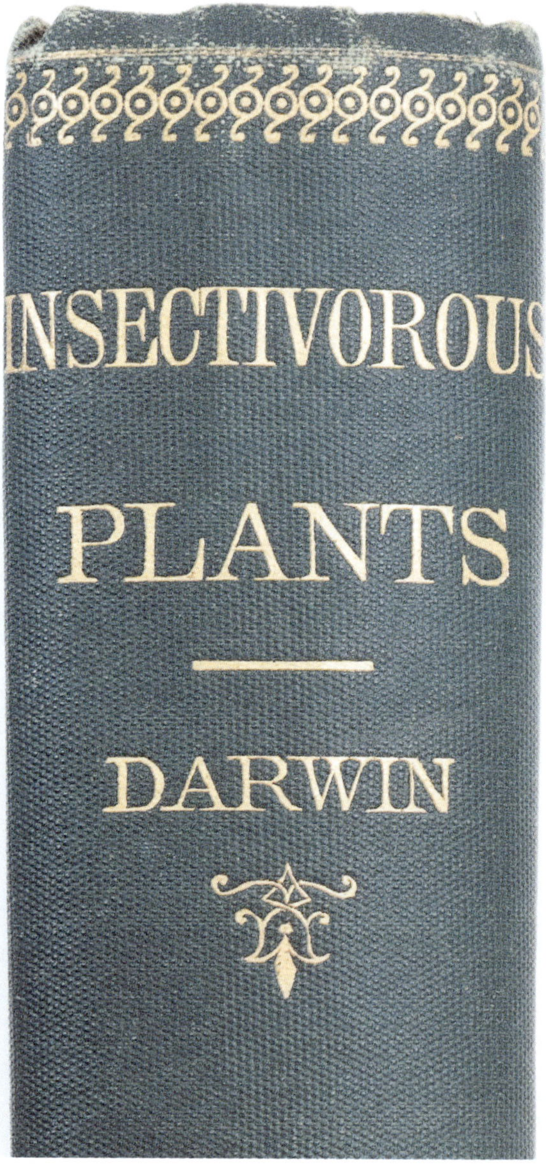
INSECTIVOROUS
PLANTS
DARWIN

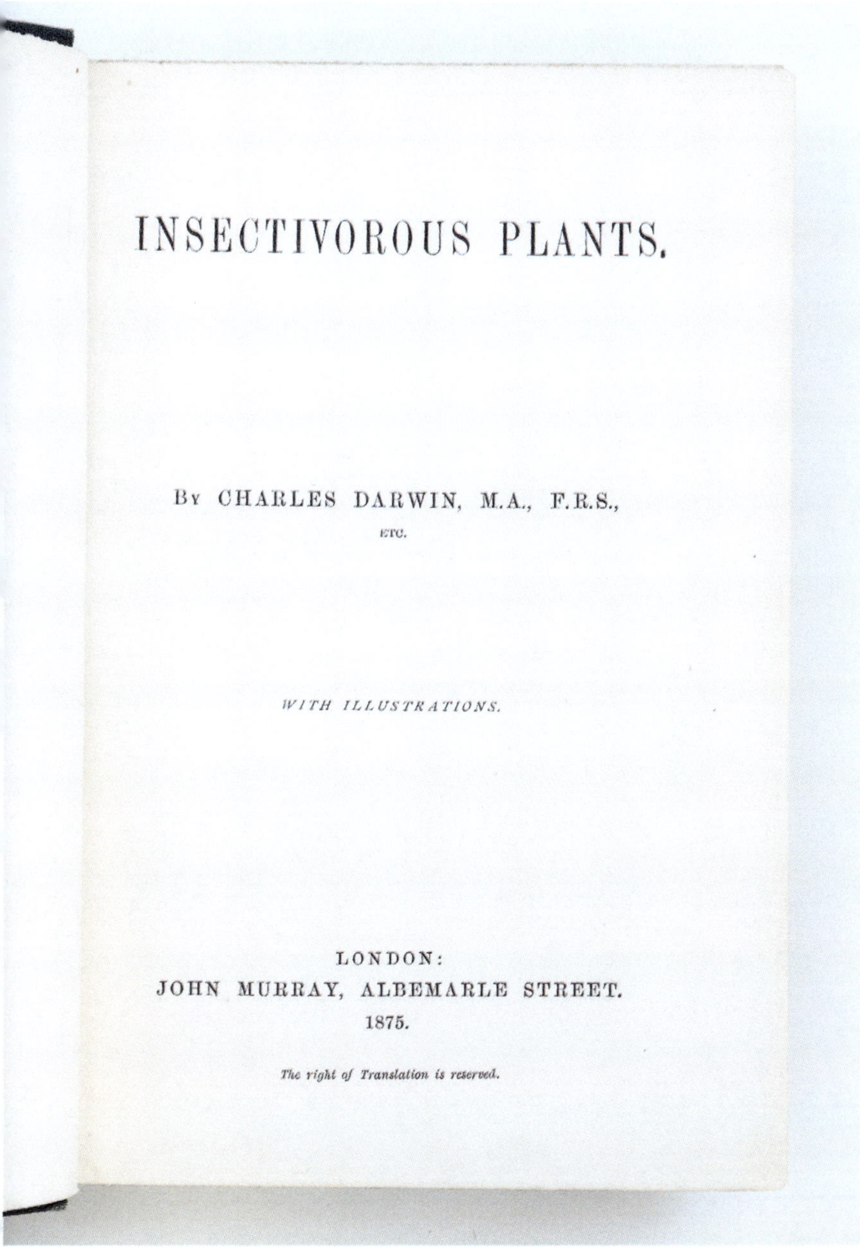

INSECTIVOROUS PLANTS.

BY CHARLES DARWIN, M.A., F.R.S.,
ETC.

WITH ILLUSTRATIONS.

LONDON:
JOHN MURRAY, ALBEMARLE STREET.
1875.

The right of Translation is reserved.

short discal tentacles of a greenish, colour. Thes little masses incessantly change their forms and posi tions, being never at rest. A single mass will ofte separate into two, which afterwards reunite. The movements are rather slow, and resemble those Amoebæ or of the white corpuscles of the blood. W

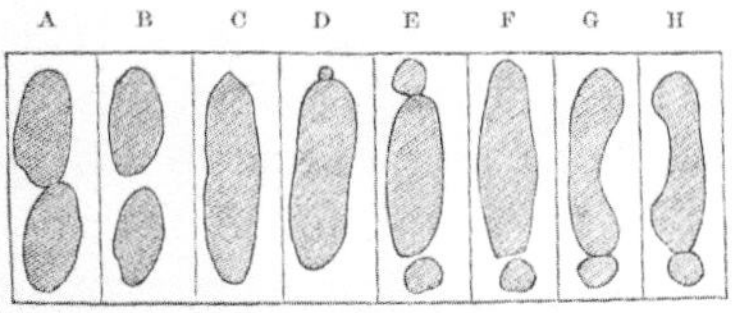

Fig. 7.

(*Drosera rotundifolia.*)

Diagram of the same cell of a tentacle, showing the various forms successively assumed by the aggregated masses of protoplasm.

may, therefore, conclude that they consist of pro plasm. If their shapes are sketched at interv of a few minutes, they are invariably seen to ha

The cell above figured was from a tentacle of a dark red leaf, which had caught a small moth, and was examined under water. As I at first thought that the movements of the masses might be due to the absorption of water, I placed a fly on a leaf, and when after 18 hrs. all the tentacles were well inflected, these were examined without being immersed in water. The cell

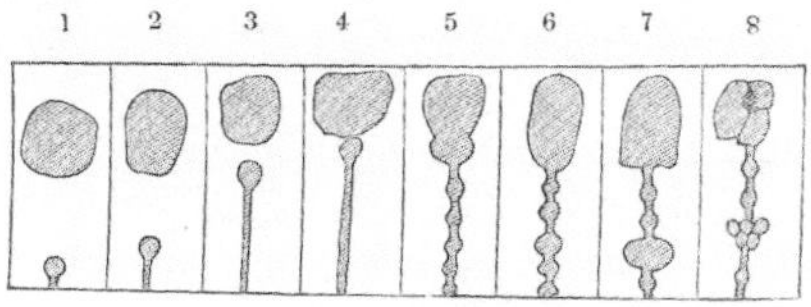

FIG. 8.

(*Drosera rotundifolia.*)

Diagram of the same cell of a tentacle, showing the various forms successively assumed by the aggregated masses of protoplasm.

here represented (fig. 8) was from this leaf, being sketched eight times in the course of 15 m. These sketches exhibit some of the more remarkable changes which the protoplasm undergoes. At first the

IBN AL-HAYTHAM, Ḥasan (965–1040).
Opticae thesaurus. Basel, 1572.

A foundational work in the science of optics and light.

In his *Book of Optics*, Arab mathematician and scientist Hasan Ibn al-Haytham (Latinized Alhazen) described the principles of optics, including the laws of refraction. He was also the first to articulate an accurate theory of vision, observing that visual perception is the result of reflected light entering the eye. His writings on optics were cited by later Western scientists, including Johannes Kepler, Isaac Newton, and Galileo. The work's woodcut frontispiece, shown here, illustrates optical phenomena — refraction, reflection, and perspective — in charming vignettes.

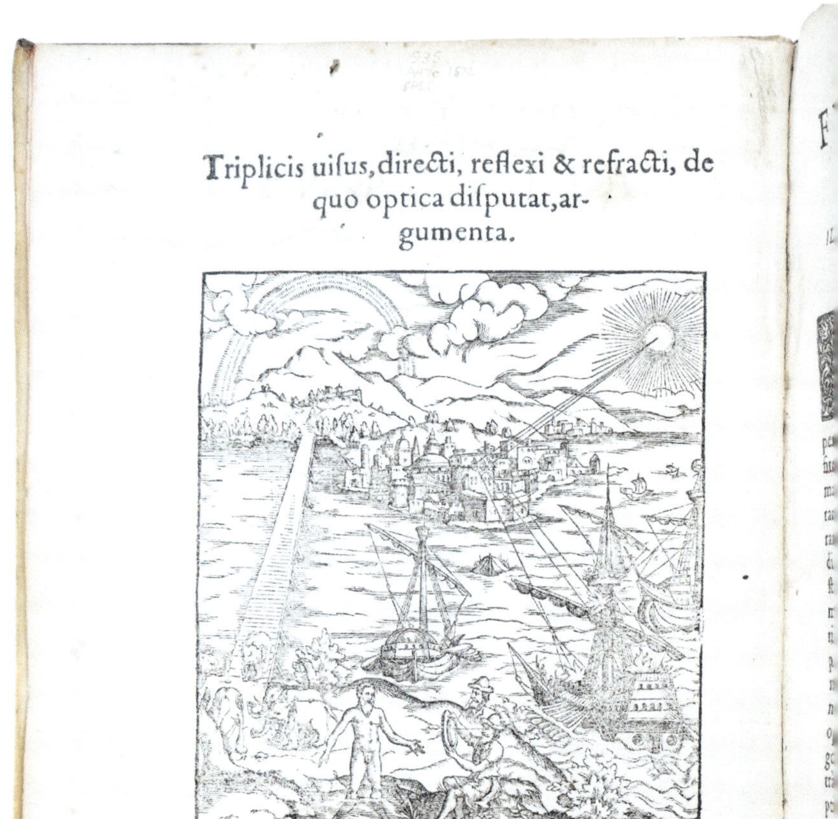

Triplicis uiſus, directi, reflexi & refracti, de
quo optica diſputat, ar-
gumenta.

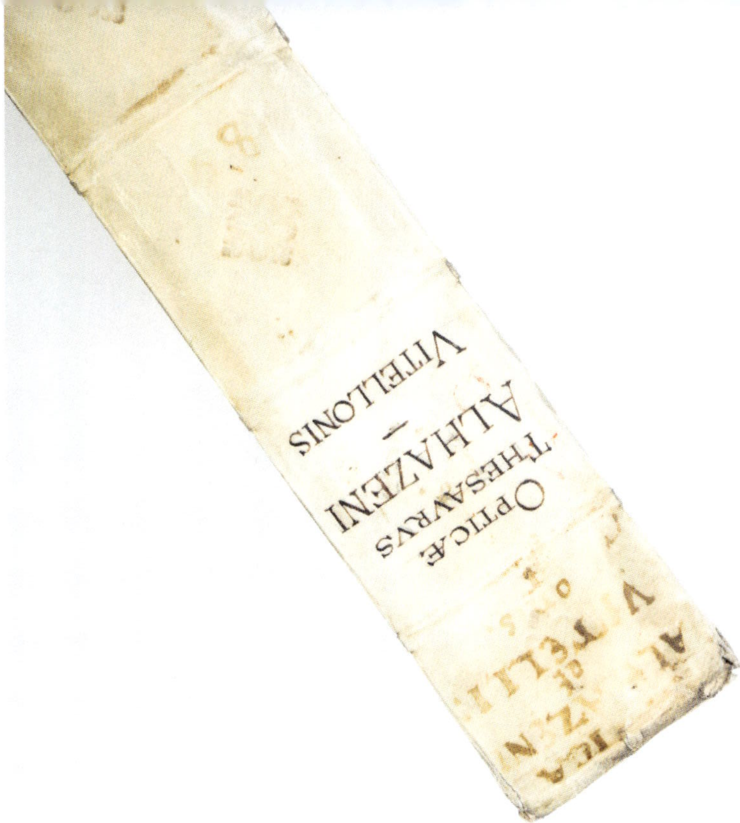
OPTICÆ
THESAVRVS
ALHAZENI
VITELLONIS

quodcunq; punctum illius arcus ducatur linea à puncto b: tenebit cū contingente illius puncti angulum obtusum ex parte e: [Nam semidiameter circuli ad rectam lineam per punctum illud, speculum tangētem educta, declinat à puncto b uersus e, & facit cum tangente angulum rectum per 18 p 3. Itaq; angulus tangentis & lineæ rectę à puncto b ad pūctum tactus, ductę, maior est recto: ideoq; per 11 d 1 obtusus:] & linea ducta à puncto a ad illud punctum, tenebit cū contingente illa angulū acutum ex parte l: [quia angulus semidiametri & tangentis rectus est per 18 p 3: & recta à puncto a ad illud punctū ducta secat circulū d e g.] Quare si ab illo puncto fieret reflexio: esset angulus acutus æqualis obtuso [per 12 n 4.] Iterū à nullo puncto arcus g l potest fieri reflexio. Sumatur enim punctum quodcunq;: & sit z: & ducatur linea a z o, secans perpendicularē in puncto o: & ducatur linea contingens circulum in puncto z: [per 17 p 3] quæ cadit necessariò inter b g, & b l: [quia punctū z est inter puncta g & l] & sit m z: & f g circulū cōtingat in puncto g. Palā ex superiorib. [18 n] quòd proportio b n ad n q, sicut b f ad f q. Eodem modo proportio b n ad n o, sicut proportio b m ad m o: Sed [per 9 ax. 8 p 5] maior est proportio b n ad n q, quā b n ad n o. Igitur [per 11 p 5] maior est proportio b f ad f q, quàm b m ad m o. Quod planè impossibile: cum [per 9 ax] b f sit minor b m, & f q maior m o. [ideoq; ratio b f ad f q minor est ratione b m ad m o, ut patet ex 8 p 5.] Restat ergo, ut à puncto z non fiat reflexio. Verùm quòd ab aliquo puncto arcus g d non fiat reflexio: sic constabit. Sumatur quodcunq; punctum: & sit t: & ducatur linea b t: & linea a t h, secans b n in pūcto h: & [per 17 p 3] ducatur cōtingens circulū in pūcto t: quę sit p t. Erit ergo ex superiorib. [18 n] ꝓportio b n ad n h, sicut b p ad p h: & b n ad n q, sicut b f ad f q: Sed [per 9 ax. 8 p 5] b n ad n h maior est, quā b n ad n q: ergo [per 11 p 5] maior est proportio b p ad p h, q̄ b f ad f q: quod planè falsum: cum [per 9 ax] b t sit maior b p, & p h maior f q. [ideoq; ratio b p ad p h minor est ratione b f ad f q, ut constat ex 8 p 5.] Restat ergo, ut à nullo puncto arcus g d fiat reflexio puncti b. Quare quodlibet punctum ab uno solo puncto speculi reflectitur ad uisum. Ergo una sola erit linea reflexionis cuiuslibet puncti uisi. Quare unica unius puncti imago. Si aūt punctum b fuerit in perpendiculari uisuali: palàm [per 11 n 4] quòd reflectetur ab uno solo puncto, per quod perpendicularis, tantū: & unica erit eius imago: & erit propter continuitatem aliorum punctorum, in loco imaginis proprio.

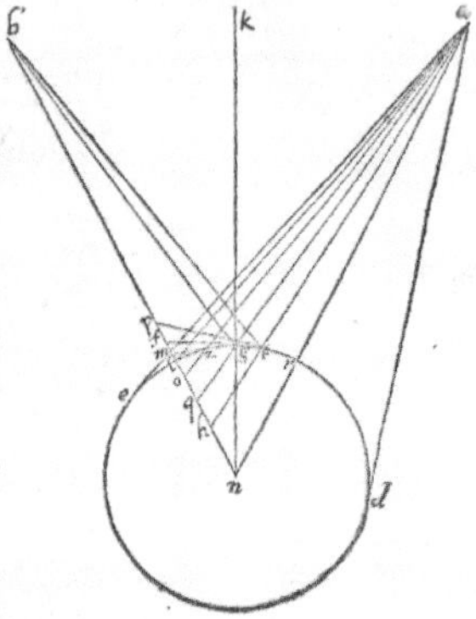

30. *Si duo perpendicularis incidentis puncta, à speculo sphærico conuexo ad unum uisum reflectantur: locus tum imaginis tum reflexionis, puncti centro speculi propinquioris erit remotior: imaginis ab eodem centro: reflexionis à uisu. 17 p 6.*

Amplius: si in aliqua diametro sumātur duo puncta ex parte centri eadem: locus imaginis centro propinquioris, erit remotior à centro sphęrę, loco imaginis puncti remotioris à cētro sphęrę. Verbi gratia dico, quòd locus imaginis puncti p, remotior est à centro, loco imaginis puncti b: & punctum reflexionis puncti p remotius ab a puncto uisus, puncto reflexionis puncti b, quod est punctum g. Dico, quòd punctum p non reflectitur, nisi ab aliquo puncto arcus g l. Palàm enim, quòd non reflectitur ab aliquo puncto arcus l e, nisi à puncto l: [sicq; per 11 n 4 reflectetur per perpendicularem b l n, nō ad uisum, in a positum] nec à puncto g: cū b reflectatur ab eo, [ad uisum scilicet a ex thesi: ideoque nullum aliud punctum, ut p, ab eodem puncto g reflectetur ad eundē uisum a p pręcedētē numerū.] Et si dicatur, qđ ab aliquo pūcto arcus g d: sit illud pūctū t:

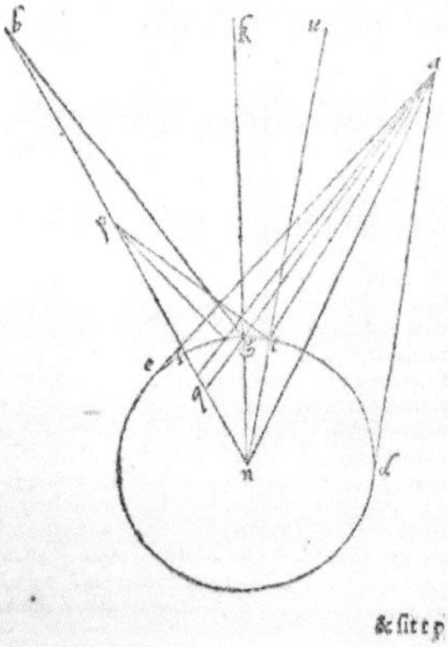

& sit t p

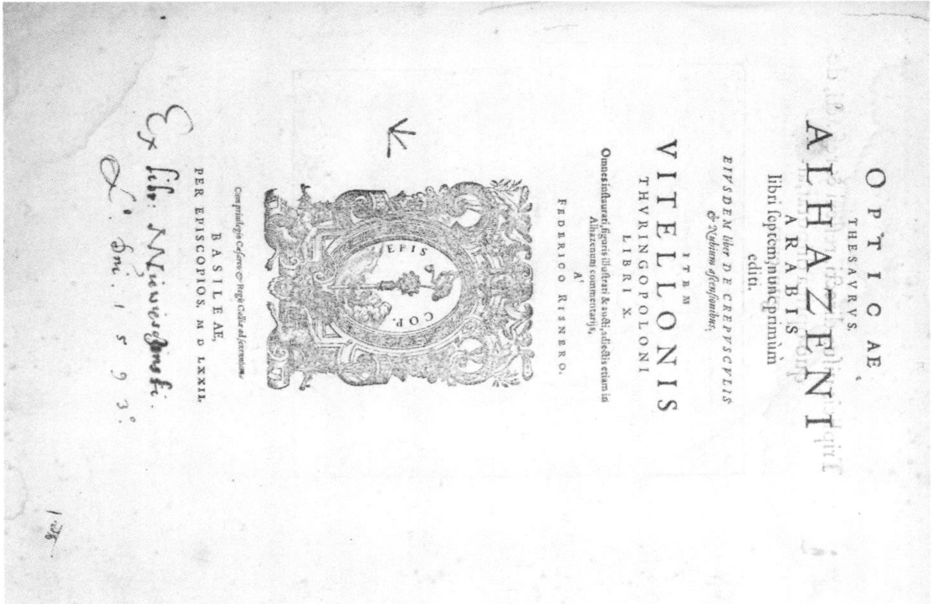
OPTICAE THESAVRVS.

ALHAZENI ARABIS libri septem, nunc primùm editi.

EIVSDEM liber DE CREPVSCVLIS & Nubium ascensionibus.

ITEM

VITELLONIS THVRINGOPOLONI LIBRI X.

Omnes instaurati, figuris illustrati & aucti, adiectis etiam in Alhazenum commentarijs,

A FEDERICO RISNERO.

Cum priuilegio Cæsareo & Regis Galliæ ad sexennium.

BASILEAE,
PER EPISCOPIOS. M D LXXII.

Ex libr: Mniewiesinski.
A°. Dni. 1593.

↑ A FULLER HISTORY of SCIENCE

The history of science was long dominated by European innovations and thinkers. This is slowly changing. Special Collections aims to curate books and objects from multiple cultures, languages, and traditions. Held in the Posner Memorial Collection, this edition of al-Haytham's *Book of Optics* — written in Arabic during the tenth-century Islamic scientific renaissance, translated into Latin, and printed in sixteenth-century Switzerland — is emblematic of the migratory and transcultural nature of scientific progress.

CURIE, Marie (1867–1934).
Traité de radioactivité. Paris, 1910.

CURIE, Pierre (1859–1906) and CURIE, Marie.
Sur une substance nouvelle radio-active, contenue dans la pechblende. Paris, 1898.

The definitive work on radioactivity.

Marie Curie's *Traité de Radioactivité* was the first comprehensive work on radioactivity, a term she coined. Expanding on her earlier discoveries of polonium and radium, Curie presents the results of her research into radioactive elements and their properties.

The *Traité* is illustrated with photographic plates documenting the effects of radioactivity on various materials. These include photographic emulsions exposed to radium's invisible rays. Produced using Curie's carefully controlled experiments, these images offered striking proof of radiation's material effects.

Published in two volumes, *Traité de Radioactivité* solidified Curie's status as a preeminent scientist. By 1910, she had already received a Nobel Prize in Physics (1903, shared with Pierre Curie and Henri Becquerel). The discoveries described in this book would contribute to her second Nobel, in Chemistry (1911). Curie's earlier treatise "on a new radioactive substance" was published with her husband and research partner, Pierre Curie, in 1898.

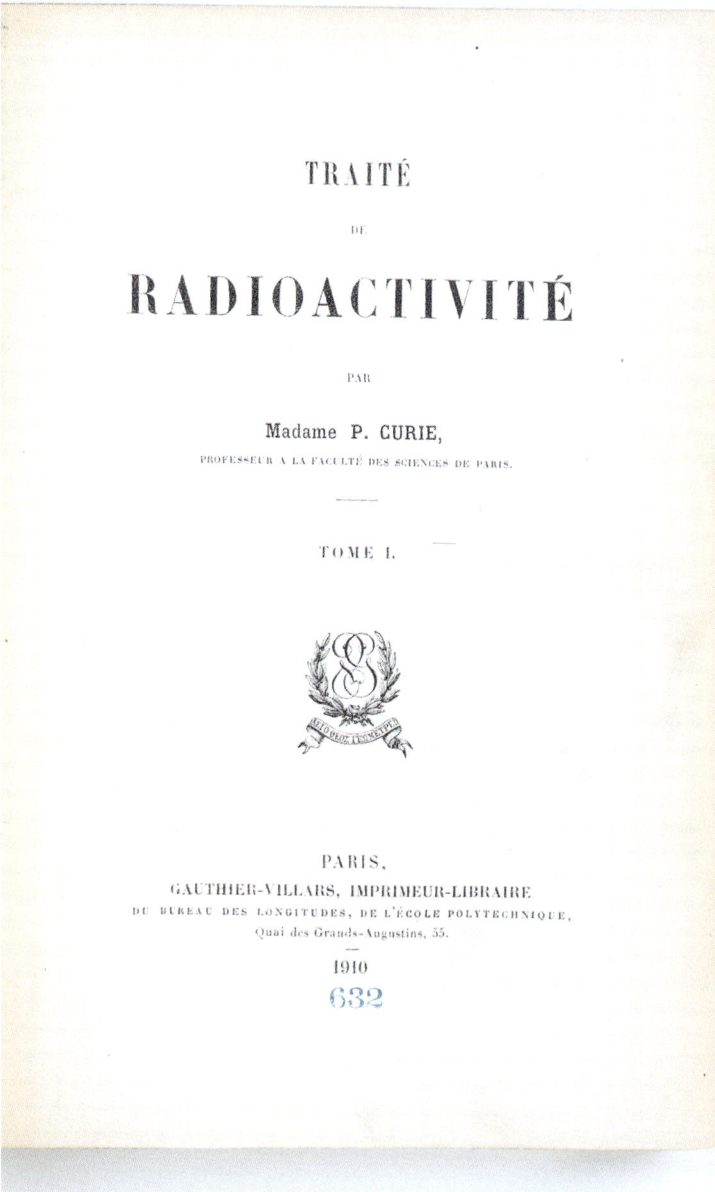

TRAITÉ

DE

RADIOACTIVITÉ

PAR

Madame P. CURIE,

PROFESSEUR A LA FACULTÉ DES SCIENCES DE PARIS.

TOME I.

PARIS,
GAUTHIER-VILLARS, IMPRIMEUR-LIBRAIRE
DU BUREAU DES LONGITUDES, DE L'ÉCOLE POLYTECHNIQUE,
Quai des Grands-Augustins, 55.

1910

Chlorure de Radium pur

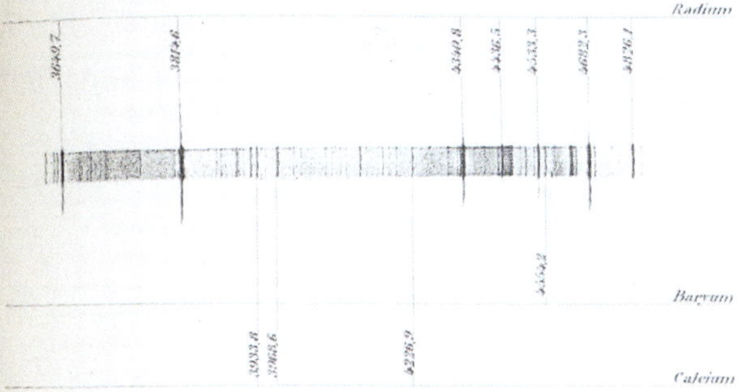

Chlorure de Radium pur
additionné de 0,6 % de chlorure de baryum

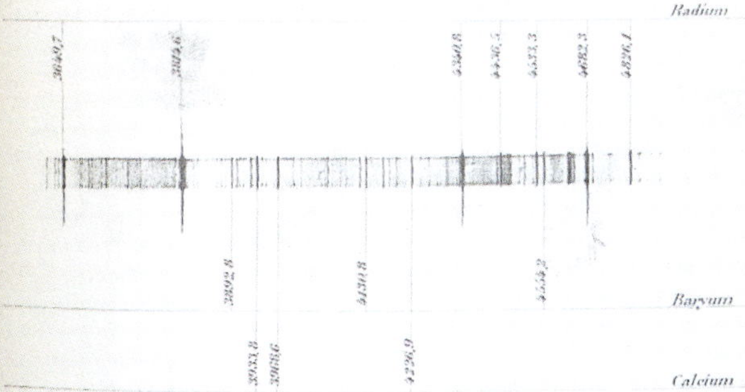

tube à potentiel très élevé, elle se charge négativement, ce qu'il est aisé de vérifier. De même, en touchant du doigt un tube de Crookes loin de la cathode, la paroi touchée devient cathode et il y a répulsion.

» Soit maintenant un tube à cathode plane centrée, de même diamètre que le tube. Les surfaces équipotentielles sont sensiblement planes et le faisceau est cylindrique. Vient-on à réduire le diamètre de la cathode, les surfaces de niveau se courbent et le faisceau est divergent. Si la cathode présente la forme d'un rectangle allongé, les rayons cathodiques doivent s'étaler en éventail dans un plan perpendiculaire à la plus grande dimension du rectangle, et c'est en effet ce qui a lieu.

» Supposons, au contraire, une cathode sphérique concave : à un vide peu avancé, les rayons émis forment un cône creux ; menons un plan tangent à ce cône, le rayon contenu dans ce plan est repoussé d'une manière prépondérante par la partie de la cathode située du même côté de ce plan que le centre. De cette dissymétrie résulte une déviation du rayon qui tend à devenir parallèle à l'axe du cône. On peut également dire que les projectiles cathodiques, rencontrant obliquement les surfaces de niveau, se comportent comme des corps pesants lancés obliquement de haut en bas. De là cet allongement bien connu du foyer cathodique, d'autant plus marqué que le vide est plus avancé et le champ, par suite plus intense, près de la cathode. Plaçant au-devant de celle-ci un diaphragme à deux trous, on a deux faisceaux concourants, rectilignes à partir du diaphragme, se coupant cependant au delà du centre de courbure de la cathode ; c'est donc surtout au voisinage de celle-ci que se produit l'inflexion des trajectoires, là précisément où le champ a son maximum d'intensité. »

PHYSICO-CHIMIE. — *Sur une substance nouvelle radio-active, contenue dans la pechblende* (¹). Note de M. **P. Curie** et de Mme **S. Curie**, présentée par M. Becquerel.

« Certains minéraux contenant de l'uranium et du thorium (pechblende, chalcolite, uranite) sont très actifs au point de vue de l'émission des rayons de Becquerel. Dans un travail antérieur, l'un de nous a montré que

(¹) Ce travail a été fait à l'École municipale de Physique et Chimie industrielles. Nous remercions tout particulièrement M. Bémont, chef des travaux de Chimie, pour les conseils et l'aide qu'il a bien voulu nous donner.

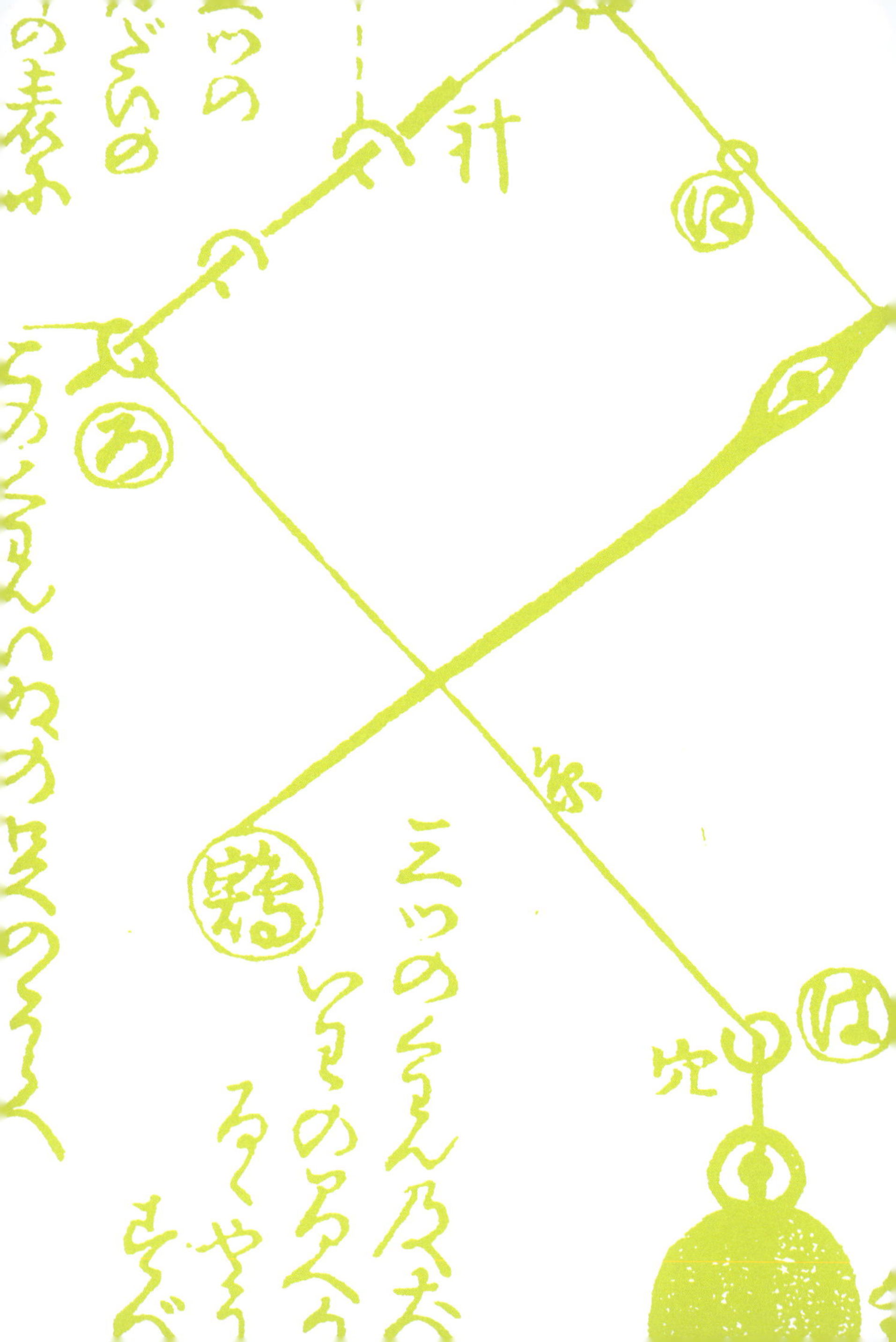

ろ
あ
鶴
穴
は

Preclariſſimus liber e
cacissimi: in artem Geo

ficiē applicatioqꝫ nō direc
linee recte rectiline⁹ angu
ſteterit duoqꝫ anguli vtrob
¶ Lineaqꝫ linee ſupſtās ei

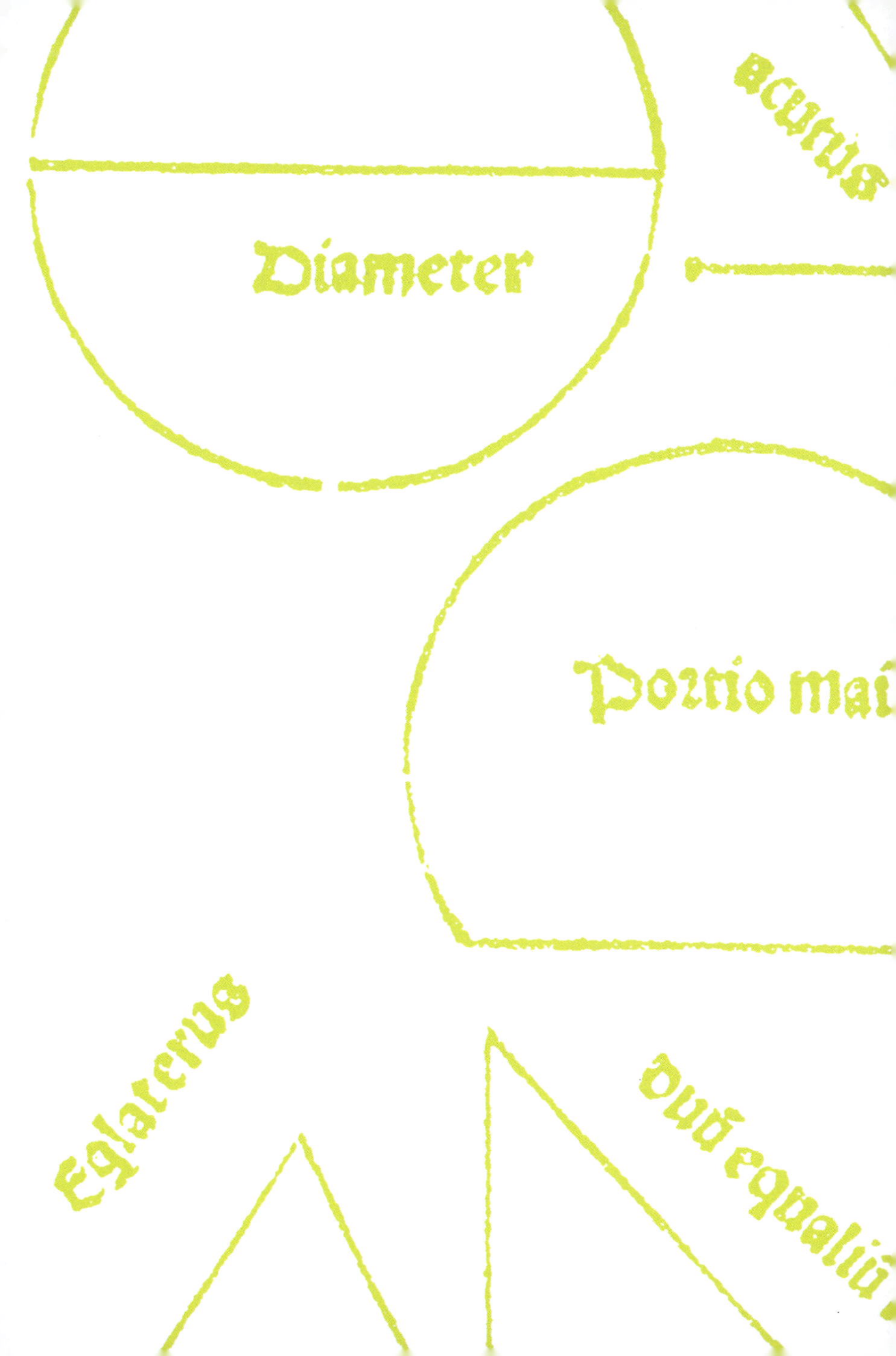
Diameter
Poꝛtio mai

PART VII

Technology & Media

The history of the book is also the history of a technology: the printed word. Described here are examples of recordkeeping and media across forty centuries, from a Babylonian clay tablet to one of the earliest works on computer-aided type design.

CUNEIFORM TABLET.
Bismaya (Adab, Iraq), circa 2350 BCE.

Ancient recordkeeping.

This is the oldest object in Special Collections: a tablet fashioned of clay bearing a cuneiform inscription in Akkadian, a Semitic language spoken in Bronze-Age Mesopotamia. The inscription is difficult to decipher, but appears to be a record for a temple in the Babylonian city of Bismaya, present-day Adab, Iraq.

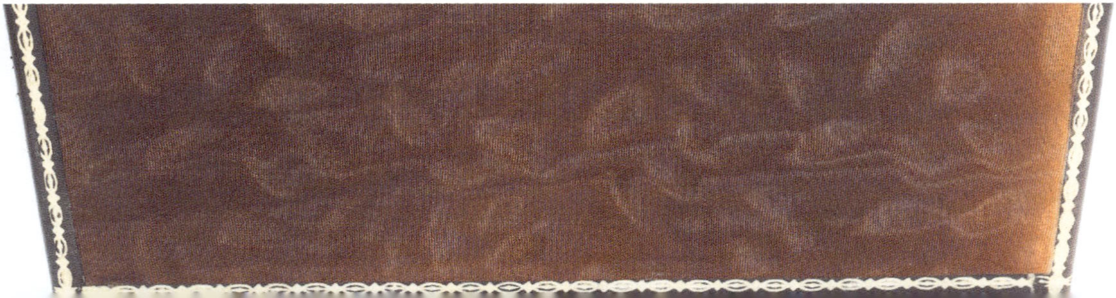

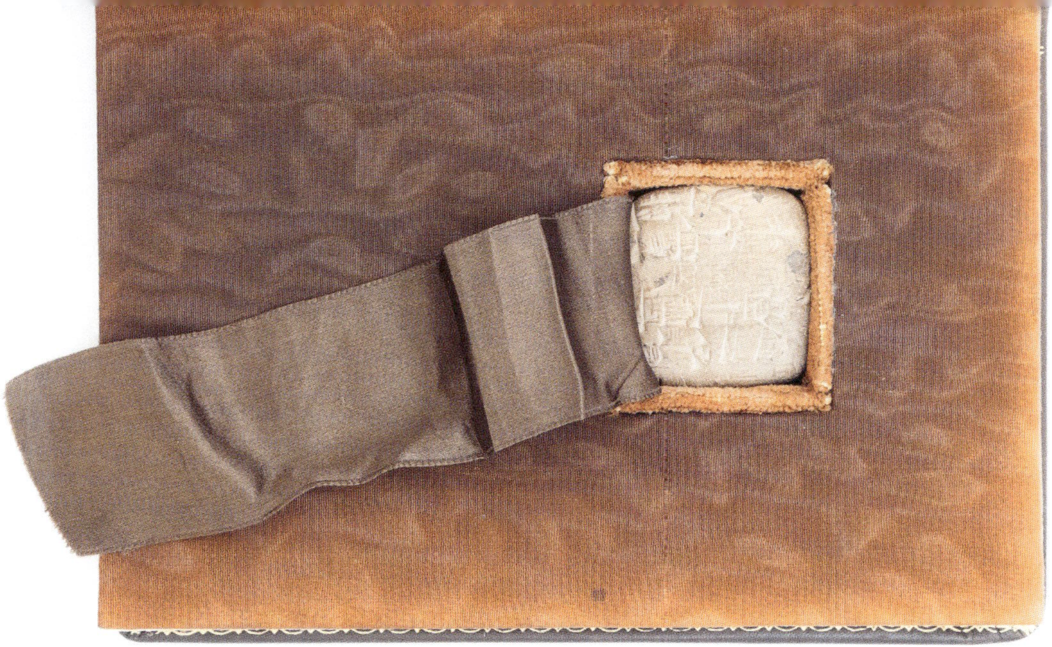

GUTENBERG BIBLE [fragment].
Mainz, circa 1455.

An original leaf from the first printed Bible.

This leaf was removed from an imperfect copy of the Gutenberg Bible and sold as a fragment. It came to Special Collections from the library of Rachel McMasters Miller Hunt, an important collector of rare botanical books who, with her husband Roy A. Hunt, funded the construction of CMU's Hunt Library. Gutenberg's innovation combined the mechanistic action of a screw press with cast-in-metal letters that were inked to print text.

The invention of the printing press and movable type was transformational: during the fifty years after Gutenberg's death, an estimated fifty thousand editions were printed, amounting to nearly half a million copies.

Printed with the Latin Vulgate text of the Bible, Special Collections' fragment is the 235th leaf of the Gutenberg Bible and carries a passage from the second Book of Ezra (Nehemiah).

ad sinistram phadaia · misahel · ⁊ mel-
chia · ⁊ asum · ⁊ aseph dana zacharia ·
⁊ mosollā. Et aperuit esdras librū co-
rā omī pplo. Sup uniūsum quippe po-
pulū eminebat. Et cū aperuisset eū: ste-
tit omīs pplus. Et benedixit esdras do-
mino deo · voce magna. Et rspōdit
omīs pplus · amē · amen: eleuās man⁹
suas. Et incuruati sunt: ⁊ adorauerūt
deū proni in terrā. Porro iesue · et ba-
ni · et serabia · yamin accub · septhai · o-
dia · maasia · celita · azarias · iozabed ·
anani · phalaia leuite · silentiū faciebāt
in pplo: ad audiēdā legē. Popul⁹ aūt
stabat in gradu suo. Et legerūt in li-
bro legis dei distincte · ⁊ apperte ad intel-
ligendū: ⁊ intellexerūt cū legeretur. Dixit
aūt neemias · ipē ē athersatha · ⁊ esdras
sacerdos ⁊ scriba · et leuite interpretātes
uniūso pplo. Dies sanctificat⁹ ē dño
deo nostro. Nolite lugere ⁊ nolite flere.
Flebat enī omīs pplus: cū audiret ver-
ba legis. Et dixit eis. Ite comedite pin-
guia ⁊ bibite mulsū: ⁊ mittite partes
eis qui nō pparauerūt sibi: qa sanctus
dies dñi ē: ⁊ nolite ɔtristari. Gaudiū
etenī dñi est fortitudo nr̄a. Leuite aūt
silentiū faciebāt in omī pplo dicentes.
Tacete: qa dies sanct⁹ est: ⁊ nolite dole-
re. Abijt itaq; omīs pplus ut comede-
ret · ⁊ biberet et mitteret partes: ⁊ faceret
leticiā magnā: qa intellexerāt verba q̄
docuerat eos. Et in die scđo ɔgregati
sunt pricipes familiarū uniūsi ppli sa-
cerdotes et leuite ad esdrā scribā: ut in-
terpretaretur eis verba legis. Et inuene-
rūt scriptū ī lege · pcepisse dñm in ma-
nu moysi · ut habitent filij isrł in taber-
naclis in die solēni mensis septimi: et
ut pdicent ⁊ diuulgēt vocē in uniūsis
urbibȝ suis · ⁊ in ihrlm dicētes. Egredi-
mini ī mōtem · et afferte frondes oliue ·
et frondes ligni pulcerrimi · frondes
mirti · ⁊ ramos palmaꝝ: et frondes li-
gni nemorosi · ut fiāt tabernacła sicut
scriptū est. Et egressus est pplus ⁊ attu-
lerūt. Feceruntq; sibi tabernacula unus-
quisq; in domate suo et in atrijs suis:
et in atrijs domus dei · ⁊ in platea por-
te aquarum: et in platea porte ephra-
im. Fecit ergo uniuersa ecclesia eorum
que redierant de captiuitate taberna-
cula: et habitauerunt in tabernaculis.
Nō enī fecerāt a diebȝ iosue filij nun-
taliter filij isrł: usq; ad diē illū. Et fuit
leticia magna nimis. Legit autē in li-
bro legis dei · p dies singulos · a die pri-
mo usq; ad diem nouissimum: et fece-
runt solennitatem septem diebus: et
ī die octauo collectā iuxta ritū. IX
In die aūt vicesimoquarto mēsis hui⁹
ɔuenerūt filij isrł in ieiunio · et ī saccis:
et hum⁹ sup eos. Et separatū est semē
filioꝝ isrł ab omni filio alienigena: et
steterūt ⁊ confitebantur peccata sua et
iniqtates patrū suoꝝ. Et ɔsurrexerūt
ad standū: ⁊ legerūt ī volumine legis
dñi dei sui quater in die: ⁊ quater con-
fitebantur · ⁊ adorabāt dñm deū suum.
Surrexerūt aūt sup gradū leuitaꝝ io-
sue ⁊ bani · et cedmihel · sennī · sabnia ·
abani · sarebias bani · chanani: et cla-
mauerūt magna voce ad dñm deum
suū. Et dixerūt leuite · iosue · ⁊ chedmiel ·
bonni · asebia · serebia · arebia · odaia · se-
bna · fataia. Surgite: benedicite dño
deo nr̄o · ab eterno usq; in eternū. Et be-
nedicāt nomini glorie tue excelso: in
omī benedictione ⁊ laude. Tu ipe dñe so-
lus: tu fecisti celū et celū celoꝝ et omnē
exercitū eoꝝ · terrā ⁊ uniūsa q̄ in ea sūt:
maria ⁊ omnia que in eis sunt. Et tu
uiuificas omīa hec: ⁊ exercit⁹ celi te ado-
rat. Tu ipse dñe deus qui elegisti abrā

iesua: nongentiseptuagintatres. Filij
emmer: millequīquagintaduo. Filij
phessur: mille ducentiquadraginta-
septem. Filij arem: milledecē et septem.
Levite. Filij iosue et cedmihel filioꝝ o-
devie: septuagintaquatuor. Cātores.
Filij sia asaph: centūquadragintao-
cto. Ianitores. Filij sellū filij ather · fi-
lij thelmon filij accub filij athita filij
sobai · centūtrigintaocto. Nathinnei.
Filij soa · filij asupha filij thebaoth ·
filij ceros filij siaa filij fadō filij leba-
na filij agaba filij selmon filij anan ·
filij geddel filij gaer filij raahia · filij
rasin filij nechoda filij iezem filij aza ·
filij phazea filij besai filij munim filij
nephusim filij bechue filij acupha fi-
lij azur filij belluth · filij maida filij
arsa filij berchos · filij sizara filij the-
ma filij nasia filij atupha filij servoꝝ
salomonis filij sothai filij sopherech ·
filij pheruda filij iahala filij dercon ·
filij gedel · filij saphata filij athil filij
focereth qui erat ortus ex asbaim filio
ammon. Omnes nathinnei · et filij ser-
vorū salomonis: trecentinōgītaduo.
Hij sunt aūt qui ascenderūt de thelma-
la thelarsa cherub adon ⁊ emmer: et
nō potuerūt indicare domū patrū su-
orū et semen suū: utrū ex isrl' essent. Fi-
lij dalaia filij tobia filij nechoda sex-
centiquadragintaduo. Et de sacerdo-
tibus filij iobia filij accos filij berzel-
lai qui accepit de filiabꝫ berzellai ga-
laaditidis uxorem: ⁊ vocatꝰ est nomē
eoꝝ. Hij quesierūt scripturā suā in cen-
su ⁊ nō invenerūt: et eiecti sunt de sacer-
dotio. Dixitq; athersatha eis ut non
mānducarent de sanctis sanctoꝝ: donec
staret sacerdos doctus ⁊ eruditus. O-
mnis multitudo quasi unus · quadra-
gītaduo milia sexcētisexaginta: absq;

servis et ancillis eoꝝ qui erāt septē mi-
lia trecentitrigintaseptē: et inter eos cā-
tores ⁊ cātatrices ducente. Equi eorū
sexcentitrigintaseptem: muli eoꝝ ducē-
tiquadragintaquinq;: cameli eoꝝ qua-
dringentitrigintaquinq;: asini sex mi-
lia septingentiviginti. Hucusq; refert
quid in commentario scriptū fuerit: ex-
in neemia historia texitur. Nōnulli aūt
de principibꝫ familiaꝝ dederūt in opꝰ.
Athersatha dedit in thesaurum auri
dragmas mille: phialas quīquagin-
ta: tunicas sacerdotales quīgentasctri-
ginta. Et de principibꝫ familiaꝝ dede-
rūt in thesaurū operis auri dragmas
viginti milia · et argenti minas duo
milia ducentas. Et qđ dedit reliquus
ppłus auri dragmas viginti milia ·
et argenti minas duo milia · ⁊ tunicas
sacerdotales · sexagītaseptē. Habitave-
rūt aūt sacerdotes ⁊ levite ⁊ ianitores
et cātores ⁊ reliquū vulgꝰ ⁊ nathinnei
⁊ omnis isrl' in civitatibꝫ suis. VIII

Et venerat mensis septimꝰ. Filij
aūt isrl' erāt in civitatibus suis.
Congregatusq; omnis ppłus q̄si vir u-
nus ad plateam que est ante portam
aquarū: et dixerūt esdre scribe ut affer-
ret librū legis moysi: quā preceperat
dn̄s israheli. Attulit ergo esdras sa-
cerdos legem corā multitudine viro-
rū ⁊ mulierū · cunctisq; qui poterāt in-
telligere: in die prima mensis septimi.
Et legit in eo aperte in platea que erat
ante portam aquarū · de mane usq;
ad mediā diem: in cōspectu virorū et
mulierū et sapientiū. Et aures omnis
ppłi erant erecte ad librum. Stetit aūt
esdras scriba sup gradū ligneum quē
fecerat ad loquēdum · et steterūt iuxta
eū mathathia ⁊ semia ⁊ ania et uria ·
et elchia et maasia ad dexterā eius · et

JACQUARD, Joseph Marie (1752–1834).
À la mémoire de J.M. Jacquard. Lyon, 1839.

A "digital" portrait of an innovator of punched-card programming.

Woven in black-and-white silk, this portrait shows Jacquard at his desk. A diminutive model of his automated loom rests nearby. The inventor is shown with a drafting compass in hand inspecting a stack of punched cards. These cards — twenty-four thousand of which would have been required to weave this image — were a precursor to early computer programs: analogous to a binary digit, or "bit," each punched hole directed the thread through the shuttling action of the loom. Similar punched cards offered a mechanism for data input and storage for Charles Babbage's Analytical Engine, programmed by Ada Lovelace (1815–1852). Lovelace wrote that "the Analytical Engine weaves algebraic patterns just as the Jacquard loom weaves flowers and leaves" (see "BERNOULLI," page 545).

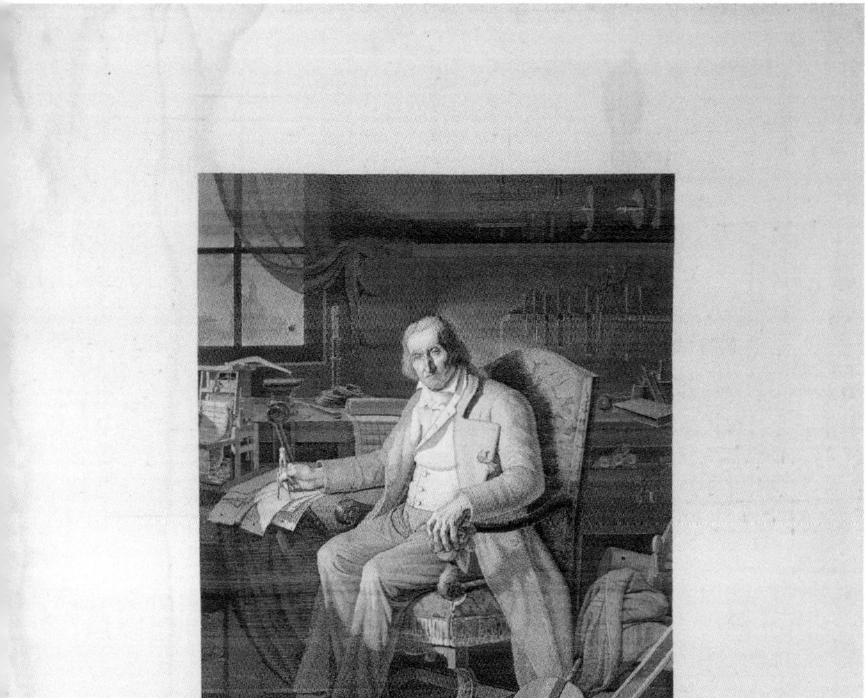
A LA MÉMOIRE DE J. M. JACQUARD.

LEIBNIZ, Gottfried Wilhelm von (1646–1716). "Nouvelle arithmétique binaire" [with:] "Explication de l'arithmétique binaire, qui se sert des seuls caracteres 0 & 1" in *Histoire de l'Academie royale des sciences.* Paris, 1720.

The invention of binary code and the mathematical basis of modern computing.

This pair of articles offered the first full account of binary arithmetic and the base-2 number system that underpins modern computing. English philosopher Francis Bacon had earlier proposed a binary system for cryptography, showing that with only two characters or symbols, "anything could signify anything." Leibniz refined binary into a far-reaching mathematical language and even proposed designs for a binary mechanical calculator. Those designs were never realized. A replica of another Leibniz calculator — this one operating on a decimal, or base-10 number system — is described above (see "LEIBNIZ," page 267).

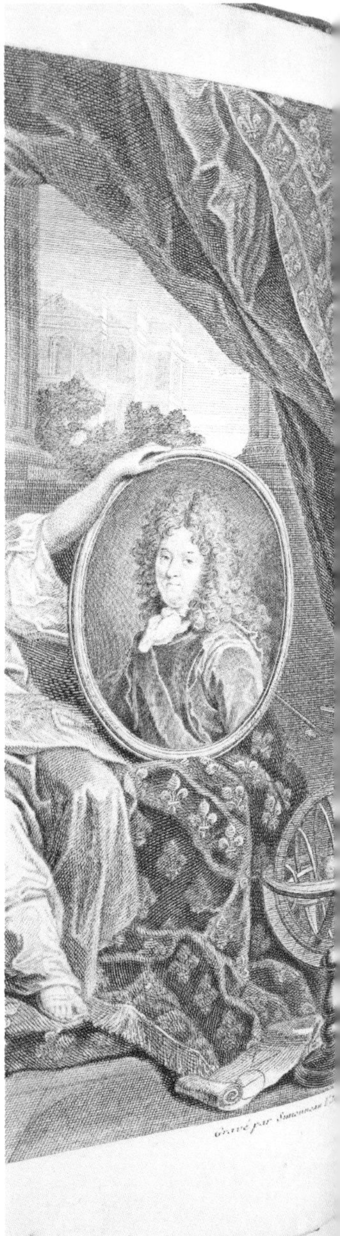
Gravé par Simonneau

HISTOIRE DE L'ACADEMIE ROYALE DES SCIENCES.

Année M. DCC. III.

Avec les Mémoires de Mathématique & de Physique, pour la même Année.

Tirés des Registres de cette Académie.

Seconde Edition, revûe, corrigée & augmentée.

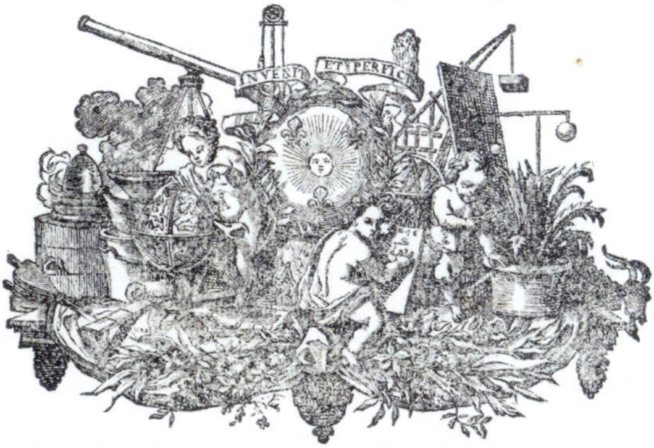

A PARIS,
Chez CHARLES-ESTIENNE HOCHEREAU,
Quay des Augustins, au Phénix.

M. DCC. XX.

AVEC APPROBATION, ET PRIVILEGE DU ROI.

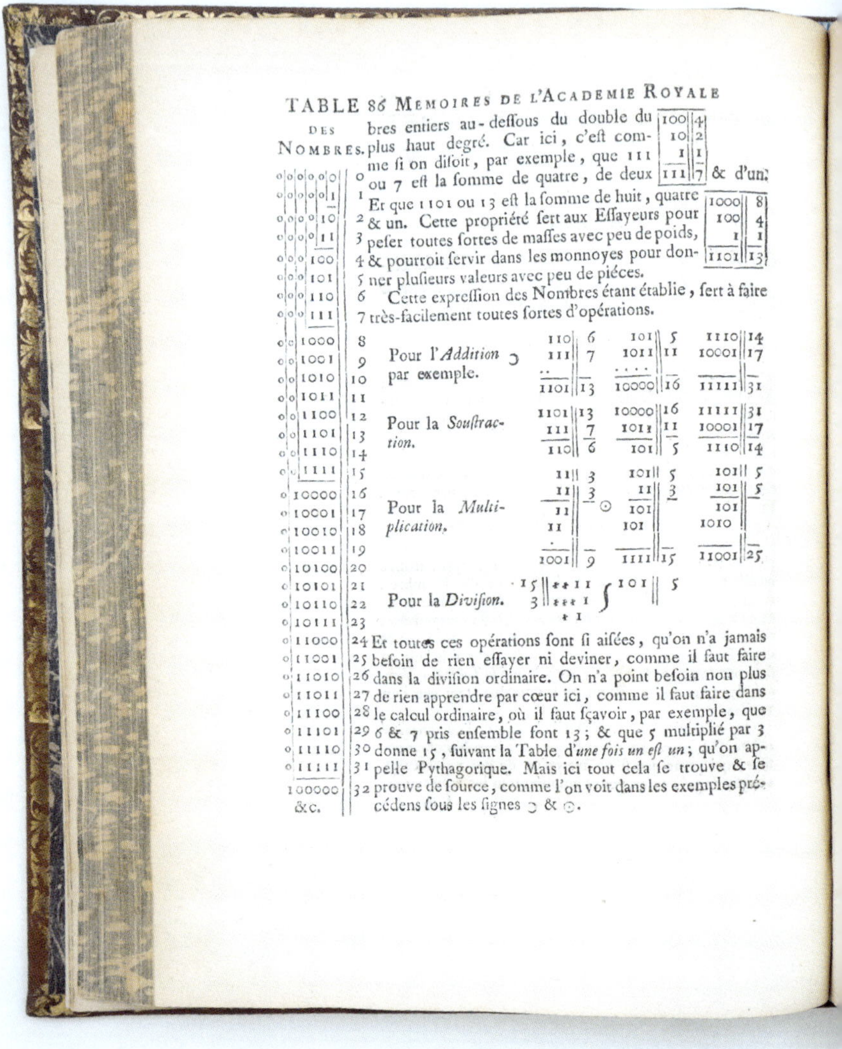

TABLE DES NOMBRES.

000000	0
000001	1
000010	2
000011	3
000100	4
000101	5
000110	6
000111	7
001000	8
001001	9
001010	10
001011	11
001100	12
001101	13
001110	14
001111	15
010000	16
010001	17
010010	18
010011	19
010100	20
010101	21
010110	22
010111	23
011000	24
011001	25
011010	26
011011	27
011100	28
011101	29
011110	30
011111	31
100000	32
&c.	

86 MEMOIRES DE L'ACADEMIE ROYALE

bres entiers au-dessous du double du plus haut degré. Car ici, c'est comme si on disoit, par exemple, que 111 ou 7 est la somme de quatre, de deux & d'un.

100	4
10	2
1	1
111	7

Et que 1101 ou 13 est la somme de huit, quatre & un. Cette propriété sert aux Essayeurs pour peser toutes sortes de masses avec peu de poids, & pourroit servir dans les monnoyes pour donner plusieurs valeurs avec peu de piéces.

1000	8
100	4
1	1
1101	13

Cette expression des Nombres étant établie, sert à faire très-facilement toutes sortes d'opérations.

Pour l'*Addition* ↄ par exemple.

110	6	101	5	1110	14
111	7	1011	11	10001	17
1101	13	10000	16	11111	31

Pour la *Soustraction.*

1101	13	10000	16	11111	31
111	7	1011	11	10001	17
110	6	101	5	1110	14

Pour la *Multiplication.* ⊙

11	3	101	5	101	5
11	3	11	3	101	5
11		101		101	
11		101		1010	
1001	9	1111	15	11001	25

Pour la *Division.*

15	**11	101	5
3	***1		
	*1		

Et toutes ces opérations sont si aisées, qu'on n'a jamais besoin de rien essayer ni deviner, comme il faut faire dans la division ordinaire. On n'a point besoin non plus de rien apprendre par cœur ici, comme il faut faire dans le calcul ordinaire, où il faut sçavoir, par exemple, que 6 & 7 pris ensemble font 13; & que 5 multiplié par 3 donne 15, suivant la Table d'*une fois un est un*; qu'on appelle Pythagorique. Mais ici tout cela se trouve & se prouve de source, comme l'on voit dans les exemples précédens sous les signes ↄ & ⊙.

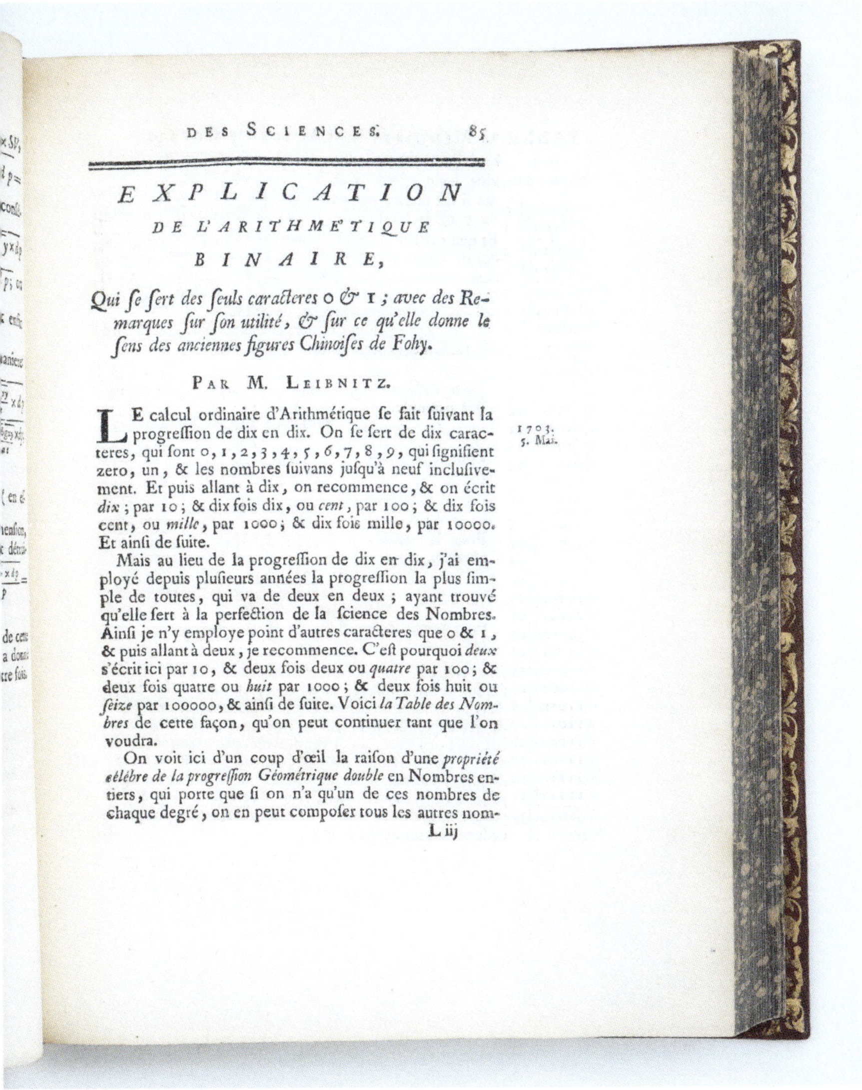

EXPLICATION DE L'ARITHMETIQUE BINAIRE,

Qui ſe ſert des ſeuls caracteres 0 *&* 1 *; avec des Remarques ſur ſon utilité, & ſur ce qu'elle donne le ſens des anciennes figures Chinoiſes de Fohy.*

PAR M. LEIBNITZ.

LE calcul ordinaire d'Arithmétique ſe fait ſuivant la progreſſion de dix en dix. On ſe ſert de dix caracteres, qui ſont 0, 1, 2, 3, 4, 5, 6, 7, 8, 9, qui ſignifient zero, un, & les nombres ſuivans juſqu'à neuf incluſivement. Et puis allant à dix, on recommence, & on écrit *dix* ; par 10 ; & dix fois dix, ou *cent*, par 100 ; & dix fois cent, ou *mille*, par 1000 ; & dix fois mille, par 10000. Et ainſi de ſuite. 1703. 5. Mai.

Mais au lieu de la progreſſion de dix en dix, j'ai employé depuis pluſieurs années la progreſſion la plus ſimple de toutes, qui va de deux en deux ; ayant trouvé qu'elle ſert à la perfection de la ſcience des Nombres. Ainſi je n'y employe point d'autres caracteres que 0 & 1, & puis allant à deux, je recommence. C'eſt pourquoi *deux* s'écrit ici par 10, & deux fois deux ou *quatre* par 100 ; & deux fois quatre ou *huit* par 1000 ; & deux fois huit ou *ſeize* par 10000, & ainſi de ſuite. Voici *la Table des Nombres* de cette façon, qu'on peut continuer tant que l'on voudra.

On voit ici d'un coup d'œil la raiſon d'une *propriété célèbre de la progreſſion Géométrique double* en Nombres entiers, qui porte que ſi on n'a qu'un de ces nombres de chaque degré, on en peut compoſer tous les autres nom-

BYRNE, Oliver (1810–1880).
The First Six Books of the Elements of Euclid.
London, 1847.

A radical experiment in visual learning.

Byrne's edition of Euclid reimagined classical geometry through primary colors and bold graphic design. Intended to make mathematical reasoning more intuitive, the book replaced symbolic abstraction with diagrams and elemental shapes. Published by William Pickering and printed in red, blue, and yellow by Charles Whittingham, it anticipated modern infographics and Bauhaus aesthetics.

Following its publication in 1847, Byrne's *Euclid* received some attention for its visual boldness but was not widely adopted. Its striking use of color — unusual and costly for the time — was admired but seen as more artistic than practical. Sales were modest, and the edition fell into obscurity.

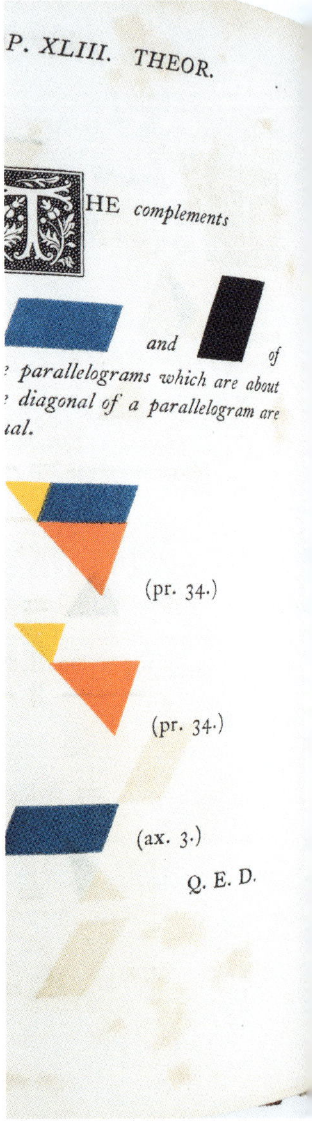

P. XLIII. THEOR.

THE *complements*

and *of*

parallelograms which are about
diagonal of a parallelogram are
ual.

(pr. 34.)

(pr. 34.)

(ax. 3.)

Q. E. D.

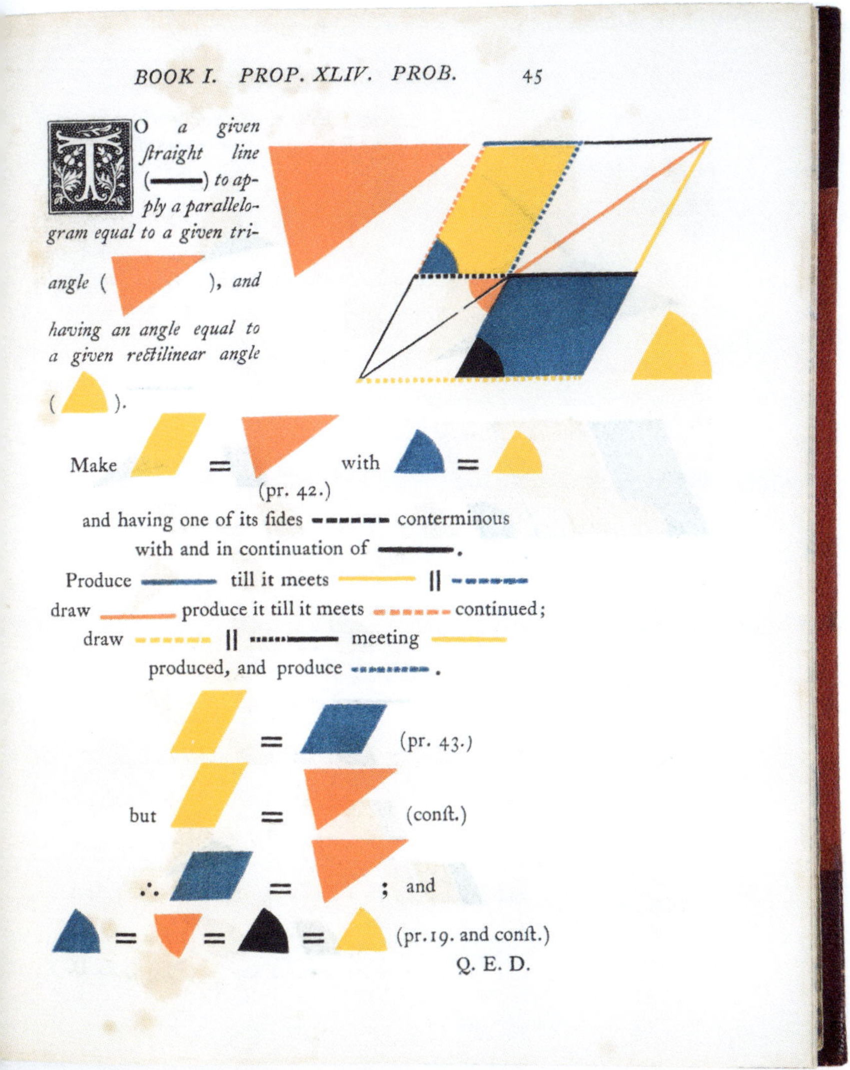

BOOK I. PROP. XLIV. PROB. 45

TO a given straight line (——) to apply a parallelogram equal to a given triangle ( ), and having an angle equal to a given rectilinear angle ( ).

Make = with = (pr. 42.)

and having one of its sides conterminous with and in continuation of .

Produce till it meets || draw produce it till it meets continued; draw || meeting produced, and produce .

= (pr. 43.)

but = (const.)

∴ = ; and

= = = (pr. 19. and const.)

Q. E. D.

KNUTH, Donald E. (1938–).
T*EX and METAFONT: New Directions in*
Typesetting. American Mathematical Society, 1979.

A revolution in digital type design.

With TEX and METAFONT, Turing award-winner Donald Knuth introduced a mathematical system for font design. Informed by Knuth's research into sixteenth- and seventeenth-century typography, METAFONT and TEX sought to make "typefaces of real beauty." LaTeX, a digital typesetting system based on TEX, remains the standard for the visual presentation of technical and mathematical writing. Special Collections' copy of the first edition was owned by CMU Professor and AI pioneer Allen Newell (1927–1992).

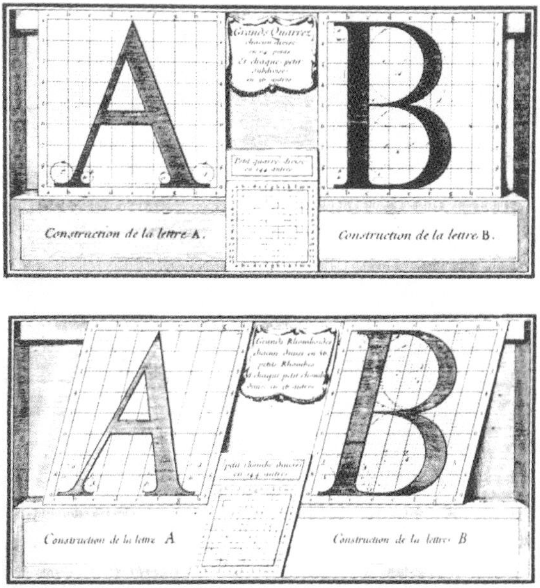

FIGURE 6. Roman and italic letters designed for Louis XIV of France [24].

METAFONT
a system for alphabet design

GENERATION OF TYPEFACES by mathematical means was first tried in the fifteenth century; it became popular in the sixteenth and seventeenth centuries; and it was abandoned (for good reason) during the eighteenth century. Perhaps the twentieth century will turn out to be the right time for this idea to make a comeback, now that mathematics has advanced and computers are able to do the calculations.

Modern printing equipment based on raster lines—in which metal "type" has been replaced by purely combinatorial patterns of zeros and ones that specify the desired position of ink in a discrete way—makes mathematics and computer science increasingly relevant to printing. We now have the ability to give a completely precise definition of letter shapes that will produce essentially equivalent results on all raster-based machines. Furthermore it is possible to define infinitely many styles of type at once; computers can "draw" new fonts of characters in seconds, so that a designer is able to perform valuable experiments that were previously unthinkable.

METAFONT is a system for the design of alphabets suited to raster-based devices that print or display text. The characters you are reading were all designed with METAFONT, in a completely precise way; and they were developed rather hastily by the author of the system, who is a rank amateur at such things. It seems clear that further work with METAFONT has the potential of producing typefaces of real beauty, so this manual has been written for people who would like to help advance the art of mathematical type design.

Essay

Old Books, New Technologies: Computational Research in Special Collections[1]

"Type," Harry Carter once remarked, "is something you can pick up and hold in your hand" (Carter 1969, 5). Pieces of type, or type "sorts," are objects cast in metal, each carrying a reversed image of a typographic letter. In the hand-press period (roughly 1450–1800), these metal sorts were placed on a printing press and inked to imprint their image (again reversed) onto paper. Made of a pliant metal alloy, early printing types were often damaged during presswork, broken or bent by the heavy action of the press, or spoiled by debris. Evidence of this damage is visible in the inked impressions left by these sorts, as figure 1 shows. Printers kept these disfigured types in use while damages accrued, providing a useful signal for bibliographers and book historians.

No two pieces of type will degrade in precisely the same way, and so damaged and distinctive types provide a kind of typographical fingerprint specific to a particular print shop. Such "fingerprints-in-type" can reveal the identities of the individual(s) responsible for a book's making. This is not a problem only in theory: recent estimates suggest that a large part of the printed record from the hand-press period remains unattributed — a result of imprints omitting the names of printers or clandestine efforts to evade restrictions on freedom of speech and the press (Tolonen et al. 2021). Cautious printers were often simply unwilling to put their names to controversial texts and ideas.

The comparison and analysis of distinctive and damaged type sorts is an established practice of analytical bibliography and has been theorized by scholars like Fredson Bowers, Peter Blayney, Adrian Weiss, and Charlton Hinman. And for good reason: "a printer's type fonts [...]," Weiss observed, "together with identifiable types, can be considered de facto proof of his identity" (1990, 96). Hinman, in *The Printing and Proof-Reading of the First Folio of Shakespeare* (1963), identified more than 600 distinctive types that he traced through fifty-five copies of the 1623 First Folio of Shakespeare's plays. Modifying and in some ways correcting Hinman's work, Blayney outstripped this mark, finding 2,131 appearances of 571 distinctive types in his study of a single play, the 1608 quarto edition of *King Lear* — this in addition to hundreds of factotums (decorative frames designed to hold individual pieces of metal type), printers' flowers, woodcut ornaments, and titling letters (Blayney 1982, 432–539).

But for all its brute success, the work of Blayney and others highlights the problem of scale in type or font analysis. While distinctive types can reveal the identities of anonymous printers, they are often hidden in plain sight, disguised in a tangle of otherwise undamaged and inconspicuous letters: any given edition printed in the hand press period might contain hundreds of thousands of individual characters.

Identifying and extracting distinctive examples from this trove is a task only the most committed bibliographers attempt. Computers and machine learning offer a solution: because type analysis is essentially a form of pattern matching, it is particularly susceptible to automation.

The Print & Probability project — an interdisciplinary team of researchers led by Professor Christopher Warren (CMU English) — updates the type analysis of Weiss, Blayney, Hinman, and others through computational methods. The project works with high-resolution digital copies of early printed books made by scanning copies held in Special Collections and other institutional repositories. Running these images through the project's machine-learning pipeline yields thousands of distinctively damaged characters. To tame this mass of evidence, the project focuses on sixteen uppercase letters (ABCDEFGHKLMNPRTW), choosing these letters, in particular, for their size and form (uppercase, robustly serifed characters seem to accrue more damages, and their added height helps to disclose minute features that might be overlooked on more compact lowercase sorts). The result of this computational sifting is a vast and strange alphabet of bent Ws, shabby Ds, bow-legged Hs, and nicked Cs (figure 2), bringing the scale and speed of the digital to a traditionally analog form of scholarly work.

This shift in the quantity (and format) of evidence available for bibliographical analysis is transformative, not least because its form — printed pages rendered as digital files — has created the opportunity for a kind of inquiry that differs from non-computational font analysis in degree if not in kind. But bibliographers have been slow to recognize the value of evidence in scanned books — a byproduct perhaps of an ingrained distrust of photographic reproduction. Any mechanical copy, the bibliographer W.W. Greg

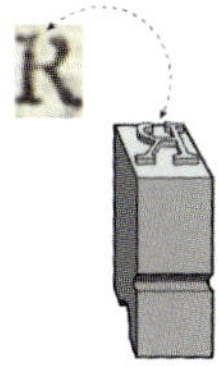

Figure 1 ← A damaged piece of type (or "type sort"): an uppercase R with a fracture through its rounded "counter." The metal sort shown in the figure is a digital reconstruction based on a real type impression in a seventeenth-century book. By permission of Taylor Berg-Kirkpatrick.

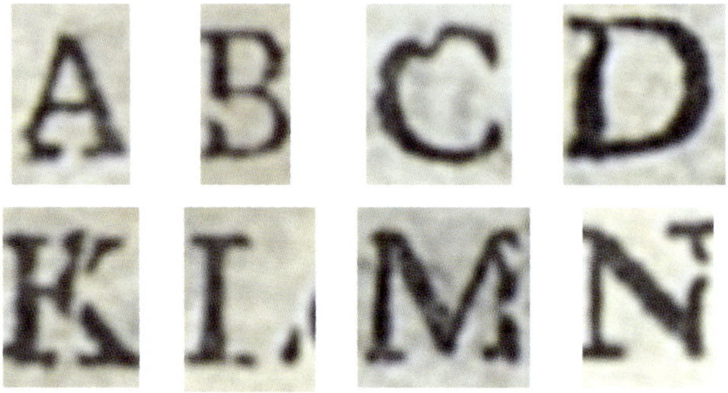

Figure 2 ↘ The end result of the Print & Probability project's computational pipeline: a strange alphabet of damaged and distinctive letters.

argued, "obscures what it reproduces" (quoted in Tanselle 1989, 25). This point of view reflects the perceived and real limitations of earlier technologies, such as microfilm and xerography, that were (and are) inadequate for the fine-grained analysis of bibliographical evidence. Yet modern high-resolution cameras make this blanket distrust of reproductions increasingly outdated.

There are signs of new thinking, however. Sarah Werner imagines a future in which scans of early printed books will be examined by "computer script" and suggests that "we are moving towards a time when digital facsimiles are going to be seen as digital objects in their own right: not as surrogates for a printed book […], but as different ways to experience that object." Werner concludes with a prediction: "for some uses, the material text might be better suited; for others, a digital image might be a better choice" (Werner 2016, 182). Modifying the point, Zachary Lesser and Whitney Trettien argue for a "bibliographic approach to the digital facsimile that acknowledges both its independence from and its close, functional relationship to the material text that it aims to reproduce" (Lesser and Trettien 2021, 418). Ryan Cordell meanwhile proposes a new type of bibliographical inquiry

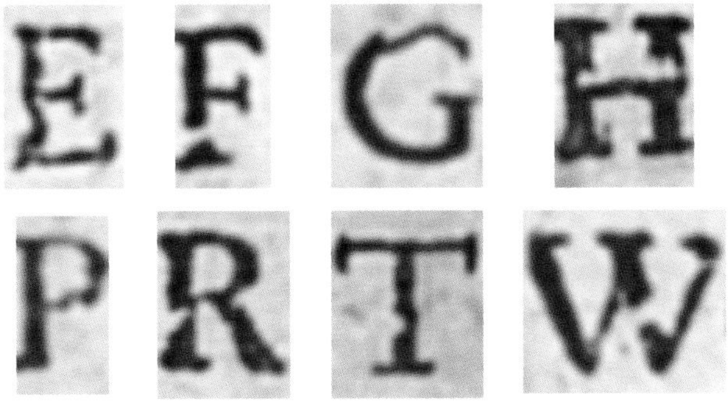

based on an "experimental approach to the digitised archive" (Cordell 2020, 419). Dubbed "speculative bibliography," Cordell's approach "weaves together textual editing and computational speculation" (529). Implicit in the writing of these scholars is the idea that a new kind of bibliography — based on digital images and text — is inevitable: new tools and methods are needed to make use of large bodies of evidence, much of it lodged in digital formats. The work of the Print & Probability team, which relies on the computational analysis of digital copies of early printed books, represents one step in this direction.

New tools are also needed to improve upon existing methods. The type and font analysis developed by twentieth-century analytical bibliographers is hampered by its difficulty, a fact even its principal theorists acknowledged. Weiss admitted that font analysis is a "devastatingly tedious task" (Weiss 1991, 185). Hinman was even more despairing, describing collation (the process of comparing copies of the same edition) as "one of the most arduous and time-consuming of all possible bibliographical activities — and often [...] one of the dullest" (Hinman 1947, 100). Hinman's position didn't change with time. Sixteen years later he wrote that collation and type analysis require "continuous attention to very small matters—among other things to the almost infinitely various 'patterns' in which hundreds of individual types appear and reappear" (Hinman 1963, 12). While Hinman was here writing about the 1623 First Folio of Shakespeare's plays, many books present comparable difficulties. Some editions contain several million letters, each at least potentially made by a distinctively damaged piece of type. Hinman's Dantean warning given at the end of his introduction to *The Printing and Proofreading* still rings true: "from some complexity, at least, the evidence itself allows no escape" (Hinman 1963, 14).

Hinman's solution was technological. The analysis of typographical evidence in the First Folio "had been made possible," Hinman wrote, "by [...] the perfecting of an optical instrument which greatly facilitates the collation of any given book" (Hinman 1963, 7). Eight years before, Hinman announced the invention of this "optical instrument" — a device eventually named the Hinman collator — in a short article: "the new collating machine makes the bibliographical exploitation of [a] vast body of material possible for the first time" (Hinman 1955, 134). Hinman's instrument-aided analysis, which he called "mechanized collation," proved influential; several Hinman collators are still functional, including one held in Special Collections (see "HINMAN COLLATOR," page 407).

There is a rich tradition, then, starting with Hinman, that uses instruments and technology to examine early printed books. In prototyping and developing instruments of its own, the Print & Probability project participates in this tradition, offering a computational analog to Hinman's "mechanized" collator — one that delivers on Hinman's promise of a tool for discerning "the almost infinitely various 'patterns' in which [...] individual types" recur.

The Print & Probability project's algorithmic pipeline operates by deconstructing a digitized page into its most basic constituent parts: individual pieces of type. First, the pipeline automatically segments page scans into their constituent columns of text. These columns of text are then segmented into their constituent lines, which are then segmented into their constituent letters. These isolated letters are classified with OCR (Optical Character Recognition) and assessed for damage against a synthetic, idealized letterform — a visual average of non-damaged characters of the same sort. The farther a letter departs from this ideal model, the higher its "damage score." A bent and fractured uppercase F, for instance, will be assigned a higher damage score compared to an uppercase F that is unbent and unfractured.

This computer-aided sifting of typographical evidence amounts to a mode of bibliographical research that the Print & Probability project has called *computational bibliography* (Warren et al. 2021). But while the field (such as it is) is still new, its methods are not: Blaise Agüera y Arcas, working with Paul Needham on Gutenberg's fifteenth-century Donat-Kalendar (DK) types, adopted a similar tack as early as 1999, naming their method "computational analytical bibliography" (Agüera y Arcas 2002, 2003). New or not, as the technology has evolved, computational methods show increasing promise. Uniquely suited to pattern recognition and variant detection, computers can spot evidence that is either imperceptible or only laboriously perceived.

The Print & Probability project has already identified the printers responsible for several anonymously published works of English literature. One article concluded that John Richardson printed the anonymous "Ornaments" edition of Thomas Hobbes's *Leviathan* (Warren et al. 2021; see "HOBBES," page 397).

Another article uncovered the identities of three printers involved in the publication of John Milton's seminal defense of a free press, *Areopagitica* (1644) (Warren et al. 2020). And a pair of papers, both published in 2023, showed that the unknown printer of the Fourth Folio of Shakespeare's plays (1685) was Robert Everingham (Warren et al. 2023; Lemley et al. 2024). With each discovery — and there are many forthcoming and yet to be made — the Print & Probability project demonstrates that old books invite scholarly work that is as technological as it is humanistic and historical.

1 This essay is an edited excerpt from Lemley et al. 2024. The research described here is richly interdisciplinary and collaborative. I would like to thank my co-authors and collaborators who together make up the Print & Probability team: Christopher N. Warren, Taylor Berg-Kirkpatrick, Nikolai Vogler, D.J. Schuldt, Laura S. DeLuca, Kari Thomas, Elizabeth Dieterich, Kartik Goyal, Max G'Sell, and John Ladd.

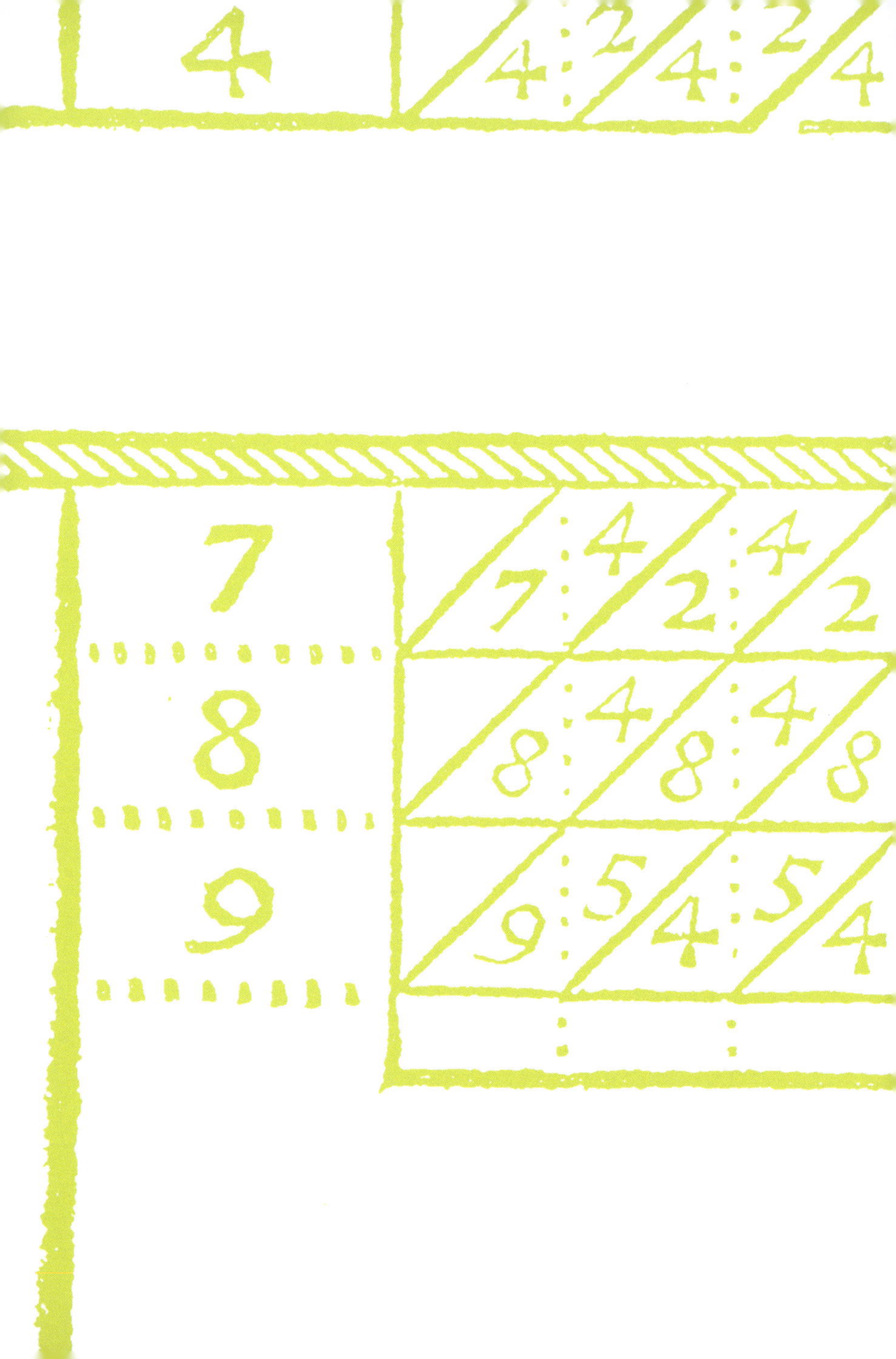

Dixaines
Unités
Q
Q
S
S

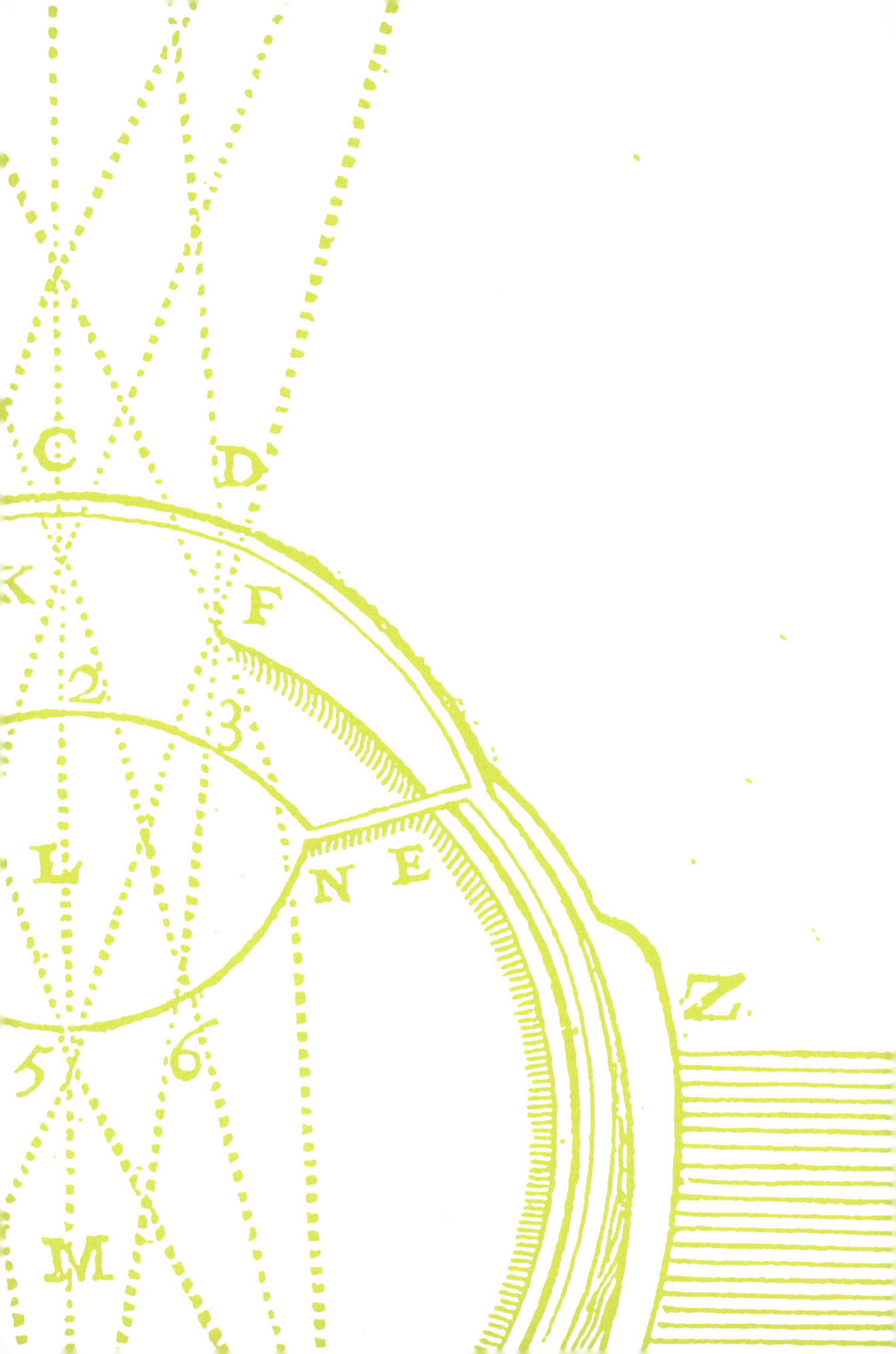
C
D
K
F
2
3
L
N
E
Z
5
6
M

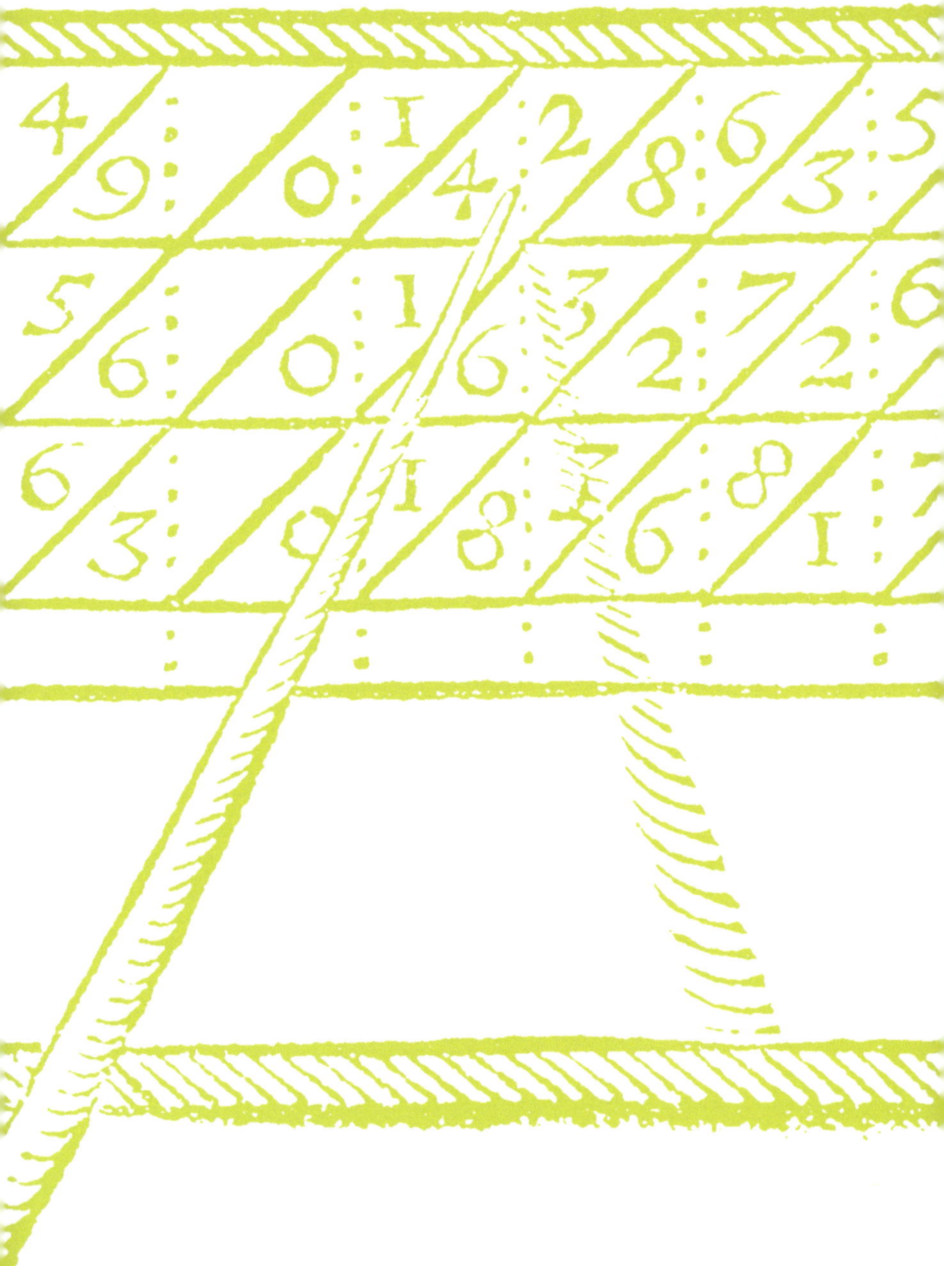

PART VIII

Old Books, New Technologies

Special Collections is more than a storehouse for rarities and historical curiosities. It is also a space for discovery and experimentation. At CMU, rare books and early technologies inspire knowledge-making and kindle the curiosity of students and scholars.

HOBBES, Thomas (1588–1679).
Leviathan: Or The Matter, Form, & Power of a Common-wealth Ecclesiastical and Civil.
London, 1651 (actually 1690)

A clandestine edition of one of the most radical works in the history of political thought.

Hobbes's *Leviathan* was so controversial that no printer involved in any of its three seventeenth-century English editions put his or her name on it. Written during the English Civil War (1642–1651), *Leviathan* argued that society would devolve into chaos without a strong central authority. This copy derives from the so-called "Ornaments" edition, named for the distinctive typographic ornament printed on its title page. For nearly four hundred years, scholars failed to identify the person responsible for its printing.

In 2021, an interdisciplinary team of literary historians, librarians, and computer scientists at CMU developed a machine learning algorithm that matched distinctively damaged typographic letters that appear in the Ornaments edition to letters that appear in other books printed by John Richardson, proving that Richardson was responsible for the edition. This is one example of how Special Collections supports and engages in cutting-edge research that is as technological as it is historical (see "Old Books, New Technologies: Computational Research in Special Collections," page 379).

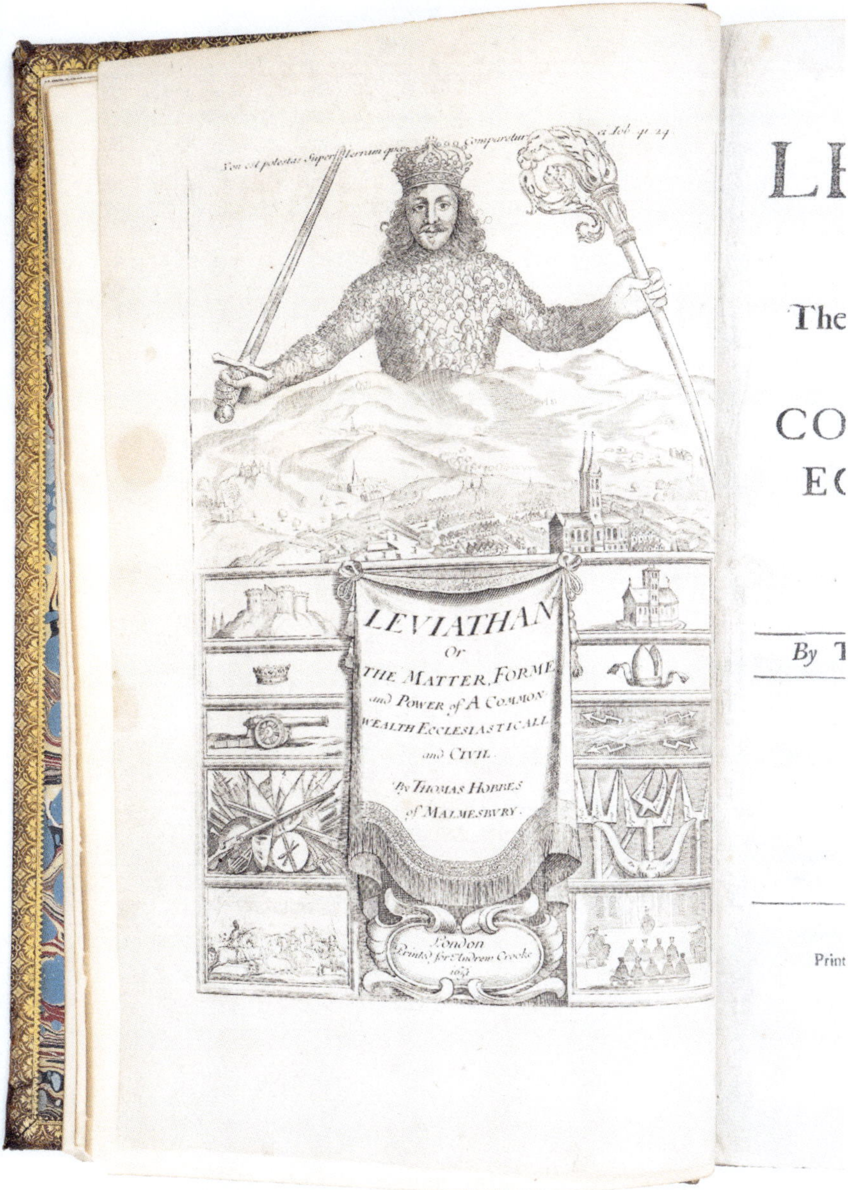
Non est potestas Super Terram quae Comparetur ei Iob. 41. 24
LEVIATHAN
Or
THE MATTER, FORME and POWER of A COMMON-WEALTH ECCLESIASTICALL and CIVIL.
By THOMAS HOBBES of MALMESBURY.
London
Printed for Andrew Crooke

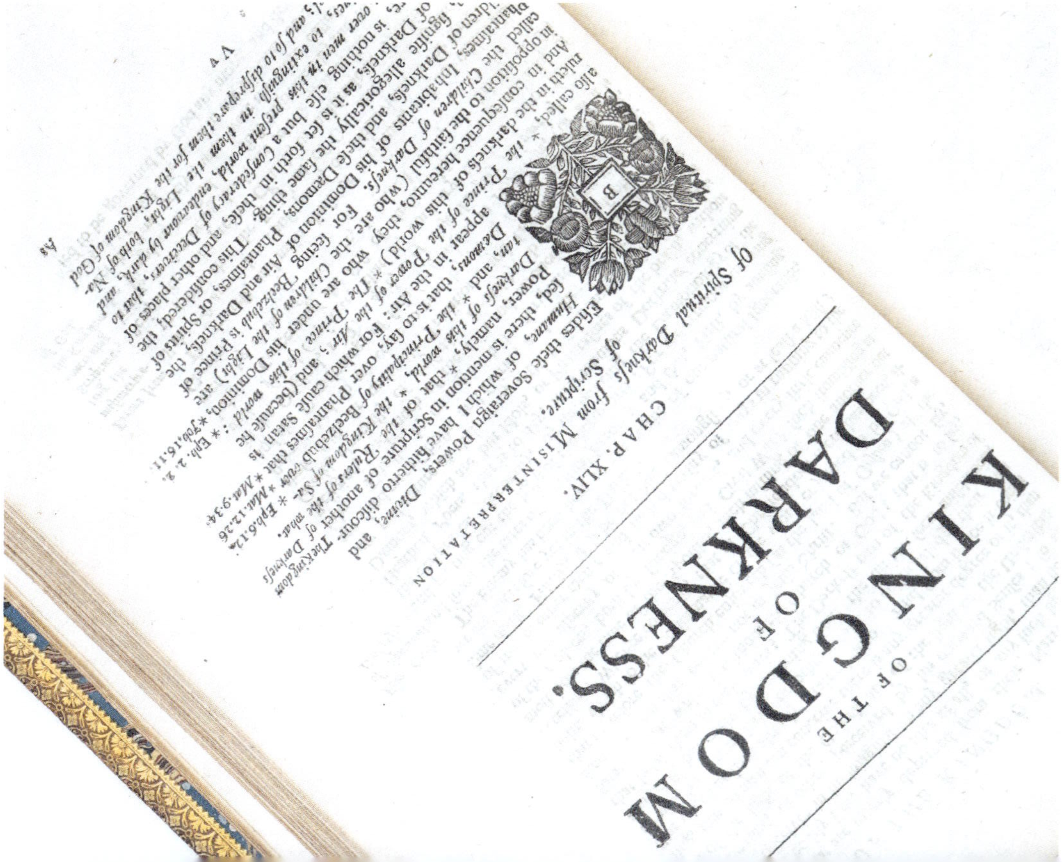

OF THE
KINGDOM
OF
DARKNESS.

CHAP. XLIV.

Of Spiritual Darkness from MISINTERPRETATION *of Scripture.*

BEsides these Soveraign Powers, *Divine*, and *Humane*, of which I have hitherto discoursed, there is mention in Scripture of another Power, namely, * that of *the Rulers of the Darkness of this world*, * *the Kingdom of Satan*, and * *the Principality of* Beelzebub over *Dæmons*, that is to say, over Phantasmes that appear in the Air: For which cause Satan is also called, * *the Prince of the Power of the Air*; and (because he ruleth in the darkness of this world) * *The Prince of this world*: And in consequence hereunto, they who are under his Dominion, in opposition to the faithful (who are the *Children of the Light*) are called the *Children of Darkness*. For seeing *Beelzebub* is Prince of Phantasmes, Inhabitants of his Dominion of Air and Darkness, the Children of Darkness, and these Dæmons, Phantasmes, or Spirits of ... signifie allegorically the same thing. This considered, the ... of Darkness, as it is set forth in these, and other places of ...re, is nothing else but a *Confederacy of Deceivers, that to ... over men in this present world, endeavour by dark, and ...nes, to extinguish in them the Light, both of Na... ...l; and so to disprepare them for the Kingdom of God*

The Kingdom of Darkness what.
* Eph.6.12.
* Mat.12.26
* Mat.9.34.
* Eph.2.2.
* Joh.16.11.

V v

As

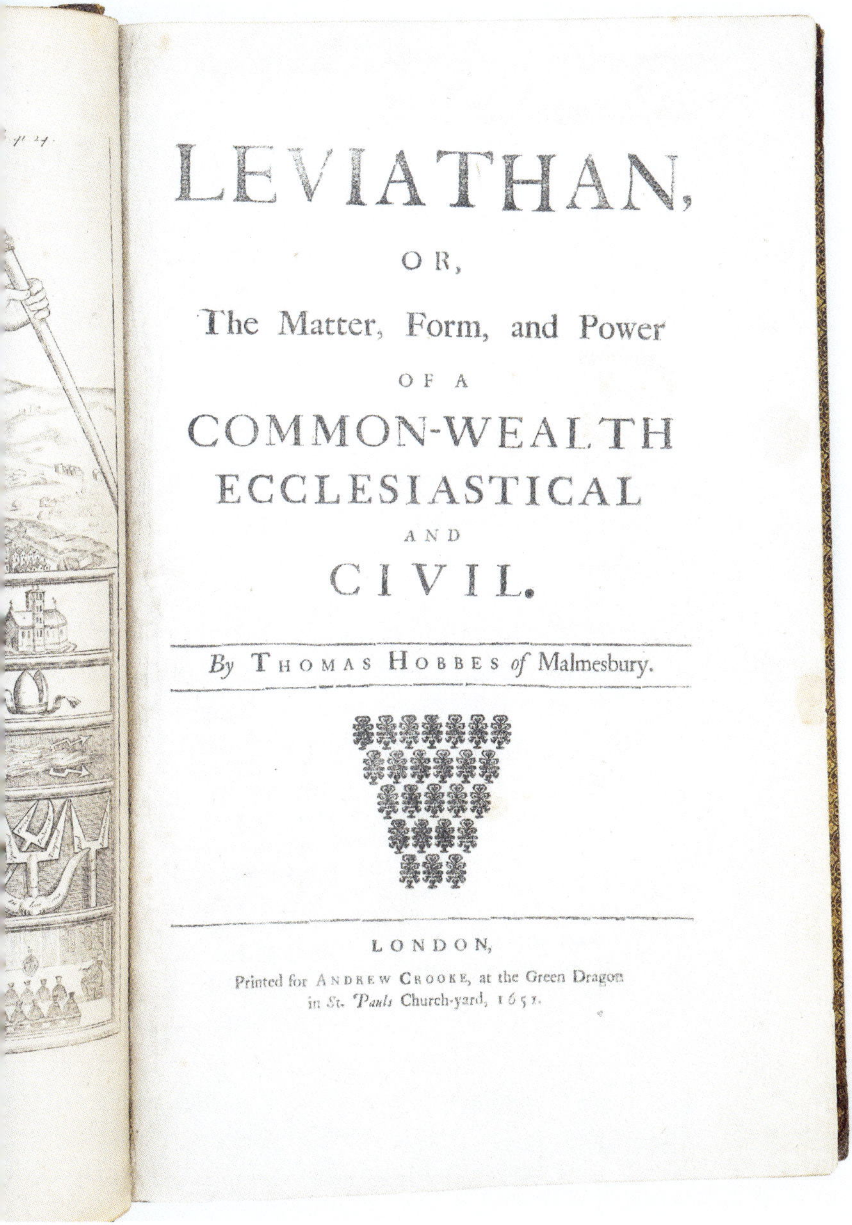

LEVIATHAN,

OR,

The Matter, Form, and Power

OF A

COMMON-WEALTH

ECCLESIASTICAL

AND

CIVIL.

By THOMAS HOBBES *of* Malmesbury.

LONDON,

Printed for ANDREW CROOKE, at the Green Dragon in *St. Pauls* Church-yard, 1651.

AT&T Picturephone, Mod II (set of 2).
Western Electric, 1970.

A midcentury precursor to Zoom and FaceTime.

The Picturephone Mod II was a technological marvel. Its 5.5" × 5" screen offered a full-motion black and white picture with 250 lines of resolution at thirty interlaced frames per second. The device also contained an innovative silicon photodiode array camera with a resolution of 0.8 megapixels; a small integrated mirror could be flipped, allowing it to transmit either the user or documents laid in front of the device. The picture tube and integrated circuits were made at the Western Electric plant in Reading, PA, and the first transmission using the Picturephone occurred in Pittsburgh. For transmission, the Picturephone required three twisted pairs of ordinary telephone cable — two pairs for video, and one for audio and signaling. The Picturephone could also "call" mainframe computers and render rudimentary graphics, with user input provided via a phone's number keypad.

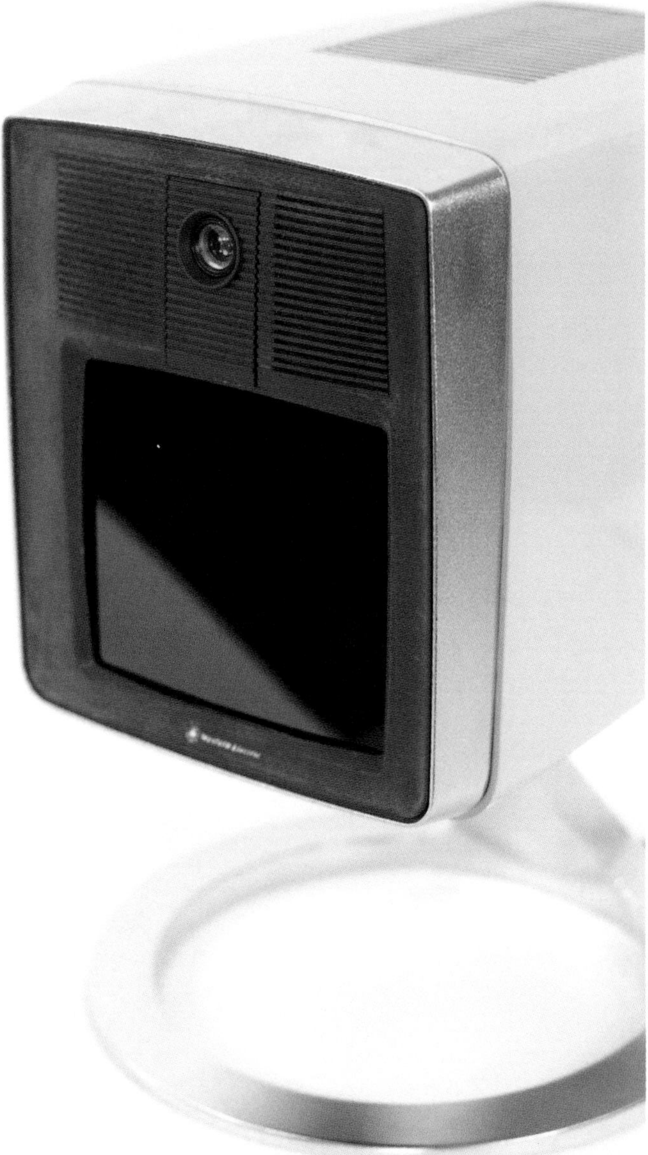

A HUMANIST'S LABORATORY

HINMAN, Charlton (1911–1977). Hinman Collator. Arthur M. Johnson Company, circa 1963.

"In appearance the machine is at once awesome and a little ludicrous..."

— Charlton Hinman

The Hinman Collator is an optical device used to compare two copies of the same book. Invented by Shakespeare scholar Charlton Hinman, the Collator works on the basis of stereoptics: two copies of the same book are placed on cradles beneath lamps that blink. In one blink, the operator sees one of the copies through the device's binocular eyepiece. In the next blink, the operator sees the other copy. This rapid blinking causes the two copies to converge, tricking the operator's mind into seeing them as one. Any printed variants — damaged type, alternate spellings, changes in page layout — are made visible by the mind's failure to resolve two subtly dissimilar images.

↓ By permission of the Folger Shakespeare Library (Digital image 48192).

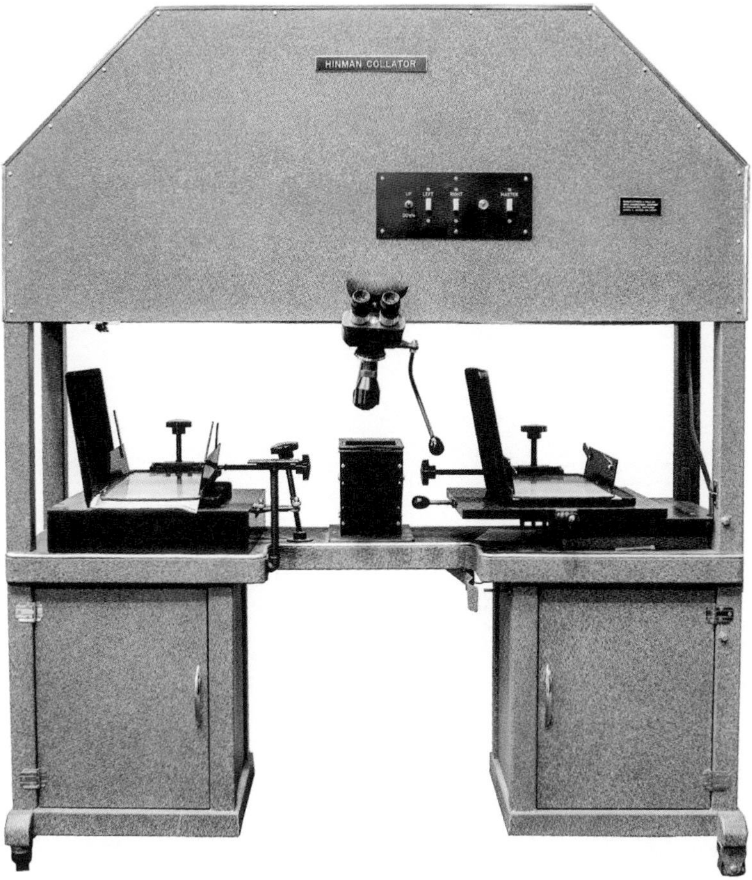
HINMAN COLLATOR
LEFT
RIGHT

On The Hinman Collator, Machine Vision & The Mechanical Origins of the Digital Humanities

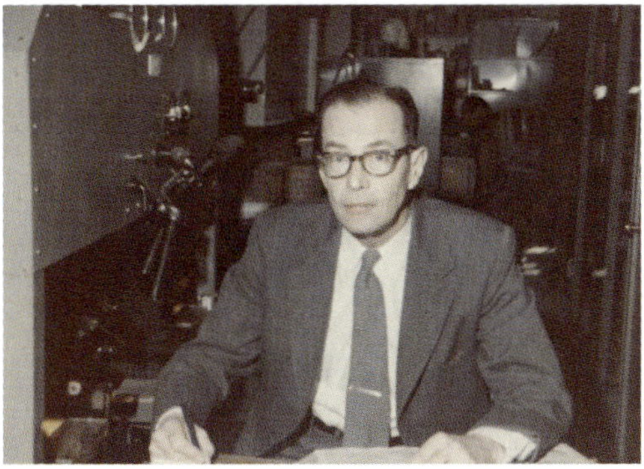

Figure 1 ↑ Charlton Hinman with the Collator in the background, undated. By permission of Kansas University Archives.

In the final decade of his life, the Shakespeare scholar and bibliographer Charlton Hinman (1911–1977) began to lose his sight. In an obituary published in *The Book Collector*, Fredson Bowers — Hinman's friend and onetime academic advisor — suggested that the scholar's near-sightedness was the result of "excessive strain over the course of years and too many hours of concentrated working the Collator" (Bowers 1977, 391). "The Collator" that Bowers refers to here is an optical-mechanical device invented by Hinman that's used to compare two copies of the same book. Since Hinman's death, the collator has taken on the name of its inventor and is now known as the Hinman Collator. Special Collections acquired a rare surviving example of the device in 2019 (see "HINMAN COLLATOR," page 407).

Scholarship in the humanities is not without its own toolkit of technologies, some ancient (the book wheel, the magnifying lens, the index) and others of more recent invention (topic modeling, digital facsimiles, the card catalog). The printed book is itself a kind of technology, one uniquely suited to the human hand and mind — perhaps explaining its persistence in the

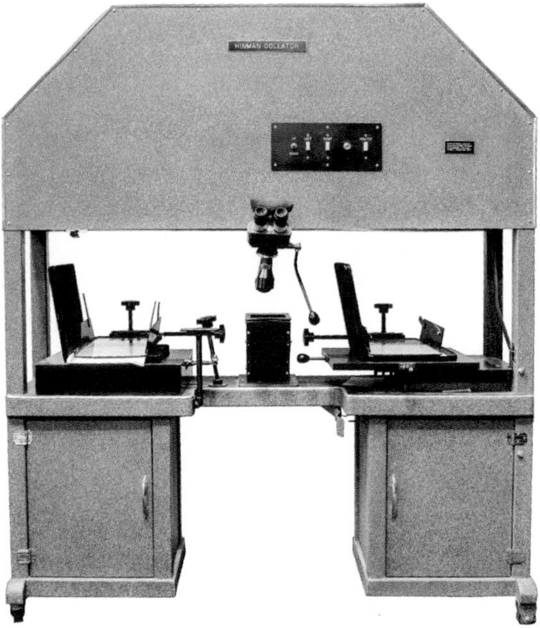

Figure 2 ↓ The Hinman Collator. By permission of the Folger Shakespeare Library (Digital image 48192).]

face of newfangled alternatives. But the strangest of these humanistic technologies, both in appearance and function, is perhaps Hinman's Collator. With its angular framework, pushbutton switches, and Bakelite nameplate, the collator betrays its origins in postwar academia. Ungainly and poorly balanced on too-small wheels, it resembles a diminutive, vinyl-upholstered ziggurat — a strange mid-century altar onto which countless scholars have placed their bookish offerings. But when in use, the collator's oddities become purposeful, and the device takes on a less unnerving guise. It is an object by turns disquieting, functional, and endearing.

The Hinman Collator enacts an optical test known as the flicker paradigm. This is a visual phenomenon by

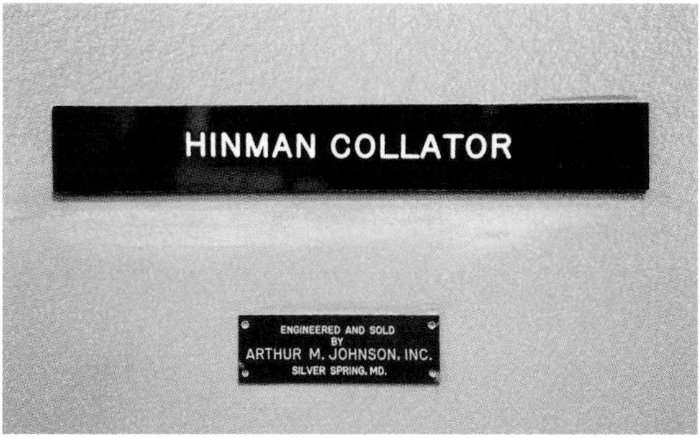

Figure 3 ↗ The Hinman Collator's nameplate. By permission of Carnegie Mellon University Libraries, Special Collections.

which the brain detects subtle differences between two alternating images. When using the collator, the operator places two copies of the same book on cradle-like platforms under lamps that blink in rapid alternation. After bringing the two copies into focus and alignment, the operator peers through a binocular eyepiece, starts the device's blinking lamps using its knee-activated switch, and sets in motion the collator's optical effect: in one blink, the operator sees one copy, and in the next, the other. This rapid blinking — A then B, B then A — causes the two copies to converge and superimpose, tricking the eye into seeing two copies as one. Any printed variants — damaged type, alternate spellings, changes in page layout — appear to shift. And the eye's failure to resolve the two images makes visible even the smallest of discrepancies.

One writer has likened the flicker of these optically disclosed variants to a "dance or wiggle," though the Hinman Collator offers little in the way of grace or choreography (Smith 2000, 131). Charlton Hinman himself admitted that "the machine is at once awesome and a little ludicrous. [...] it is ungainly looking at any time; but it presents an especially odd spectacle when in use" (Hinman 1955, 132). Gesturing again

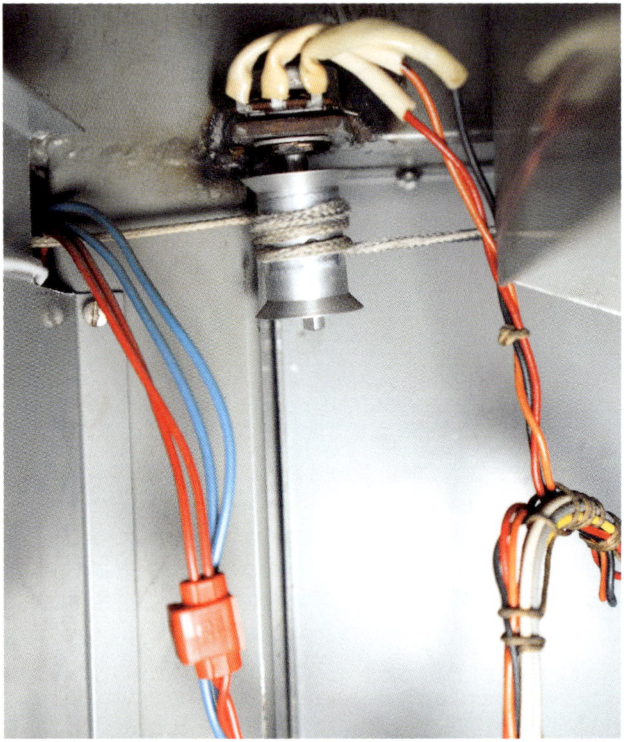

Figure 4 ↓ The inner workings of the Hinman Collator's on-switch: an aluminum barrel wrapped in nylon cord. By permission of Carnegie Mellon University Libraries, Special Collections.

toward the device's strange aspect, Hinman noted that "Rube Goldberg is inevitably suggested" (132). The collator's inner workings bear out the comparison: the device's on-switch, a dial-like potentiometer that tracks rotational distance to set the speed of its blink, resembles (or even *is* — it's hard to tell) an emptied spool of thread. In place of the absent thread is a nylon cord wrapped around the spool's aluminum barrel and tied to a loosely coiled spring which lends the switch its squeaky recoil. The device also contains a kind of analog memory bank, again recalling the inscrutable mechanics of a Rube Goldberg machine: behind a cabinet door near the user's knee is a counter labeled with uppercase letters reading, "HOURS." The

Hinman Collator held in Special Collections records 5,118 hours of on-time, or more than six months of continuous collating.

Machine vision is now the stuff of algorithms and computer-speak, which makes the Hinman Collator's cog-and-wheel plainness all the more endearing: it is an object of nostalgia as much as it is one of scholarly pragmatism. Even the heavy suffix of the device's name seems to recall an earlier and simpler relationship with technology: the agent noun, collat*or*, personifies the machine, evoking a thinking, breathing collaborator instead of an instrument or tool. It has a sort of inelegant charisma that immaterial algorithms and code lack.

But what is meant when one claims to collate? What action does a collator perform? The etymology of *collate* is tangled, deriving from the past participle of the Latin verb *conferre* ("bring together, collect, gather, contribute, connect"). But in English — and particularly in the overlapping fields of rare books and bibliographical analysis — it has taken on a narrower meaning. To *collate* is to compare, but more strictly it means to place two copies of a book side-by-side in order to reveal something about the tangled and error-prone world of the premodern print shop. *Collate* can also imply a kind of confirmation: a process by which the completeness of a book is verified and recorded. This is the meaning of the word most often intended when it's used by antiquarian booksellers and auction houses, whose interest in completeness understandably takes precedence over the scrutiny of minute variants in the text (see James and Pratt 2019, 25–47).

Linked here by QR code is a video that approximates the stereoptical effect of the Hinman Collator's blink. It shows a passage in Act Four of Shakespeare's *Othello* in two different printings. In the later printing, "set sining" is revised to "sat singing" in the first line. Between the two versions, or "states," the typesetter has amended the phrase, replacing the *e* in *set* with an *a* and adding a *g* between the *n* and *i* in *sining*. Both readings, it turns out, are wrong: pointing to the frequent unreliability of premodern printing, the line should read, "The poore Soule sat *sighing*, by a Sicamour [sic] tree," as is clear from the context of Desdemona's lament.

← QR code linking to a video demonstration of the Hinman Collator.

By enabling the detection of variants like this one, the Hinman Collator allows scholars to judge which printing of a line is closest to what Shakespeare actually wrote. In this example of an error in the printed text of *Othello*, Hinman's machine opens a window onto a moment of orthographic and interpretive failure; its rapid blinking permits us to witness a seventeenth-century typesetter at work, attempting to make sense of an ambiguous or smudged line in his manuscript copy.

Meanwhile, in the variant states of Shakespeare's engraved portrait — also shown in the video linked above — we find more evidence of alteration. The changes revealed by the collator's blink represent the engraver's effort to modify the image to more accurately depict Shakespeare both "after the life" and after life (Shakespeare had been dead seven years when the plate was made; the image, more ghostly remnant than embodied face, was evidently made without a model). The two states of the portrait are distinguished by lines added to darken the shadow cast by Shakespeare's head onto his collar. This cross-hatching lends the later state more depth and color. But Shakespeare's beard was also made thicker in the later state, making it appear in the video as though the Bard's face flutters from clean-shaven to ruggedly stubbled and back again.

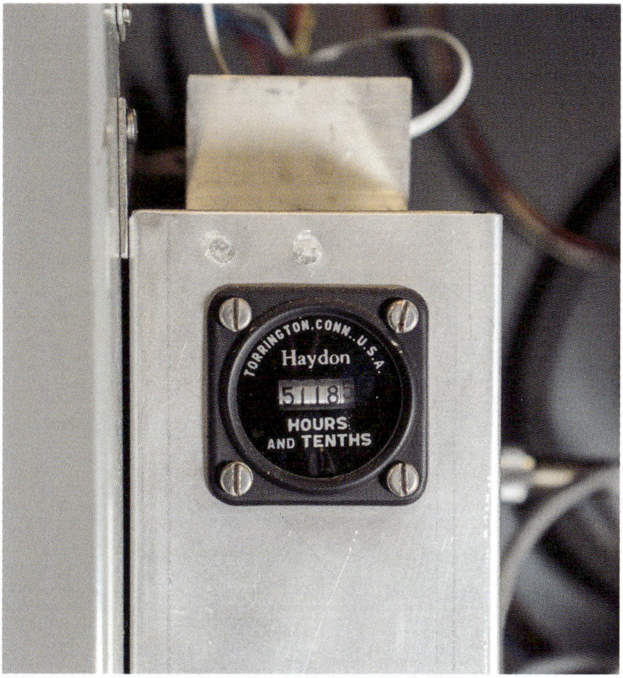

Figure 5 ↑ The Hinman Collator's mechanical counter, recording 5,118.5 hours of on-time. PBy permission of Carnegie Mellon University Libraries, Special Collections.

For all its clockwork strangeness, the story of the Hinman Collator's invention is less ludicrous than it is practical and commercial. After a prototype emerged in the 1950s (Hinman was an inveterate tinkerer and had been a Navy cryptanalyst assigned to a code-breaking unit during the Second World War), Hinman contracted with Arthur M. Johnson, a retired Navy engineer, to fabricate collators to order (Smith 2000). The collator was briefly in demand, viewed by curators and influential professors of literature as a required piece of kit for research with rare books and prints. Seeing its potential (Hinman had published a string of articles hinting at the collator's astonishing alacrity), literary scholars, bibliographers, museum curators, and textual editors who had access to the device set about tracking and explaining variants and

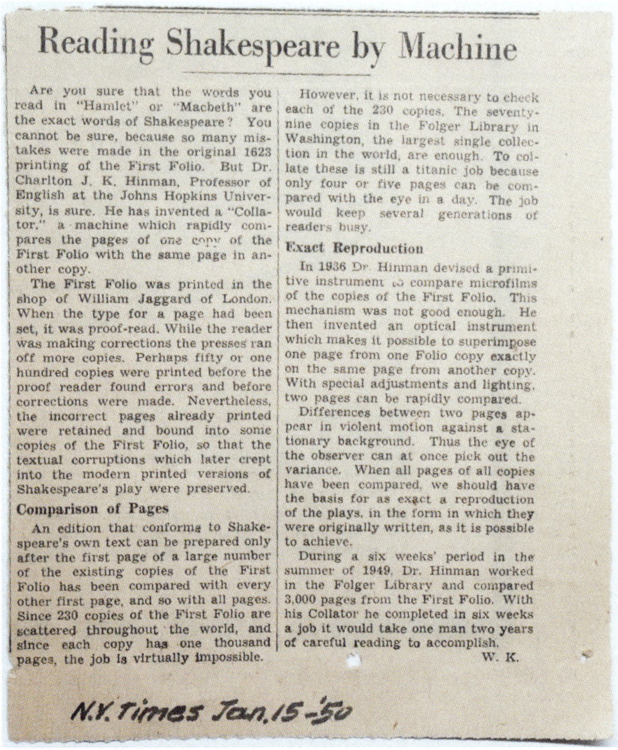

Reading Shakespeare by Machine

Are you sure that the words you read in "Hamlet" or "Macbeth" are the exact words of Shakespeare? You cannot be sure, because so many mistakes were made in the original 1623 printing of the First Folio. But Dr. Charlton J. K. Hinman, Professor of English at the Johns Hopkins University, is sure. He has invented a "Collator," a machine which rapidly compares the pages of one copy of the First Folio with the same page in another copy.

The First Folio was printed in the shop of William Jaggard of London. When the type for a page had been set, it was proof-read. While the reader was making corrections the presses ran off more copies. Perhaps fifty or one hundred copies were printed before the proof reader found errors and before corrections were made. Nevertheless, the incorrect pages already printed were retained and bound into some copies of the First Folio, so that the textual corruptions which later crept into the modern printed versions of Shakespeare's play were preserved.

Comparison of Pages

An edition that conforms to Shakespeare's own text can be prepared only after the first page of a large number of the existing copies of the First Folio has been compared with every other first page, and so with all pages. Since 230 copies of the First Folio are scattered throughout the world, and since each copy has one thousand pages, the job is virtually impossible.

However, it is not necessary to check each of the 230 copies. The seventy-nine copies in the Folger Library in Washington, the largest single collection in the world, are enough. To collate these is still a titanic job because only four or five pages can be compared with the eye in a day. The job would keep several generations of readers busy.

Exact Reproduction

In 1936 Dr. Hinman devised a primitive instrument to compare microfilms of the copies of the First Folio. This mechanism was not good enough. He then invented an optical instrument which makes it possible to superimpose one page from one Folio copy exactly on the same page from another copy. With special adjustments and lighting, two pages can be rapidly compared.

Differences between two pages appear in violent motion against a stationary background. Thus the eye of the observer can at once pick out the variance. When all pages of all copies have been compared, we should have the basis for as exact a reproduction of the plays, in the form in which they were originally written, as it is possible to achieve.

During a six weeks' period in the summer of 1949, Dr. Hinman worked in the Folger Library and compared 3,000 pages from the First Folio. With his Collator he completed in six weeks a job it would take one man two years of careful reading to accomplish.

W. K.

N.Y. Times Jan. 15 -'50

Figure 6 ↓ "Reading Shakespeare by Machine," The New York Times, 15 January 1950. By permission of Posner Memorial Collection (Collector's Files), Carnegie Mellon University Libraries, Special Collections.

typographic anomalies in early printed books. The collator allowed these scholars to attain previously unimaginable accuracy in their work.

Hinman used the Collator in his own exhaustive analysis of the 1623 First Folio of Shakespeare's plays — a project that led to the publication of a two-volume study, *The Printing and Proofreading of the First Folio of Shakespeare* (Oxford, 1963). The story of Hinman's optical-mechanical analysis of the Folio is a long one: he eventually examined fifty-five copies of the Folio over many months with the help of his device at the Folger Shakespeare Library in Washington,

D.C. Beyond cataloging and tracing six hundred-odd distinctive types through the Folio's nine hundred pages, his findings revealed a spate of previously unremarked details about the Folio's printing. His method meanwhile reshaped the field of analytical bibliography.

It's difficult to say precisely what the collator signaled to those not involved in its scholarly use. An article that appeared in the New York Times in 1950 hints at both the era's breathless embrace of a machine future and amusement at the lengths to which tweedy scholars might go for ever diminishing returns: "Reading Shakespeare by Machine," the headline reads.

But Hinman's collator did find a kind of traction and wandered beyond the world of Shakespearean textual scholarship and typographical detective work. A report by Elizabeth Mongan — Curator of Graphic Arts at the National Gallery of Art — submit ted in support of a grant proposal to fund the collator's manufacture notes that her "particular interest was to test [the machine's] usefulness in respect to print," including "engravings and etching by Schongauer, Durer and Rembrandt." Having examined prints by each of these artists (unnervingly, Mongan used the phrase "in the machine," implying a kind of mechanical ingestion or assimilation), she marveled at its "remarkable possibilities," reporting that it "disclosed undescribed differences of lines, cross hatching, etc." The device, Mongan concluded, represented a "vast time saver in that it greatly speeds up the arduous work of meticulous comparison." [1]

Hinman's instrument-aided method, which he called "mechanized collation," proved influential: though Hinman's was the first, several optical-mechanical collators followed, including the Lindstrand Comparator, Hailey's Comet, and the McCleod Portable Comparator — all named for their slightly madcap inventors. More recently, PocketHinman, a smartphone app, brings the Hinman's blink to the scale of a handheld screen, allowing users to overlay digital photographs of variant states of a text for comparison. Each of these solutions is meant to identify a common feature of early modern presswork: errors.

In seventeenth-century print shops, including the one that produced Shakespeare's First Folio, mistakes in typesetting were corrected during printing. But rather than discard the early, error-riddled sheets, uncorrected copies were bound up and sold, meaning that editions of early printed books almost always comprise a jumble of sheets, some in their corrected state, some not. Determining what Shakespeare actually wrote, Hinman realized, would thus require examining several dozen copies of the First Folio, the only reliable witness for nearly half of Shakespeare's plays. As Hinman explained in 1953, "we cannot hope for the best possible edition of the text of Shakespeare until all of the variants in the First Folio [...] have been discovered and recorded" (Hinman 1953, 288). The Hinman Collator made this painstaking search for variants practicable.

What fascinates here, though, rather than the device's historical applications, is the thing itself. According to a census published in 2002, there are just over forty-one Hinman Collators that improbably survive. Some of these are in various states of disrepair and decrepitude. Still, this is an unexpected number — the survival of forty-one collators is a reminder of the brute endurance of these machines: only fifty-nine were manufactured, the last sold in 1979 (Smith 2002, 139). The Collator held in Special Collections came to Pittsburgh by a circuitous route. Long kept in storage at the University of Wisconsin's surplus warehouse in Milwaukee, it was removed from the Shakespeare

Research Collection in the University of Wisconsin's Golda Meir Library — its home since the University purchased it from Arthur M. Johnson in 1964. Special Collections acquired it for the cost of freight, and after some logistical wrangling, it arrived on the back of an open trailer, shrouded in industrial-strength cling film.

⋆ ⋆ ⋆

Made obsolete by algorithms that detect printed variants at superhuman speeds, the Hinman Collator is no longer a useful object. Or, more accurately, it is no longer widely used. It remains useful in a way that computer vision applications are not, due to the simple fact that with some basic instruction, anyone can sit at the Hinman Collator and bring it to whirring life. Its function and controls are self-explanatory, in spite of the device's initially off-putting appearance. Yet the collator has long since given up any claim to be cutting-edge. It is an ancestral machine, a technological memento mori. Even so, its invention could be said to mark the birth of the digital humanities, an honor usually accorded to the *Index Thomisticus* — a punch-card concordance of the works of the medieval theologian Thomas Aquinas begun in the 1940s. But where the *Index* represented Big Data *avant la lettre*, the Hinman Collator offered a kind of microscopic prosthesis. It was (and is) unique in being less a means of tabulating data than an instrument designed to augment the human eye. It made possible the detection of evidence that was otherwise difficult to discern. In this respect, it is more like Galileo's telescope than, say, the relational database: it divulged new worlds.

The Hinman Collator foreshadowed the many tools and technologies that now inform research in the humanities. The Print & Probability project, an interdisciplinary team of researchers led by Professor Christopher Warren (CMU Department of English), was awarded an NEH grant in 2021 to develop computational tools and methods for detecting new evidence in early printed books (see "Old Books, New Technologies: Computational Research in Special Collections," page 379). The project team has created machine learning algorithms that can automatically assess and categorize distinctively damaged types in seventeenth-century books — a task Hinman once described as "one of the most arduous and time-consuming of all possible bibliographical activities" (Hinman 1947, 100). The Print & Probability project's tools and approach amount to a new field, "computational bibliography" — an algorithmic successor to Hinman's mechanical approach (Agüera y Arcas 2002, 2003; Warren et al. 2020, 2021, 2023; Lemley et al. 2024).

1 Mongan's report is in the Charlton Hinman Papers, Kenneth Spencer Research Library Archival Collections, University of Kansas, MS 69.

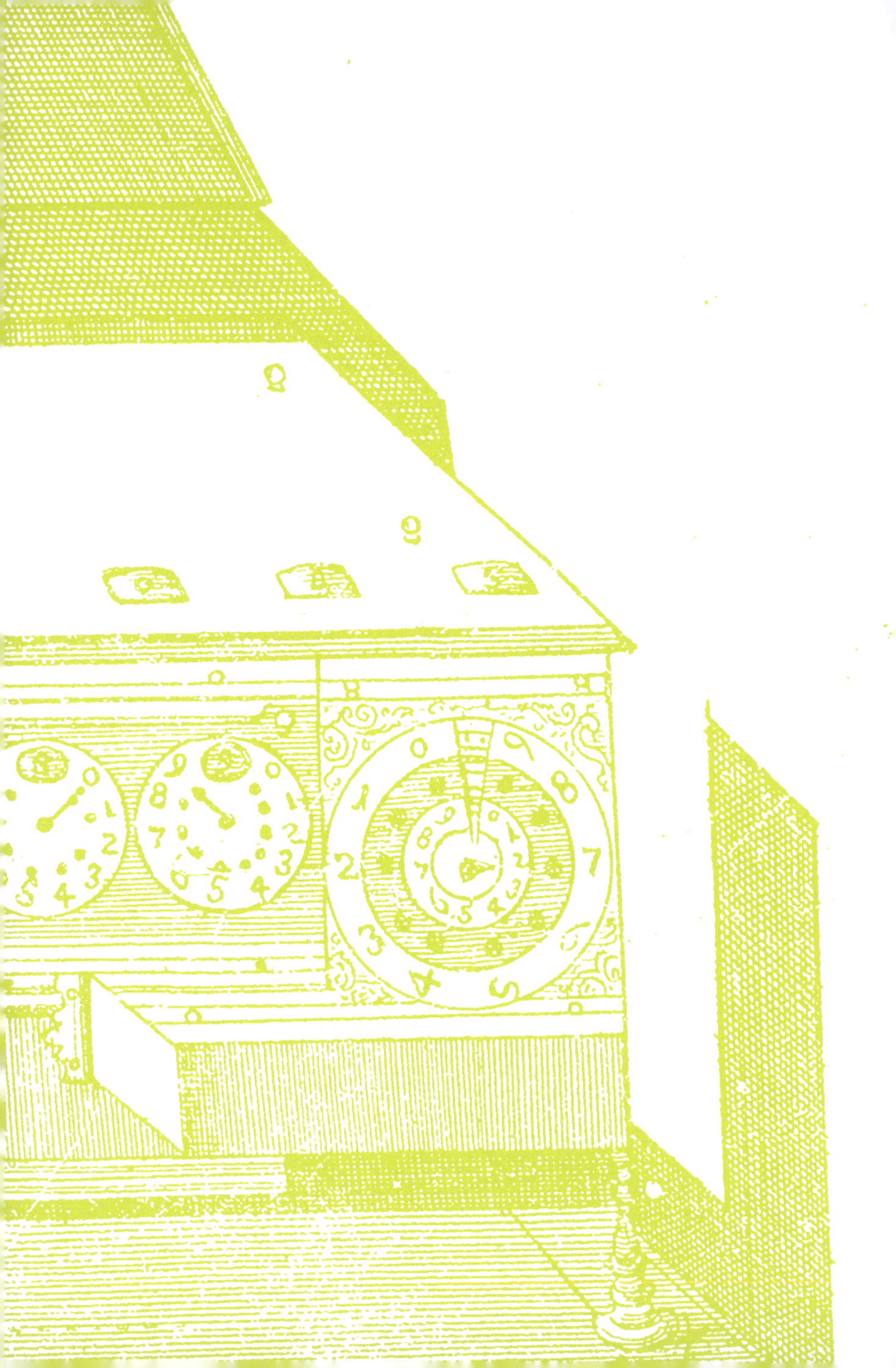

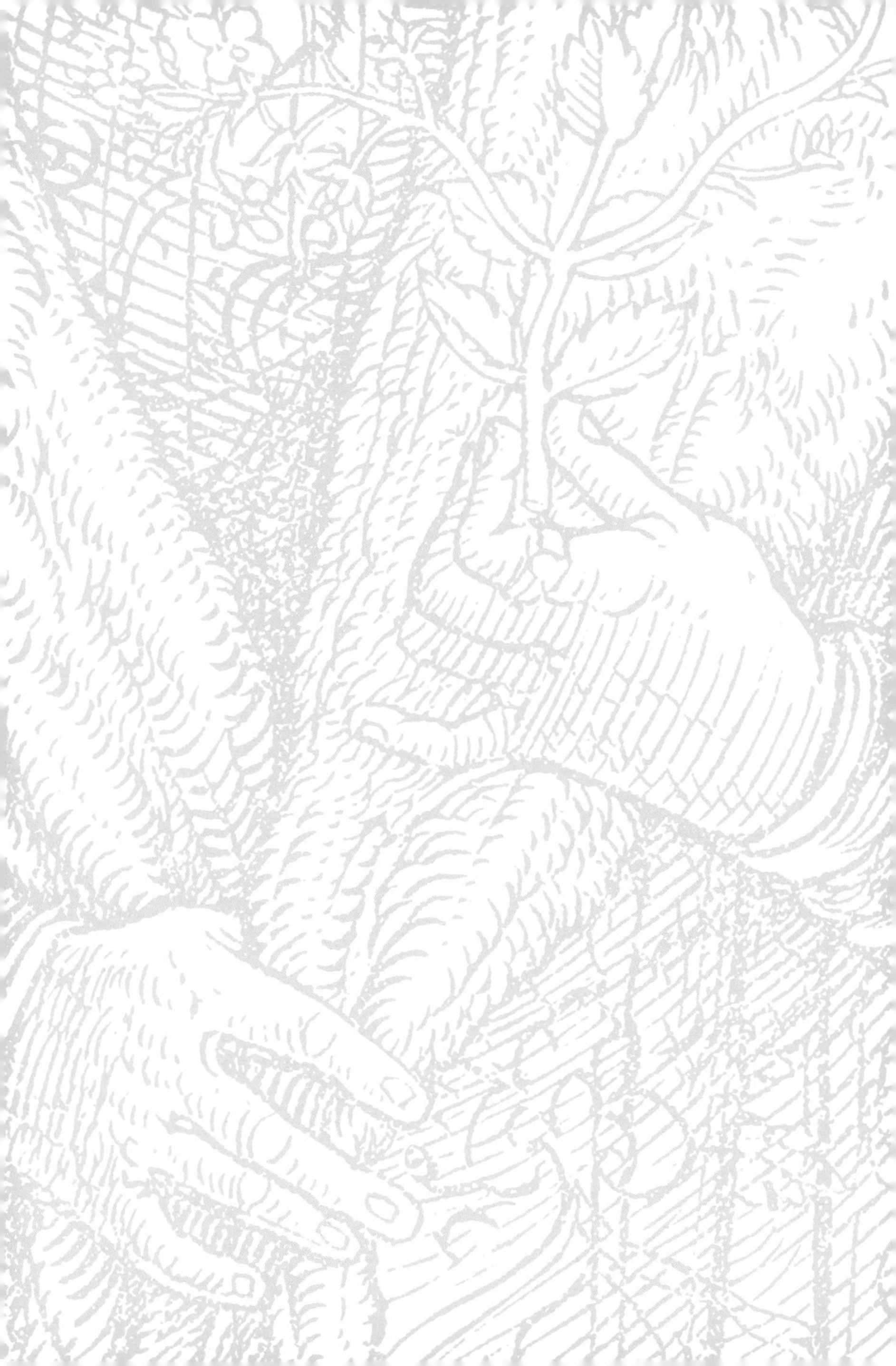

3933,8
3968,6
4130,8
4226,9
4340,8
4436,5

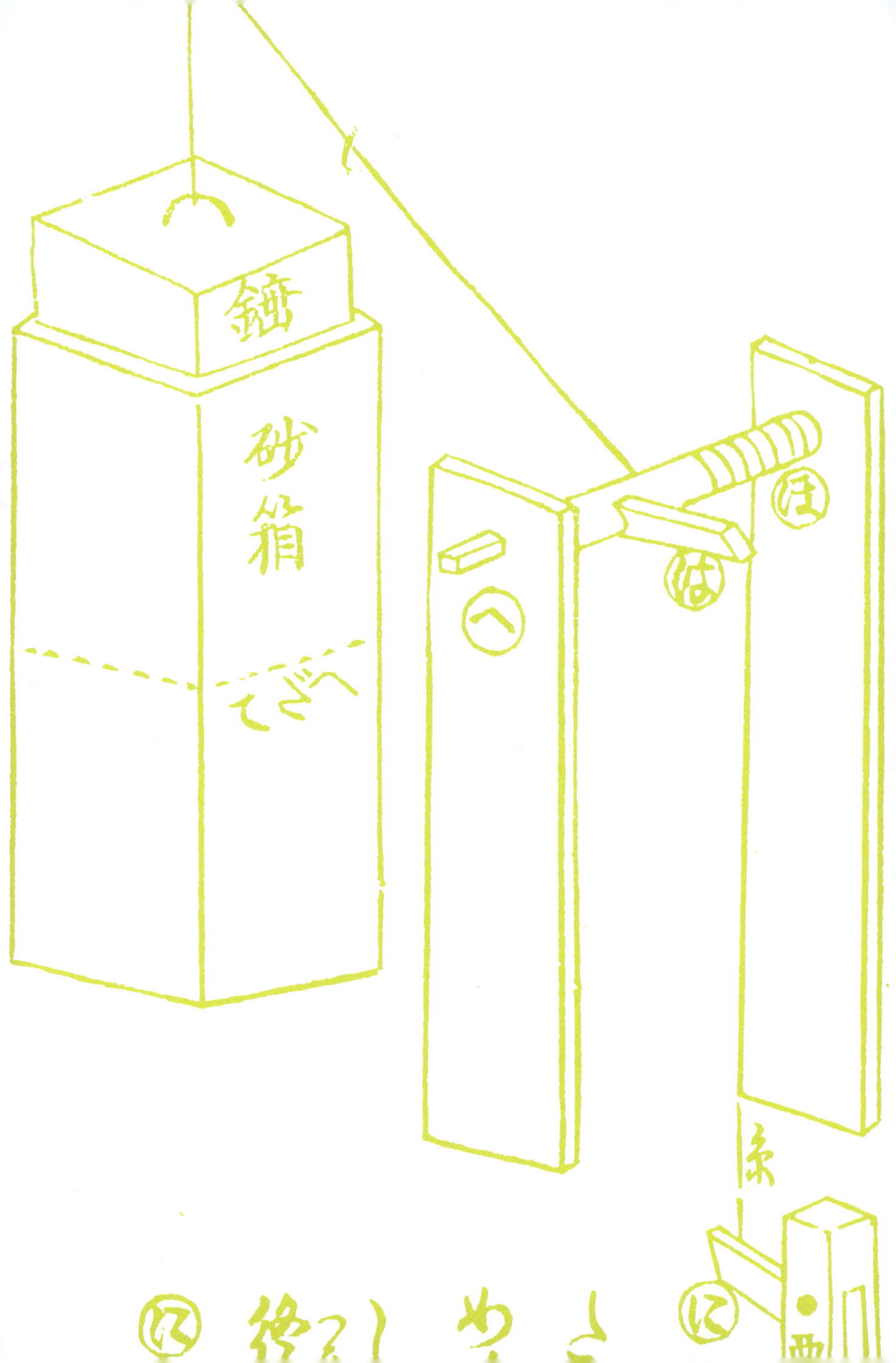
錘
砂箱
てざへ
へ
は
ほ
糸
に

PART IX

Miscellanea

Materials in Special Collections cross themes and disciplines. While its strengths include the history of science, technology, and the evolution of computation and cryptography, it also houses significant works in English literature, mathematics, botany, children's literature, and art — including an important collection of AI-generated artwork. This part presents a curated sample of items that lack a single unifying theme.

EUCLID (4th century BCE).
Elementa geometriae. Venice, 1482.

The first printed work to include mathematical diagrams and Euclid's first appearance in print.

Euclid's *Elementa Geometriae* ("The Elements of Geometry") is of profound importance in the history of mathematics and the mathematical sciences. Published in Venice in 1482 — only thirty years after the invention of the printing press — this edition introduced diagrams that represented Euclidean principles and axioms in the form of visual proofs. As Euclid's first appearance in print, this edition established a model for mathematical texts and influenced the teaching and development of geometry in the Renaissance and after.

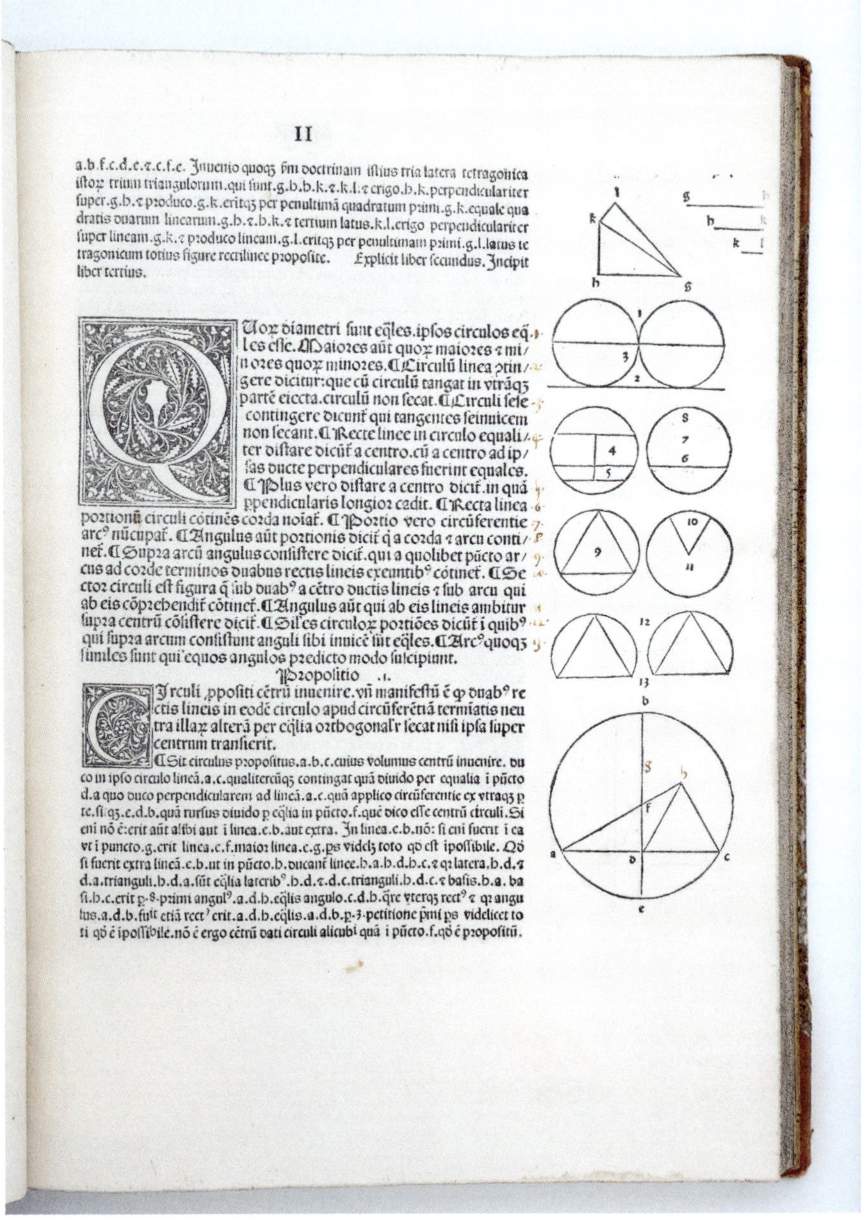

II

a.b.f.c.d.e.ꝛ.c.f.e. Inuenio quoqȝ ſm doctrinam iſtius tria latera tetragonica iſtoꝝ trium triangulorum.qui ſunt.g.h.h.k.ꝛ.k.l.ꝛ erigo.h.k.perpendiculariter ſuper.g.h.ꝛ produco.g.k.eritqȝ per penultimã quadratum primi.g.k.equale quadratis duarum linearum.g.h.ꝛ.h.k.ꝛ tertium latus.k.l.erigo perpendiculariter ſuper lineam.g.k.ꝛ produco lineam.g.l.eritqȝ per penultimam primi.g.l.latus tetragonicum totius figure rectilinee propoſite. Explicit liber ſecundus. Incipit liber tertius.

Quoꝝ diametri ſunt eq̃les.ipſos circulos eq̃les eſſe. Maiores aũt quoꝝ maiores ꝛ minores quoꝝ minores. ¶ Circulũ linea ꝯtingere dicitur: que cũ circulũ tangat in vtrãqȝ partẽ eiecta.circulũ non ſecat. ¶ Circuli ſeſe contingere dicunt̃ qui tangentes ſeinuicem non ſecant. ¶ Recte linee in circulo equaliter diſtare dicũt̃ a centro.cũ a centro ad ipſas ducte perpendiculares fuerint equales. ¶ Plus vero diſtare a centro dicit̃.in quã ꝑpendicularis longior cadit. ¶ Recta linea portionũ circuli cõtinẽs corda noiat̃. ¶ Portio vero circũferentie arc⁹ nũcupat̃. ¶ Angulus aũt portionis dicit̃ q̃ a corda ꝛ arcu continet̃. ¶ Supra arcũ angulus conſiſtere dicit̃.qui a quolibet pũcto arcus ad corde terminos duabus rectis lineis exeuntib⁹ cõtinet̃. ¶ Sector circuli eſt figura q̃ ſub duab⁹ a cẽtro ductis lineis ꝛ ſub arcu qui ab eis cõprehendit̃ cõtinet̃. ¶ Angulus aũt qui ab eis lineis ambitur ſupra centrũ cõſiſtere dicit̃. ¶ Siĺes circuloꝝ portiões dicũt̃ i quib⁹ qui ſupra arcum conſiſtunt anguli ſibi inuicẽ ſũt eq̃les. ¶ Arc⁹ quoqȝ ſimiles ſunt qui equos angulos predicto modo ſuſcipiunt.

Propoſitio .1.

Circuli ꝓpoſiti cẽtrũ inuenire.vñ manifeſtũ ẽ ꝙ duab⁹ rectis lineis in eodẽ circulo apud circũferẽtiã termĩatis neutra illaꝝ alterã per eq̃lia orthogonal'r ſecat niſi ipſa ſuper centrum tranſierit.

¶ Sit circulus propoſitus.a.b.c.cuius volumus centrũ inuenire. duco in ipſo circulo lineã.a.c.qualitercũqȝ contingat quã diuido per equalia i pũcto d.a quo duco perpendicularem ad lineã.a.c.quã applico circũferentie ex vtraqȝ ꝑte.ſitqȝ.e.d.b.quã rurſus diuido ꝑ eq̃lia in pũcto.f.quẽ dico eſſe centrũ circuli.Si eni nõ ẽ:erit aũt alibi aut i linea.e.b.aut extra. In linea.e.b.nõ: ſi eni fuerit i ea vt i puncto.g.erit linea.c.f.maior linea.c.g.ps videlȝ toto qḋ eſt ipoſſibile. Qḋ ſi fuerit extra lineã.e.b.ut in pũcto.h.ducant̃ linee.h.a.h.d.h.c.ꝛ q² latera.h.d.ꝛ d.a.trianguli.h.d.a.ſũt eq̃lia laterib⁹.h.d.ꝛ.d.c.trianguli.h.d.c.ꝛ baſis.h.a. baſi.h.c.erit ꝑ.8.primi angul⁹.a.d.h.eq̃lis angulo.c.d.h.q̃re vterqȝ rect⁹ ꝛ q² angulus.a.d.b.fuit etiã rect⁹ erit.a.d.h.eq̃lis.a.d.b.ꝑ.3.petitione p̃mi ps videlicet toti qḋ ẽ ipoſſibile.nõ ẽ ergo cẽtrũ dati circuli alicubi quã i pũcto.f.qḋ ẽ propoſitũ.

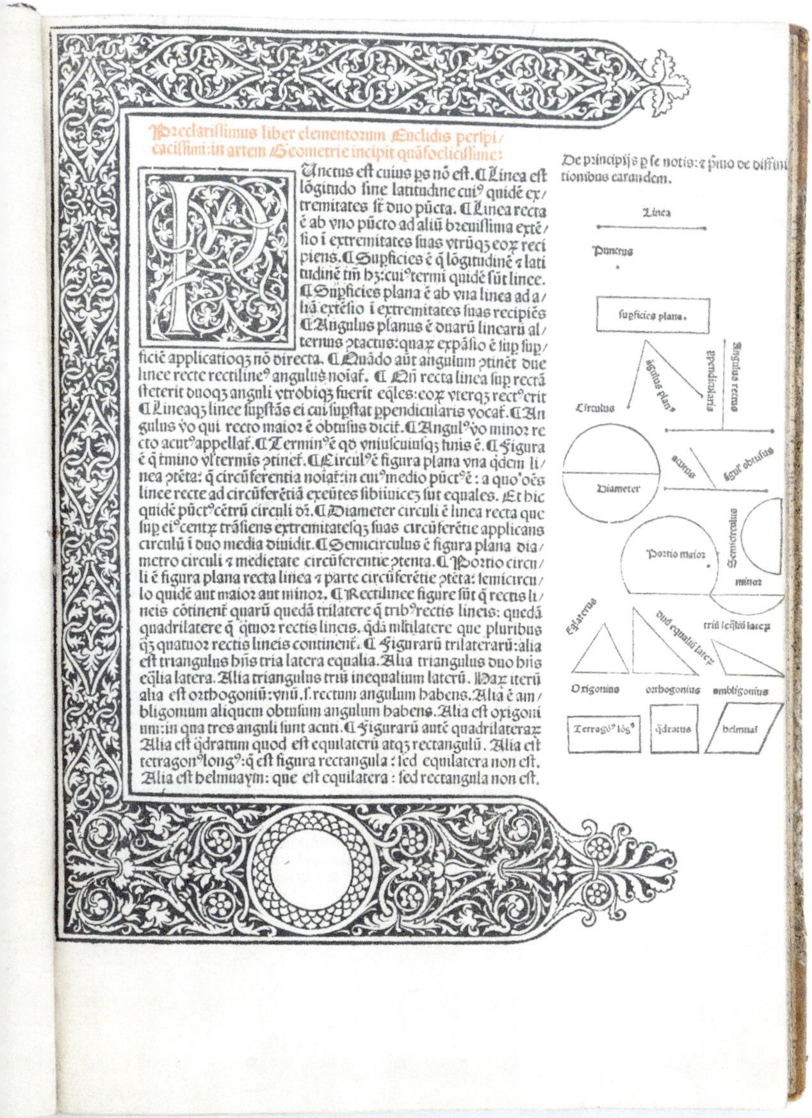

Preclarissimus liber elementorum Euclidis perspi/
cacissimi: in artem Geometrie incipit quãfoelicissime:

Punctus est cuius ps nõ est. ℂ Linea est lõgitudo sine latitudine cui⁹ quidẽ ex/tremitates sũt duo pũcta. ℂ Linea recta ẽ ab vno pũcto ad aliũ breuissima extẽ/sio i extremitates suas vtrũqꝫ eoꝝ reci/piens. ℂ Supficies ẽ q̃ lõgitudinẽ ⁊ lati/tudinẽ tm̃ hꝫ: cui⁹ termi quidẽ sũt linee. ℂ Supficies plana ẽ ab vna linea ad a/liã extẽsio i extremitates suas recipiẽs ℂ Angulus planus ẽ duarũ linearũ al/ternus ꝯtactus: quaꝝ expãsio ẽ sup sup/ficiẽ applicatioqꝫ nõ directa. ℂ Quãdo aũt angulum ꝯtinẽt due linee recte rectiline⁹ angulus noiat. ℂ Qñ recta linea sup rectã steterit duoqꝫ anguli vtrobiqꝫ fuerit eq̃les: eoꝝ vterqꝫ rect⁹ erit ℂ Lineaqꝫ linee supstãs ei cui supstat ppendicularis vocat. ℂ An/gulus vo qui recto maior ẽ obtusus dicit. ℂ Angul⁹ vo minor re/cto acut⁹ appellat. ℂ Termin⁹ ẽ qd vniuscuiusqꝫ finis ẽ. ℂ Figura ẽ q̃ tmino vl terminis ꝯtinet. ℂ Circul⁹ ẽ figura plana vna q̃dem li/nea ꝯtẽta: q̃ circũferentia noiat: in cui⁹ medio pũct⁹ ẽ: a quo oẽs linee recte ad circũferẽtiã exeũtes sibiiuicẽ sũt equales. Et hic quidẽ pũct⁹ cẽtrũ circuli dr̃. ℂ Diameter circuli ẽ linea recta que sup ei⁹ centꝝ trãsiens extremitatesqꝫ suas circũferẽtie applicans circulũ i duo media diuidit. ℂ Semicirculus ẽ figura plana dia/metro circuli ⁊ medietate circũferentie ꝯtenta. ℂ Portio circu/li ẽ figura plana recta linea ⁊ parte circũferẽtie ꝯtẽta: semicircu/lo quidẽ aut maior aut minor. ℂ Rectilinee figure sũt q̃ rectis li/neis cõtinent quarũ quedã trilatere q̃ trib⁹ rectis lineis: quedã quadrilatere q̃ q̃tuor rectis lineis. q̃dã mltilatere que pluribus q̃ꝫ quatuor rectis lineis continent. ℂ Figurarũ trilaterarũ: alia est triangulus hñs tria latera equalia. Alia triangulus duo hñs eq̃lia latera. Alia triangulus triũ inequalium laterũ. Haꝝ iterũ alia est orthogoniũ: vnũ .s. rectum angulum habens. Alia ẽ am/bligonium aliquem obtusum angulum habens. Alia est oxigoni/um: in qua tres anguli sunt acuti. ℂ Figurarũ autẽ quadrilateraꝝ Alia est q̃dratum quod est equilaterũ atqꝫ rectangulũ. Alia est tetragon⁹ long⁹: q̃ est figura rectangula: sed equilatera non est. Alia est helmuaym: que est equilatera: sed rectangula non est.

De principijs p se notis: ⁊ pmo de diffini/tionibus earandem.

THE FOUR FOLIOS

SHAKESPEARE, William (1564–1616).
Mr. William Shakespeares Comedies, Histories, & Tragedies. London, 1623.

The First Folio of Shakespeare's plays.

The First Folio aimed to be a definitive version of Shakespeare. Its editors, Shakespeare's friends and actors John Heminge and Henry Condell, claimed in their preface to have gathered Shakespeare's plays and printed them "absolute in their numbers, as he conceived them."

As the prefatory "Address to the Readers" makes clear, Heminge and Condell's motive was to ensure the accurate printing of Shakespeare's plays: while previous printings had offered "stolen, and surreptitious copies," the Folio presented the plays, "cured, and perfect of their limbs" — a metaphor that suggests the posthumous care of Shakespeare's literary body.

Gifted to Carnegie Mellon by Charles J. Rosenbloom in 1974, Carnegie Mellon's First Folio was once held in the collection of Marshall C. Lefferts (1848–1928).

"Folio"

The word folio describes the format or size of a book. A folio is a large book made up of sheets of paper folded once lengthwise, producing two leaves and four pages. Smaller book formats include quarto (two folds, four leaves) and octavo (three folds, eight leaves).

MR WILLIAM

SHAKESPEARES

COMEDIES,
HISTORIES, &
TRAGEDIES.

Publiſhed according to the True Originall Copies.

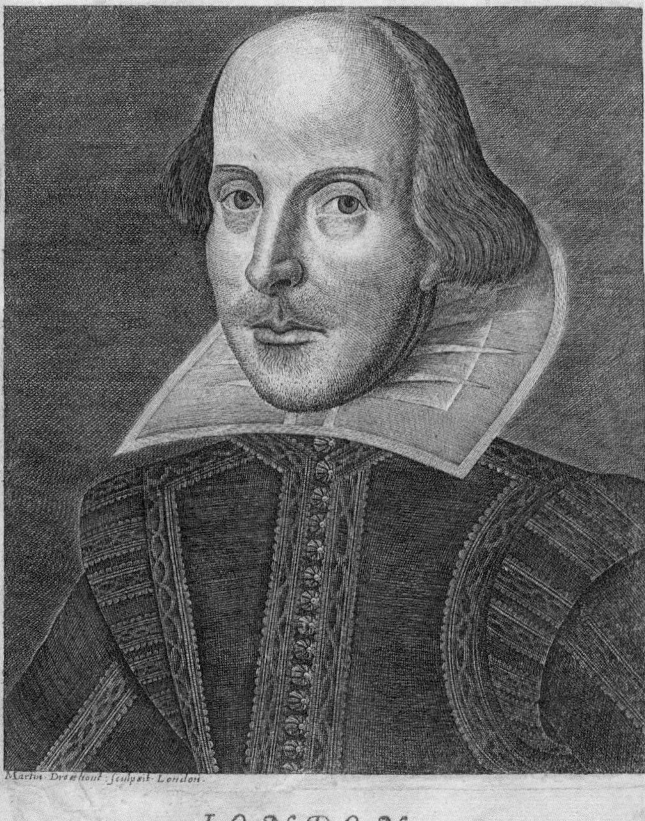

LONDON

Printed by Iſaac Iaggard, and Ed. Blount. 1623.

THE FOUR FOLIOS

SHAKESPEARE, William (1564–1616). *Mr. William Shakespeares Comedies, Histories, & Tragedies.* London, 1632.

John Milton's first appearance in print.

The Second Folio, a page-for-page reprint of the First, was published only nine years after the First Folio's appearance. This short interval suggests that the First Folio sold well and that demand for Shakespeare's plays in folio format was sustained into the 1630s. Wealthy buyers drove the demand; while Shakespeare's plays attracted audiences that spanned economic and social classes, the Folios were high-priced luxury goods.

One difference in the Second Folio is a short poem printed at the beginning of the book ("What neede my Shakespeare..."). Though unsigned, this poem is now known to be the work of John Milton (1608–1674), author of the biblical epic *Paradise Lost* and a major figure in the English literary canon. The poem, which praises Shakespeare and extols the Folio's role in preserving the plays for posterity, was Milton's first appearance in print. Milton's poem suggests that Shakespeare does not need a monument or tomb — the Folios function as their own "lasting Monument."

M^R. WILLIAM SHAKESPEARES COMEDIES, HISTORIES, and TRAGEDIES.

Publiſhed according to the true Originall Copies.

The ſecond Jmpreſsion.

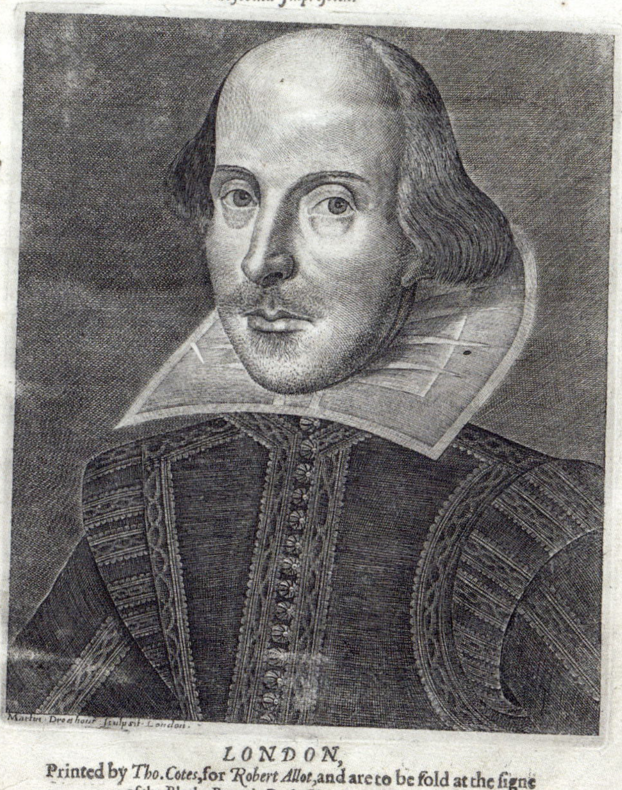

LONDON,

Printed by *Tho. Cotes*, for *Robert Allot*, and are to be ſold at the ſigne of the Blacke Beare in Pauls Church-yard. 1632.

THE FOUR FOLIOS

SHAKESPEARE, William (1564–1616). *Mr. William Shakespeares Comedies, Histories, & Tragedies.* London, 1663.

The rarest of the four Shakespeare Folios.

The Third Folio exists in two forms, or "issues." One issue, dated 1663, reprints the contents of the Second Folio (1632). The second issue bears a 1664 date on its title page and contains seven plays that do not appear in the First or Second Folios. Six of these plays are apocryphal; only *Pericles* is now accepted to be by Shakespeare. Scholars have argued that the addition of these plays was intended to make the Third Folio more appealing to potential buyers.

The Third Folio is also the rarest of the four Folios due to a 1666 fire that destroyed much of London, including the bookselling district around St Paul's Cathedral. Only 180 copies of the Third Folio survive, while 228 copies of the First Folio are extant.

"Edition Binding"

Before the introduction of "edition binding"— a method by which all copies of a particular print run are bound in the same or similar style — books were kept in sheets or only loosely stitched together until sold. The buyer would then have the sheets bound in a style of their choosing. Both of Carnegie Mellon's copies of the Third Folio are in their original bindings and exhibit a strikingly similar appearance: plain calf leather with a double line of embossed rules outlining the edge of each board and a mottled pattern of red decorating the edges of the text block. Separated for nearly four centuries, these books, perhaps bound in the same workshop in London and from the same stock of materials (animal hide, paper, twine, and glue), have been reunited by chance in Pittsburgh.

Mr. WILLIAM SHAKESPEAR'S

Comedies, Hiſtories, and Tragedies.

Publiſhed according to the true Original Copies.

The third Impreſſion.

And unto this Impreſſion is added ſeven Playes, never before Printed in Folio.

viz.

Pericles Prince of *Tyre.*
The *London Prodigall.*
The Hiſtory of *Thomas* Ld. *Cromwell.*
Sir *John Oldcaſtle* Lord *Cobham.*
The *Puritan Widow.*
A *York-shire* Tragedy.
The Tragedy of *Locrine.*

LONDON, Printed for *P. C.* 1664.

THE FOUR FOLIOS

SHAKESPEARE, William (1564–1616).
Mr. William Shakespeares Comedies, Histories, & Tragedies. London, 1685.

"Hamlet: Rpince of Denmark."

The Fourth Folio contains several oddities. Seventeen of its sheets (sixty-eight pages total) were reprinted as late as 1700 — fifteen years after their first appearance. The reason for this reprinting remains obscure. Scholars have suggested that the Fourth Folio's printer failed to deliver the required quantity, leaving the bookseller who inherited unsold copies in 1700 short-sheeted. Or maybe the missing sheets were damaged or lost while in the bookseller's warehouse, requiring replacement. The Fourth Folio also contains one of the most glaring typesetting errors in the history of English literature: the printed subtitle for Hamlet reads, "RPINCE OF DENMARK."

In 2023, an interdisciplinary team of researchers at Carnegie Mellon attributed the printing of part of the Fourth Folio to Robert Everingham. Using machine learning and computer vision techniques, the team was able to match distinctive typography in the Fourth Folio to typography in other books printed by Everingham. Prior to the team's work, the identities of the Fourth Folio's printers were unknown (see "Old Books, New Technologies: Computational Research in Special Collections," page 379).

THE TRAGEDY OF HAMLET RPINCE of DENMARK.

Actus Primus. Scæna Prima.

Enter Barnardo *and* Franciſco, *two* Centinels.

Bar. WHo's there?

Fran. Nay anſwer me: Stand and unfold your ſelf.

Bar. Long live the King.

Fran. Barnardo.

Bar. He.

Fran. You come moſt chearfully upon your hour.

Bar. 'Tis now ſtruck twelve, get thee to Bed, *Franciſco.*

Fran. For this relief much thanks: 'tis bitter cold,
And I am ſick at heart.

Bar. Have you had quiet Guard.

Fran. Not a Mouſe ſtirring.

Bar. Well, good Night. If you do meet *Horatio* and *Marcellus*, the Rivals of my Watch, bid them make haſt.

Enter Horatio *and* Marcellus.

Fran. I think I hear them. Stand, who's there?

Hor. Friends to this ground.

Mar. And Liege-men to the *Dane.*

Fran. Give you good Night.

Mar. O farewel, honeſt Souldier, who hath reliev'd you?

Fran. Barnardo has my place: give you good Night.

[*Exit* Franciſco.

Mar. Holla, *Barnardo.*

Bar. Say, what is *Horatio* there?

Hor. A piece of him.

Bar. Welcome, *Horatio*, welcome, good *Marcellus.*

Mar. What, has this thing appear'd again to Night.

Bar. I have ſeen nothing.

Mar. Horatio ſays, 'tis but our phantaſie,
And will not let belief take hold of him
Touching this dreaded ſight, twice ſeen of us,
Therefore I have intreated him along
With us, to watch the minutes of this Night,
That if again this Apparition come,
He may approve our Eyes, and ſpeak to it.

Hor. Tuſh, tuſh, 'twill not appear.

Bar. Sit down a while,
And let us once again aſſail your Ears,
That are ſo fortified againſt our ſtory,
What we two Nights have ſeen.

Hor. Well, ſit we down,
And let us hear *Barnardo* ſpeak of this.

Bar. Laſt night of all,
When yon ſame Star, that's weſtward from the Pole
Had made his courſe t'illume that part of Heaven.
Where now it burns, *Marcellus* and my ſelf,
The Bell then beating one.

Mar. Peace, break thee off:

Enter the Ghoſt.

Look where it comes again.

Bar. In the ſame figure like the King that's dead.

Mar. Thou art a Scholar, ſpeak to it, *Horatio.*

Bar. Looks it not like the King? Mark it, *Horatio.*

Hor. Moſt like: It harrows me with fear and wonder.

Bar. It would be ſpoke to.

Mar. Queſtion it, *Horatio.*

Hor. What art thou that uſurp'ſt this time of Night,
Together with that fair and warlike form
In which the Majeſty of buried *Denmark*
Did ſometimes march: By Heaven I charge thee ſpeak.

Mar. It is offended.

Bar. See, it ſtalks away.

Hor. Stay: ſpeak; ſpeak: I charge thee, ſpeak.

[*Exit* Ghoſt.

Mar. 'Tis gone, and will not anſwer.

Bar. How now, *Horatio*? You tremble and look pale:
Is not this ſomething more than fantaſie?
What think you on't?

Hor. Before my God I might not this believe
Without the ſenſible and true avouch
Of mine own Eyes.

Mar. Is it not like the King.

Hor. As thou art to thy ſelf,
Such was the Armour he had on,
When th'ambitious *Norway* combated:
So frown'd he once, when in an angry parle
He ſmote the ſledded Poleaxe on the Ice.
Tis ſtrange.

Mar. Thus 'twice before, and juſt at this ſame hour,
With Martial ſtalk, hath he gone by our Watch.

Hor. In what particular thought to work, I know not:
But in the groſs and ſcope of my opinion,

This

EINSTEIN, Albert (1879–1955).
"Die Grundlage der allgemeinen Relativitätstheorie" in *Annalen der Physik*, IV. Leipzig, 1916.

EINSTEIN, Albert (1879–1955).
Die Grundlage der allgemeinen Relativitätstheorie. Leipzig, 1916.

The first appearance of Einstein's theory of relativity in print.

Einstein's paper outlined general relativity, revolutionizing our concept of space, time, and gravity. Published in the journal *Annalen der Physik* (*Annals of Physics*), this first printing introduced the idea that massive objects — stars, galaxies, and black holes — warp spacetime. Einstein's work on relativity theory contributed to his Nobel Prize in physics in 1921.

A signed copy of the first standalone printing of the theory of relativity.

This separate printing followed the first appearance of Einstein's theory of relativity in the *Annalen der Physik* in 1916 (see previous entry). The later printing incorporates revisions to the text and an introduction written by Einstein. Special Collections' copy, held in the Posner Memorial Collection, is signed by Albert Einstein on the title page.

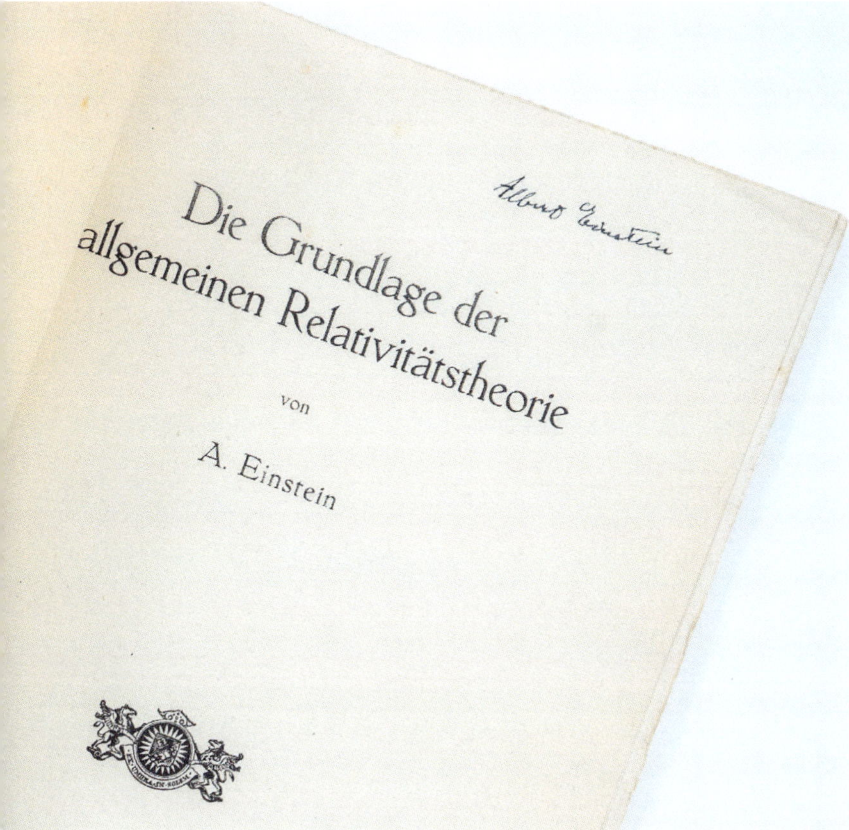
Albert Einstein
Die Grundlage der
allgemeinen Relativitätstheorie
von
A. Einstein

MELVILLE, Herman (1819–1891).
The Whale. London, 1851.

A defining work of American literature.

Special Collections holds a first edition of Melville's whaling epic, published in London under the title *The Whale* one month before its American release as *Moby-Dick*. This edition differs significantly from the American text, omitting Melville's Epilogue and softening or excising language deemed profane or inappropriate for publication.

Based on Melville's reading of an account of the killing of the "Mocha Whale" — an albino sperm whale rumored to have attacked whaling ships — the novel describes Ahab's monomaniacal pursuit of Moby Dick.

Gifted to CMU Libraries by Charles J. Rosenbloom, Special Collections' copy survives in its original binding: the spines of all three of its volumes are stamped with a gilt and embossed image of a whale.

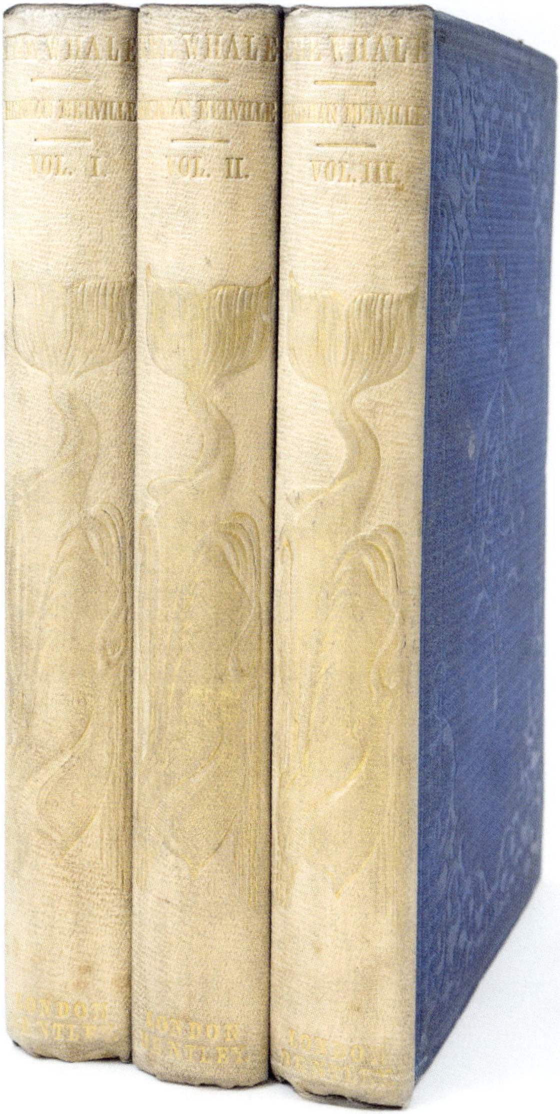
VOL. I
VOL. II.
VOL. III.
LONDON
BENTLEY.
LONDON

MENDEL, Gregor Johann (1822–1884).
“Versuche über Pflanzen-Hybriden”
in *Verhandlungen des naturforschenden Vereines in Brünn, IV.* Brünn, 1865.

The origin of modern genetics.

Gregor Mendel was an Augustinian monk and scientist whose experiments with pea plants (*Pisum sativum*) established the fundamental laws of heredity and laid the groundwork for modern genetics. Published in 1866, Mendel's "Experiments on Plant Hybridization" was overlooked in its time: Hugo de Vries (1848–1935), Carl Correns (1864–1933), and Erich von Tschermak (1871–1962) independently confirmed his findings in 1900, three decades after their first publication.

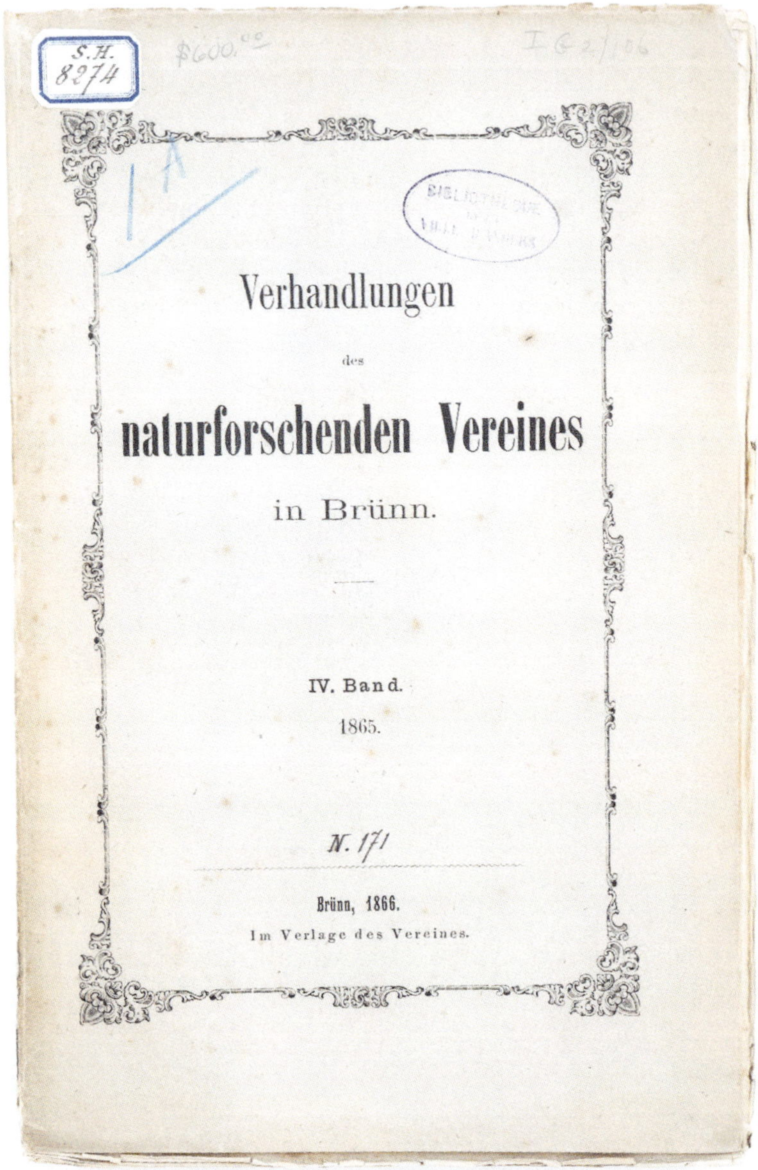
S.H.
8274
Verhandlungen
des
naturforschenden Vereines
in Brünn.
IV. Band.
1865.
N. 171
Brünn, 1866.
Im Verlage des Vereines.

Versuche über Pflanzen-Hybriden.

Von

Gregor Mendel.

(Vorgelegt in den Sitzungen vom 8. Februar und 8. März 1865.)

Einleitende Bemerkungen.

Künstliche Befruchtungen, welche an Zierpflanzen desshalb vorgenommen wurden, um neue Farben-Varianten zu erzielen, waren die Veranlassung zu den Versuchen, die her besprochen werden sollen. Die auffallende Regelmässigkeit, mit welcher dieselben Hybridformen immer wiederkehrten, so oft die Befruchtung zwischen gleichen Arten geschah, gab die Anregung zu weiteren Experimenten, deren Aufgabe es war, die Entwicklung der Hybriden in ihren Nachkommen zu verfolgen.

Dieser Aufgabe haben sorgfältige Beobachter, wie Kölreuter, Gärtner, Herbert, Lecocq, Wichura u. a. einen Theil ihres Lebens mit unermüdlicher Ausdauer geopfert. Namentlich hat Gärtner in seinem Werke „die Bastarderzeugung im Pflanzenreiche" sehr schätzbare Beobachtungen niedergelegt, und in neuester Zeit wurden von Wichura gründliche Untersuchungen über die Bastarde der Weiden veröffentlicht. Wenn es noch nicht gelungen ist, ein allgemein giltiges Gesetz für die Bildung und Entwicklung der Hybriden aufzustellen, so kann das Niemanden Wunder nehmen, der den Umfang der Aufgabe kennt und die Schwierigkeiten zu würdigen weiss, mit denen Versuche dieser Art zu kämpfen haben. Eine endgiltige Entscheidung kann erst dann erfolgen, bis Detail-Versuche aus den verschiedensten Pflanzen-Familien vorliegen. Wer die Ar-

1*

BRONTË, Charlotte (1816–1855).
Jane Eyre: An Autobiography. London, 1847.

A landmark of Victorian literature, published under a pseudonym.

Printed and sold under the masculine pseudonym "Currer Bell," Charlotte Brontë's *Jane Eyre* blended Gothic romance with social critique. This first edition, released in three volumes — known as "three-decker" format — marked the emergence of a powerful literary voice. Today, *Jane Eyre* remains a landmark of realist fiction. Special Collections' copy was donated to CMU libraries by Charles J. Rosenbloom in 1974. It survives in its original publisher's cloth binding.

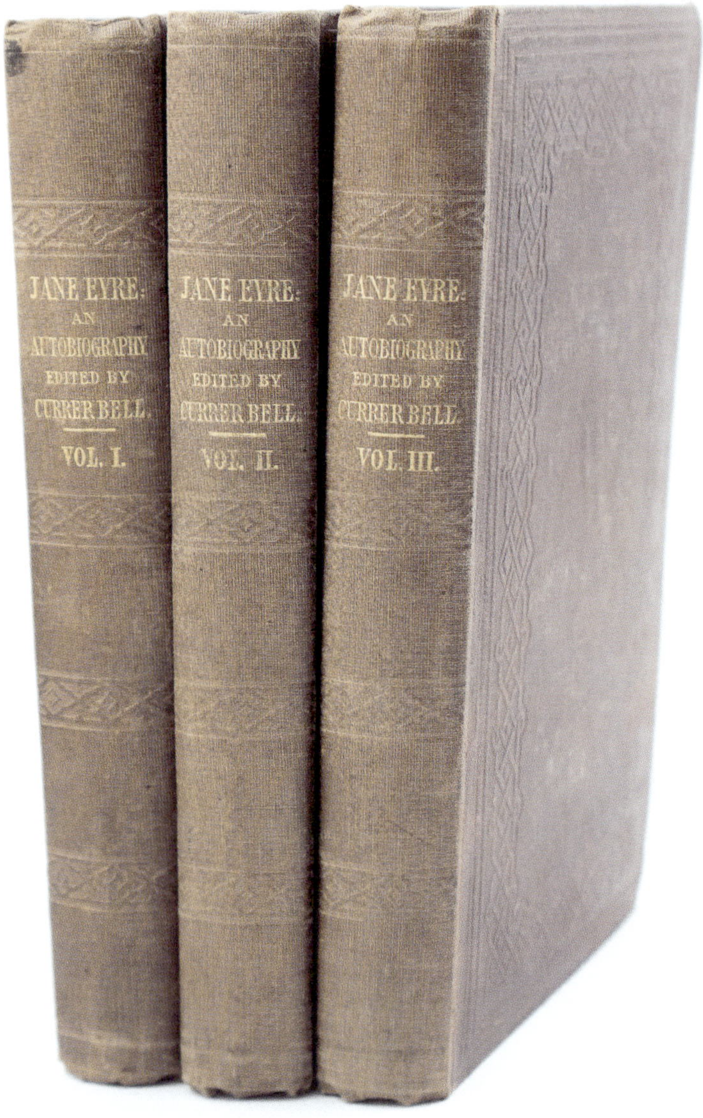
JANE EYRE:
AN
AUTOBIOGRAPHY
EDITED BY
CURRER BELL.
VOL. I.
JANE EYRE:
AN
AUTOBIOGRAPHY
EDITED BY
CURRER BELL.
VOL. II.
JANE EYRE:
AN
AUTOBIOGRAPHY
EDITED BY
CURRER BELL.
VOL. III.

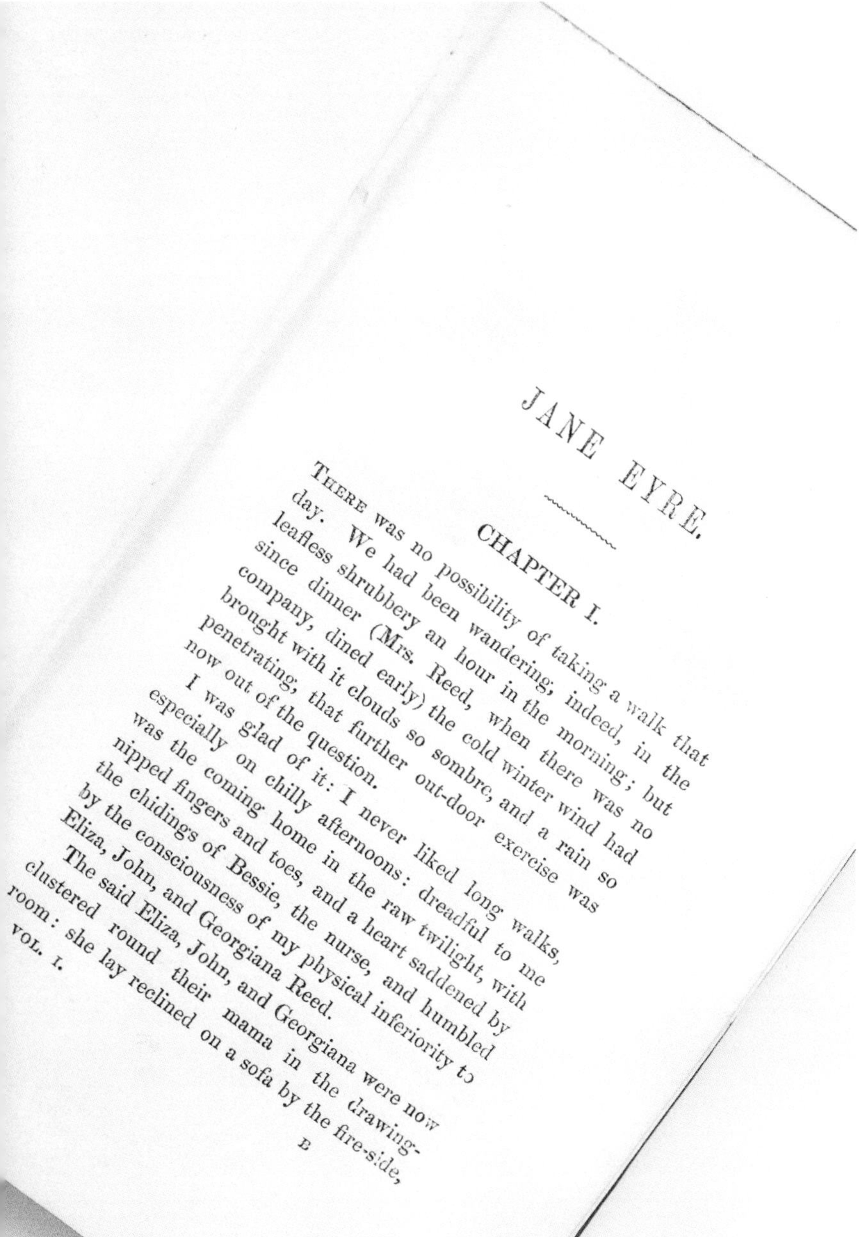

JANE EYRE.

CHAPTER I.

THERE was no possibility of taking a walk that day. We had been wandering, indeed, in the leafless shrubbery an hour in the morning; but since dinner (Mrs. Reed, when there was no company, dined early) the cold winter wind had brought with it clouds so sombre, and a rain so penetrating, that further out-door exercise was now out of the question.

I was glad of it: I never liked long walks, especially on chilly afternoons: dreadful to me was the coming home in the raw twilight, with nipped fingers and toes, and a heart saddened by the chidings of Bessie, the nurse, and humbled by the consciousness of my physical inferiority to Eliza, John, and Georgiana Reed.

The said Eliza, John, and Georgiana were now clustered round their mama in the drawing-room: she lay reclined on a sofa by the fire-side,

VOL. I. B

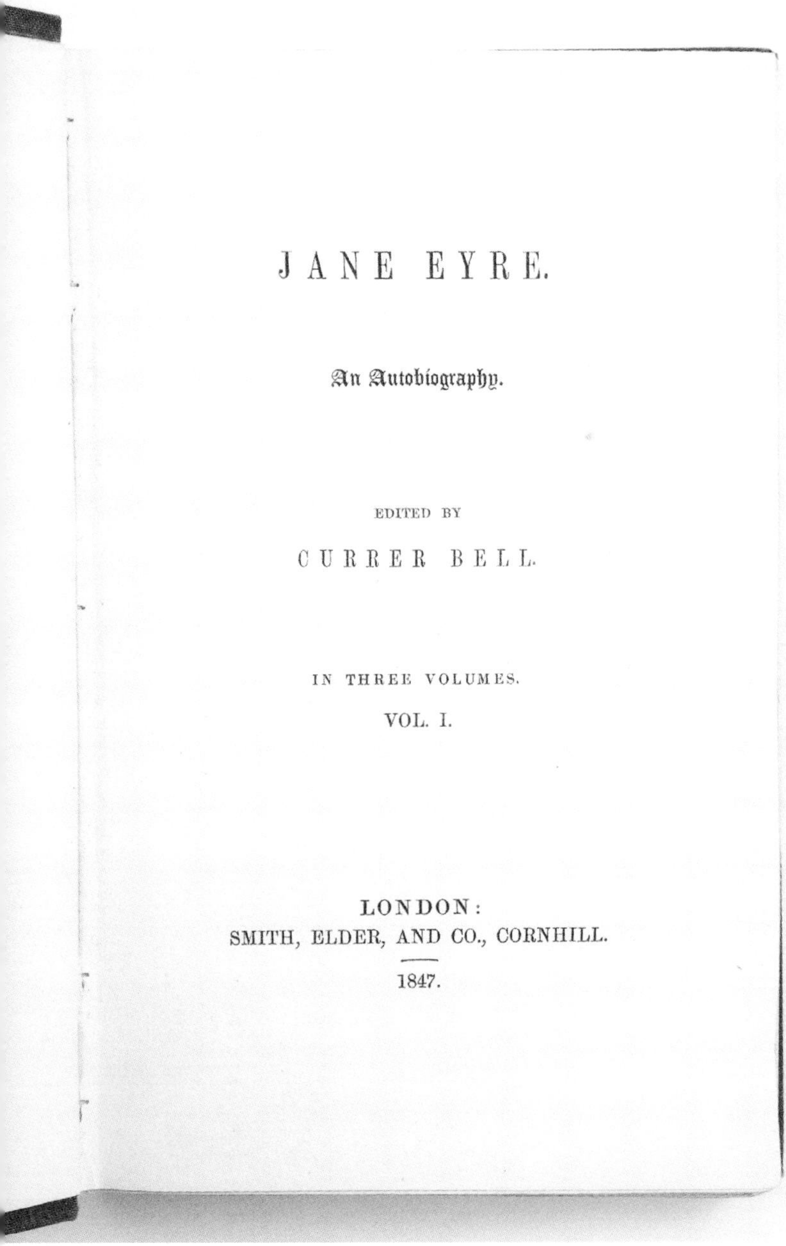

JANE EYRE.

An Autobiography.

EDITED BY

CURRER BELL.

IN THREE VOLUMES.

VOL. I.

LONDON:
SMITH, ELDER, AND CO., CORNHILL.

1847.

VESALIUS, Andreas (1514–1564).
De humani corporis fabrica libri septem.
Basel, 1543.

A revolution in anatomical science.

Published in the same year as Copernicus' *De Revolutionibus*, Vesalius's *De Humani Corporis Fabrica* ("On the Fabric of the Human Body") had a comparably transformational effect on medical theory. Divided into anatomical sections — bones, muscles, the digestive, nervous, reproductive, and venous systems — the *Fabrica* is an odd mixture of learning and levity: large woodcut letters that open the chapters of Vesalius's text show childlike allegorical figures, or "putti," engaged in surgical tasks. One shows a group dissecting a pig.

The work's frontispiece — an elaborately staged woodcut of Vesalius giving an anatomy lesson — also contains at least one joke: in the foreground are a dog and a monkey, seemingly out of place among medical students and hangers-on. Both animals are a reminder that before Vesalius, anatomical demonstrations involved the vivisection of monkeys and dogs. Look close enough, and the dog appears relieved to find itself in the audience rather than on the operating table. The monkey puts on a more enigmatic expression.

Published when Vesalius was only twenty-nine, the work overturned centuries of Galenic (named for Galen, a first-century CE Roman physician) medicine and anatomy. Rejecting classical authority, Vesalius advocated for direct observation by means of dissection. The *Fabrica* is best known for its full-page woodcut illustrations of the human body. The images depict flayed human figures in dramatic poses against classical landscapes. Created by an unknown artist, the woodcuts are commonly attributed to Jan van Calcar, a student of Titian.

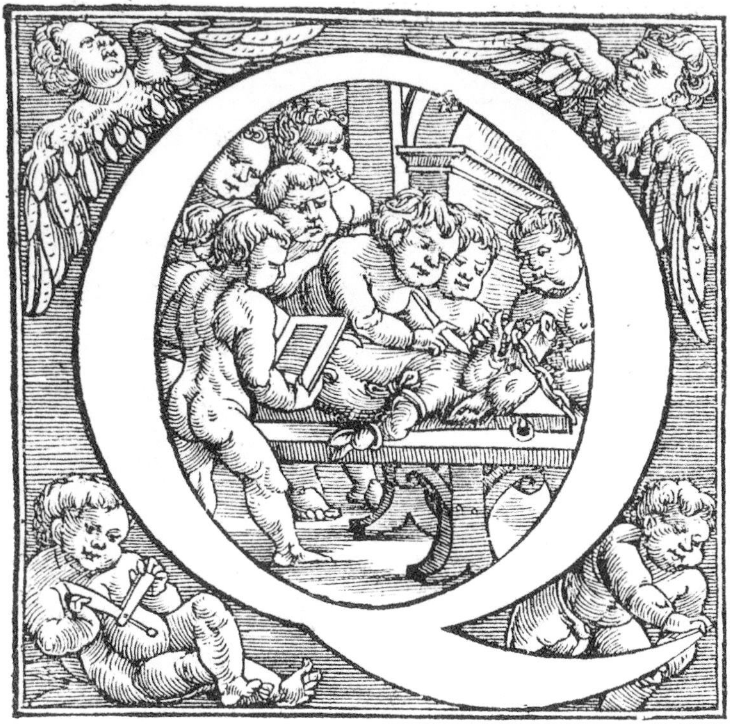

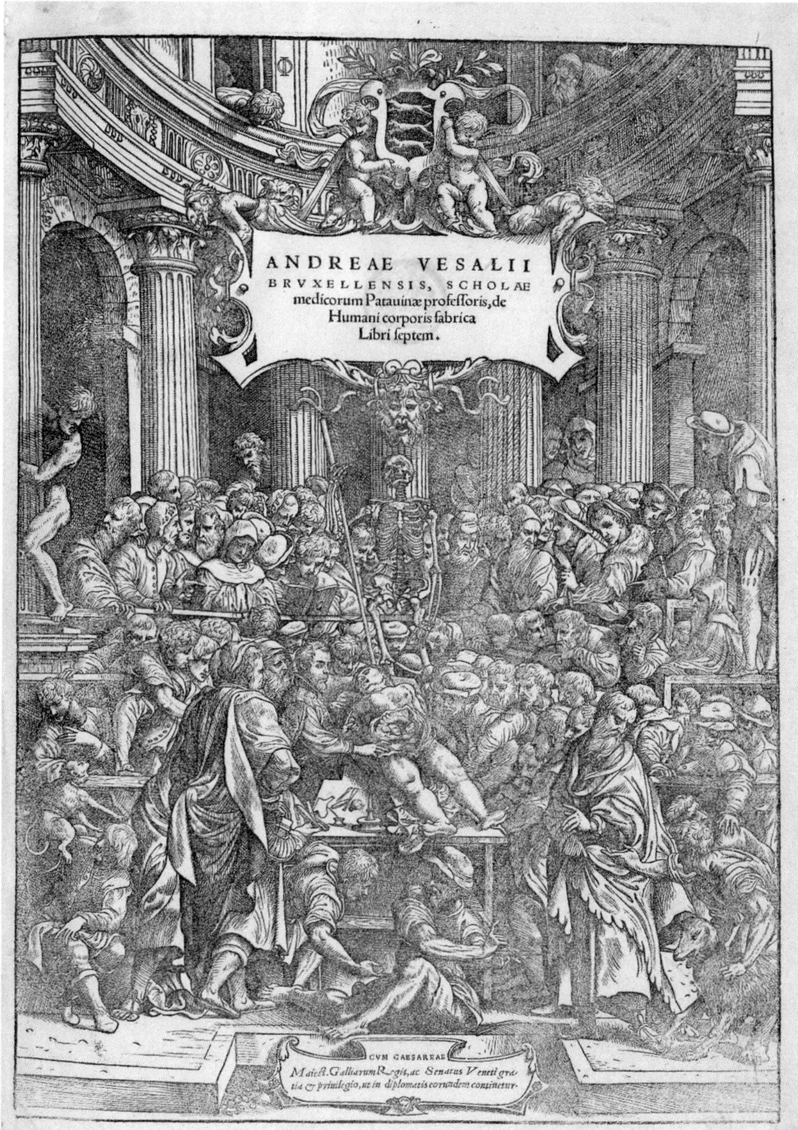
ANDREAE VESALII
BRVXELLENSIS, SCHOLAE
medicorum Patauinæ professoris, de
Humani corporis fabrica
Libri septem.
CVM CAESAREAE
Maiest. Galliarum Regis, ac Senatus Veneti gratia & priuilegio, ut in diplomatis eorundem continetur.
BASILEAE.

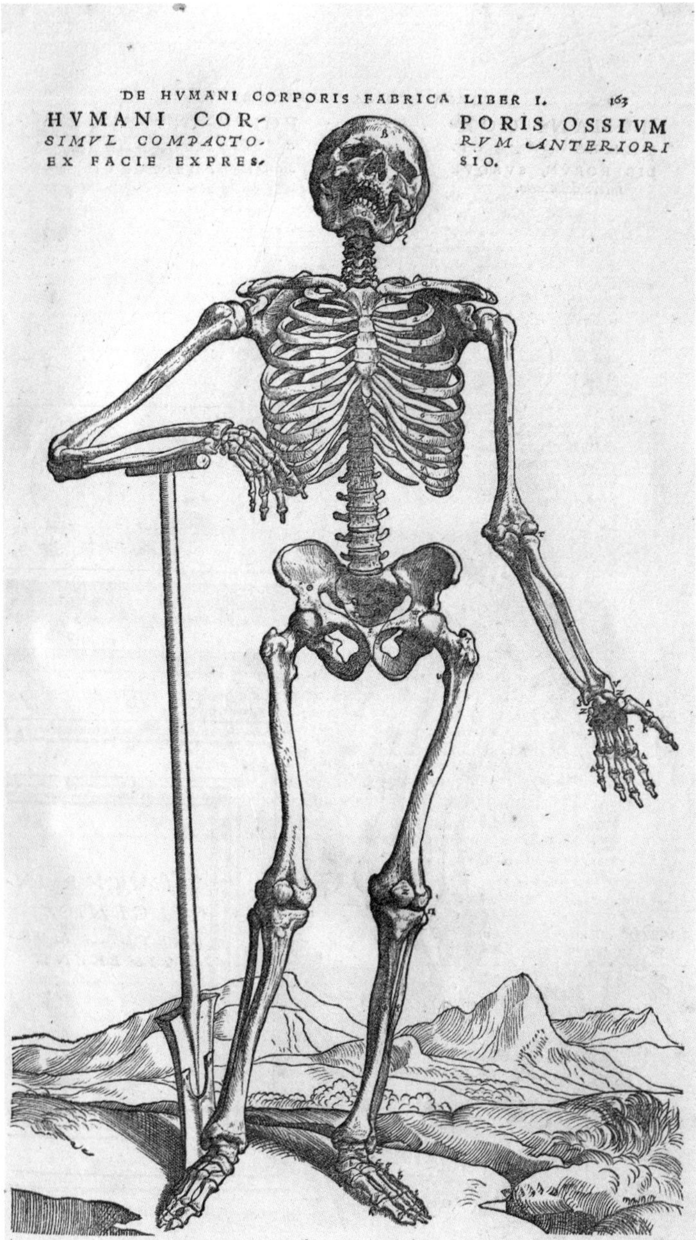
DE HVMANI CORPORIS FABRICA LIBER I. 163
HVMANI COR-
SIMVL COMPACTO-
EX FACIE EXPRES-
PORIS OSSIVM
RVM ANTERIORI
SIO.

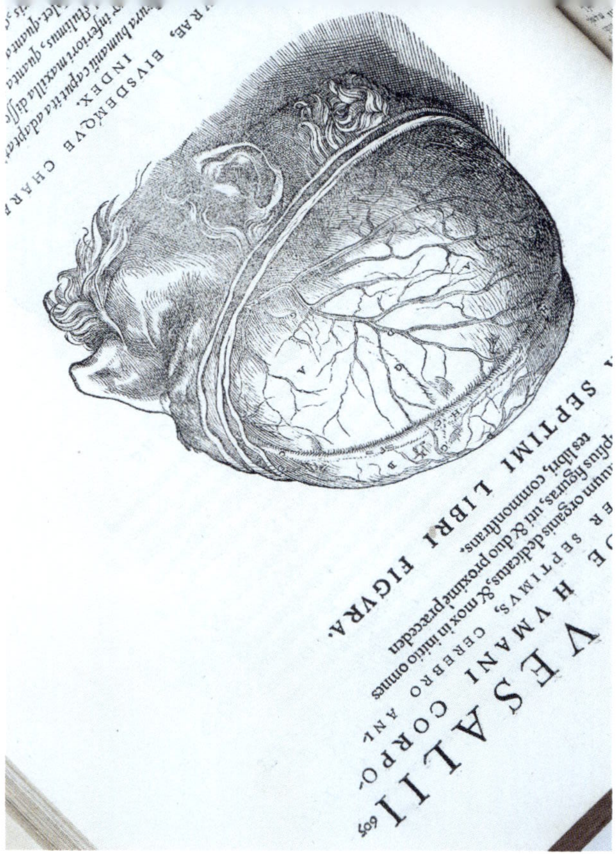

DVODECIMA SEPTIMI LIBRI FIGVRA.

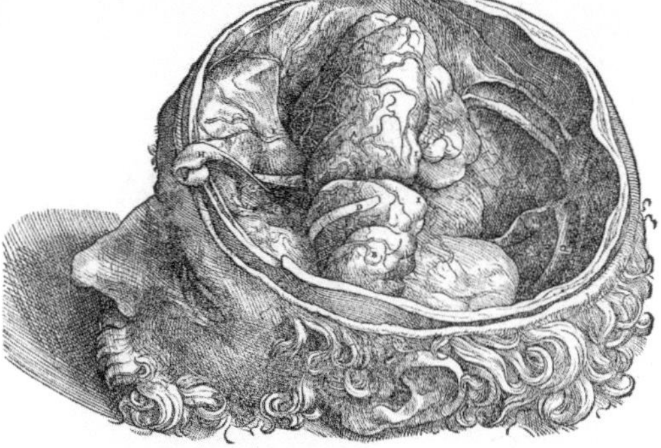

DVODECIMAE FIGVRAE, EIVSDEMQVE CHAracterum Index.

HAC figura caput à sinistro latere exprimitur, dextro nonnihil subleuato. Ex caluariæ autem amplitudine cerebellum hic euulsimus, ea tantùm cerebri relicta portione, quæ in octaua & nona figuris adhuc seruabatur. Verùm illa cerebri portio hic in sua non consistit sede, sed eleuata est, & in posteriora ab ipsa caluariæ basi aliquousque reflexa, ut tandem in conspectum uenirent cerebri processus forma neruis non absimiles, et olfactus organo subministrantes: quorum sinister è sua sede cum ipso cerebro eleuatus est, dextro adhuc duræ cerebri membranæ quæ octauo capitis ossi obducitur, connexo.

A,A *Cerebri pars dextra.*

B,B *Sinistra cerebri pars, tenui membrana adhuc quemadmodum & dextra pars obuoluta.*

C *Dextrum olfactus organum in sua sede adhuc seruatum.*

D *Sinistrum olfactus organum, unà cum cerebro in posteriora reflexum.*

E *Sinus, cui sinistrum olfactus incumbit organum. eaq́; regio duræ membranæ hic indicatur, quæ odoratus nomine hac in sede multis, sed exiguis foraminibus peruia cernitur.*

F *Vena caluariam petentium sexta, hic in duram cerebri membranam aliquot surculis excurrens.*

G *Septum sinus olfactus organis incisos interdiuidens.*

H *Duræ cerebri membranæ partis portio, quæ dextram cerebri partem à sinistra dirimit.*

I,I *Cerebrum paßim excauatæ sibi in caluaria amplitudini respondens, hic quoque ad frontis oßis regionem sibi insculptis cauernis congruit, & leuibus tuberibus prominet, quæ alij mamillares cerebri uocarunt processus.*

K *Vena caluariam petentium quinta, cui iter præbet foramen secundo neruorum pari exsculptum.*

L *L, ut & subsequentes characteres, in umbra occipitij cauitatis latitat, & parui etiam est mo-*
M. *menti. quandoquidem L caluariæ sinum indicat, cui dextra cerebelli pars congruit. M sedem,*
N.O. *quæ mediam cerebelli partem respicit. N sinum, in quem sinistra cerebelli pars reponitur. O ue-*
P.Q. *rò dextrum seu primum sinum indicat duræ membranæ. P autem tertium, Q secundum, seu sinistrum.*

BRAHE, Tycho (1546–1601).
Astronomiæ instauratæ mechanica.
Nuremberg, 1602.

A binding with a hidden astronomical pun?

The *Mechanica* illustrates the instruments used in Brahe's astronomical observatory on the Danish island of Hven. Published a year after Brahe's death in 1602, the *Mechanica* was almost immediately out of date: Galileo would turn his telescope skyward a mere seven years after its printing. Brahe's instruments, by contrast, were exclusively non-optical: mural quadrants, sextants, and armillary spheres.

Special Collections' copy of Brahe's *Mechanica* is bound in a thirteenth-century manuscript fragment of a hymn to the Virgin Mary, *Ave Maris Stella* ("Hail, Star of the Sea"). The orientation of the fragmentary leaf (the incipit of the hymn is mounted on the book's front board) suggests that the choice was one of reverent intention: the binding both marks the work's starry subject and offers its readers a piece of star-themed Marian devotion.

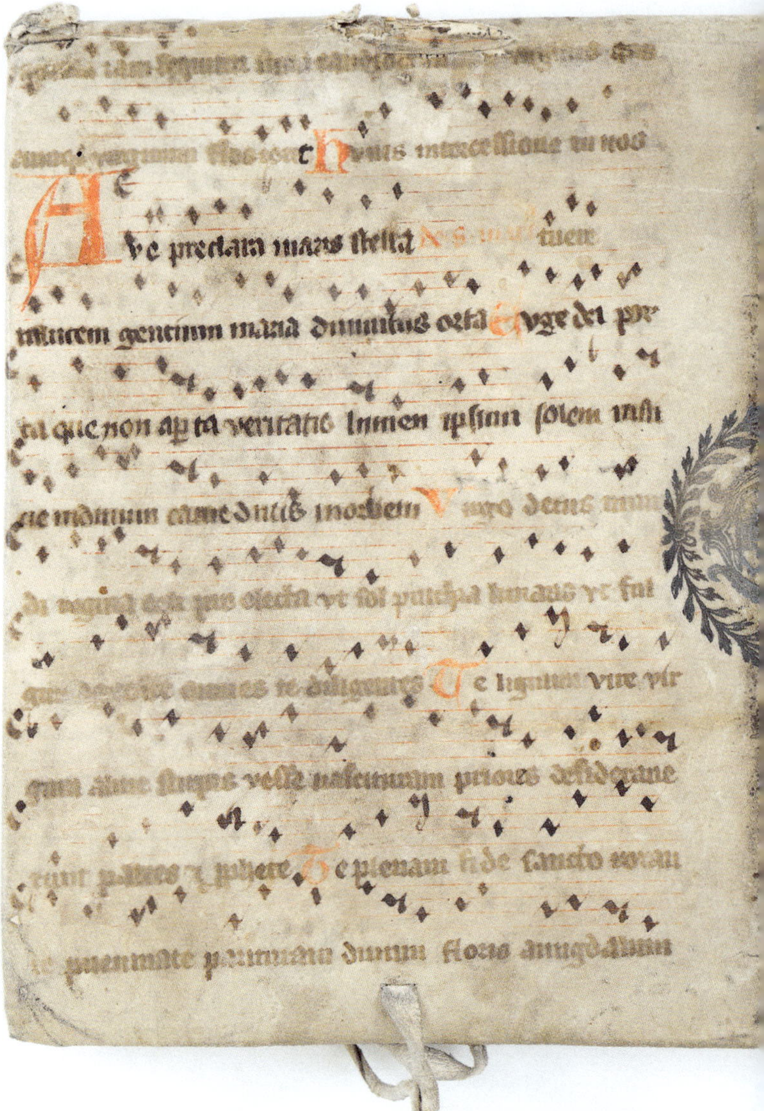
Ave preclara maris stella

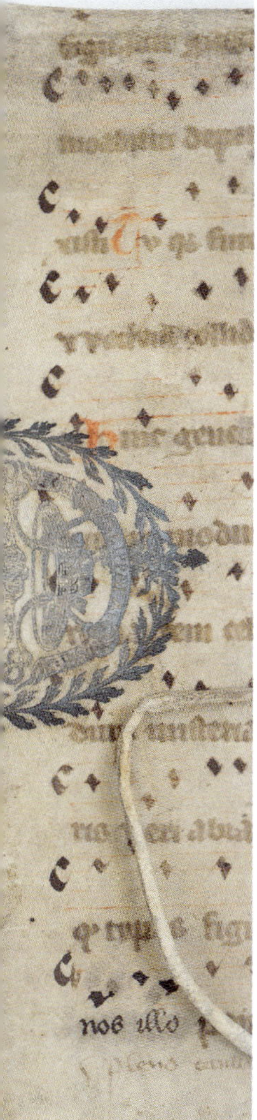

BOOLE, George (1815–1864).
An Investigation of the Laws of Thought, on Which Are Founded the Mathematical Theories of Logic and Probabilities.
London, 1854.

The mathematical origin of computational logic and digital circuits.

George Boole, a self-taught mathematician, invented the system of logic that now informs all digital computing. Boolean logic uses logic gates — the basic components of digital circuits — and Boole's logical operators (AND, OR, NOT, NAND, NOR, XOR, XNOR) to process information. Special Collections' copy is a first edition of Boole's fullest treatment of his system of logic.

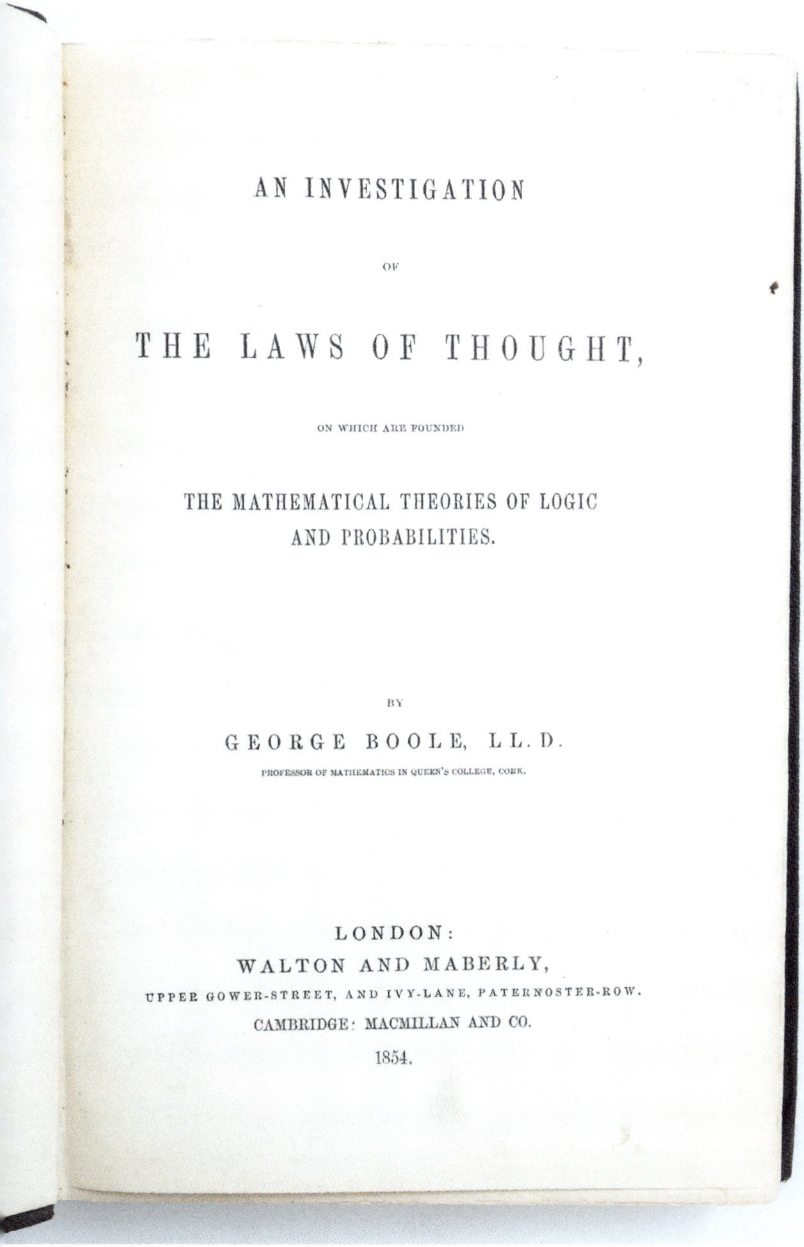

AN INVESTIGATION

OF

THE LAWS OF THOUGHT,

ON WHICH ARE FOUNDED

THE MATHEMATICAL THEORIES OF LOGIC
AND PROBABILITIES.

BY

GEORGE BOOLE, LL.D.

PROFESSOR OF MATHEMATICS IN QUEEN'S COLLEGE, CORK.

LONDON:
WALTON AND MABERLY,
UPPER GOWER-STREET, AND IVY-LANE, PATERNOSTER-ROW.
CAMBRIDGE: MACMILLAN AND CO.
1854.

CARROLL, Lewis (1832–1898).
Alice's Adventures in Wonderland.
New York, 1866.

Logic and nonsense in Wonderland.

Alice's Adventures in Wonderland is both an important work of children's literature and a playful demonstration of its author's innovations in mathematics and symbolic logic. Through the book's often absurd narrative and word games, Lewis Carroll — the pen name of Oxford mathematician Charles Lutwidge Dodgson — revels in puzzles and paradoxes that reflect his interest in recreational mathematics, decision making, and logic. Special Collections holds one copy each of the first American and first British editions. The collection also holds a first edition of Carroll's sequel to *Alice's Adventures in Wonderland: Through the Looking-Glass, and What Alice Found There* (1872).

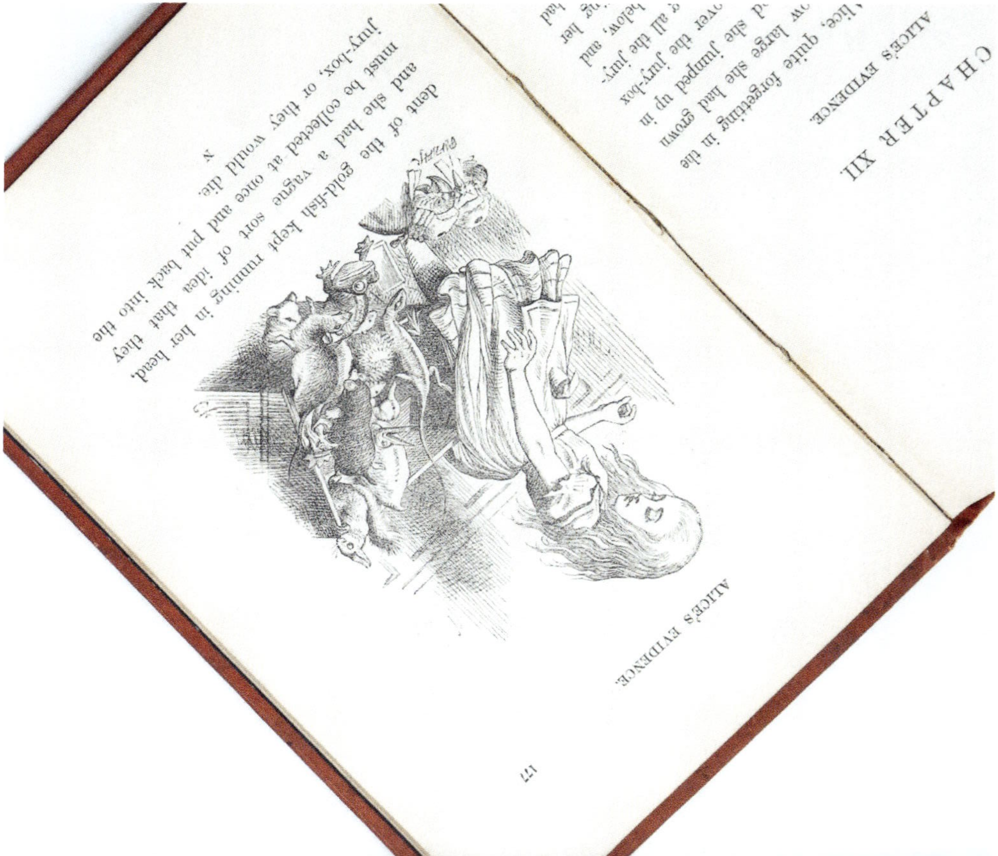

CHAPTER XII.

ALICE'S EVIDENCE.

lice, quite forgetting in the
ow large she had grown
d she jumped up in
over the jury-box
g all the jury-
below, and
ng her
had

ALICE'S EVIDENCE. 177

dent of the gold-fish kept running in her head,
and she had a vague sort of idea that they
must be collected at once and put back into the
jury-box, or they would die.

N

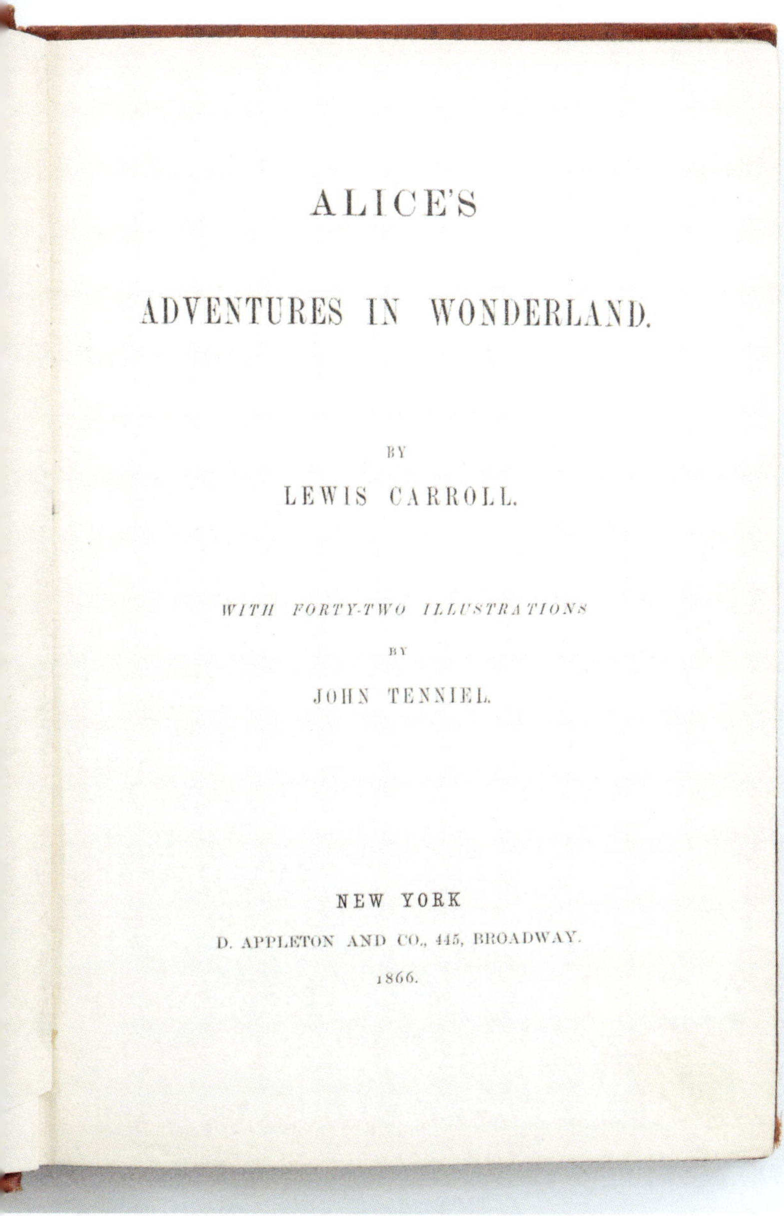

ALICE'S

ADVENTURES IN WONDERLAND.

BY

LEWIS CARROLL.

WITH FORTY-TWO ILLUSTRATIONS

BY

JOHN TENNIEL.

NEW YORK

D. APPLETON AND CO., 445, BROADWAY.

1866.

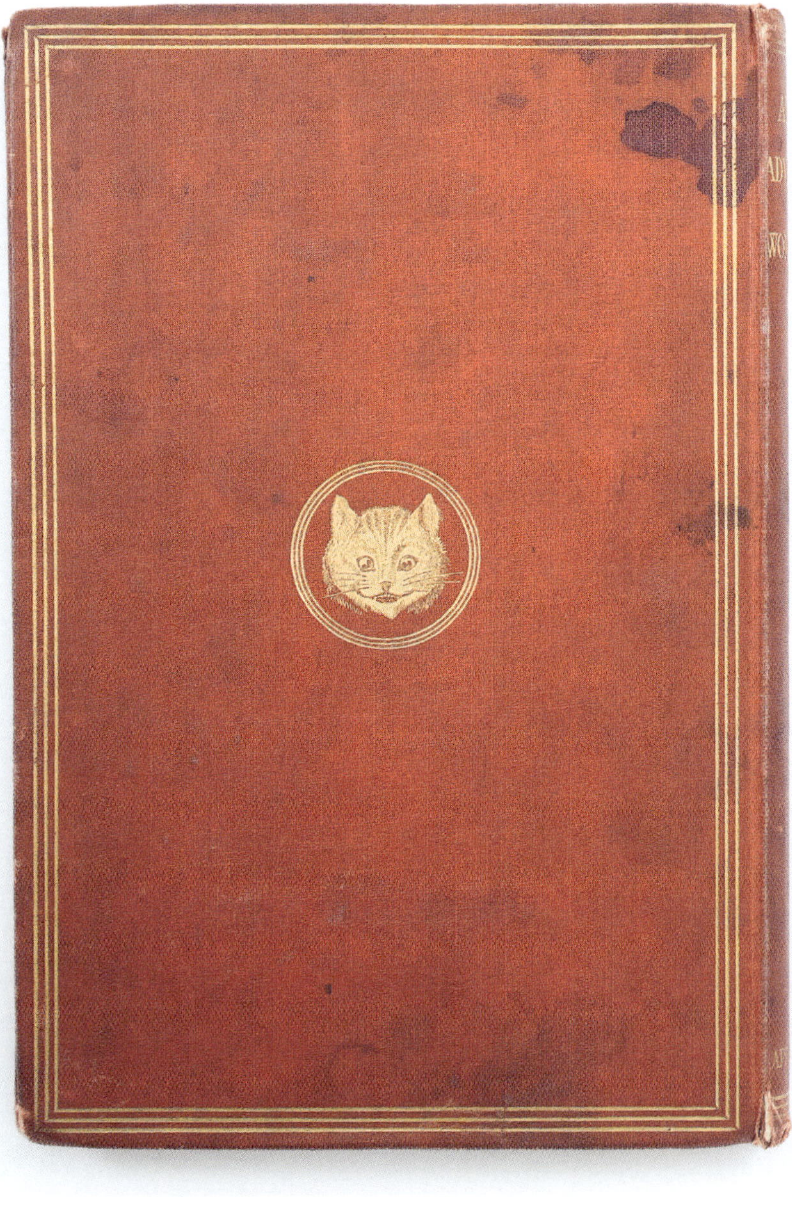

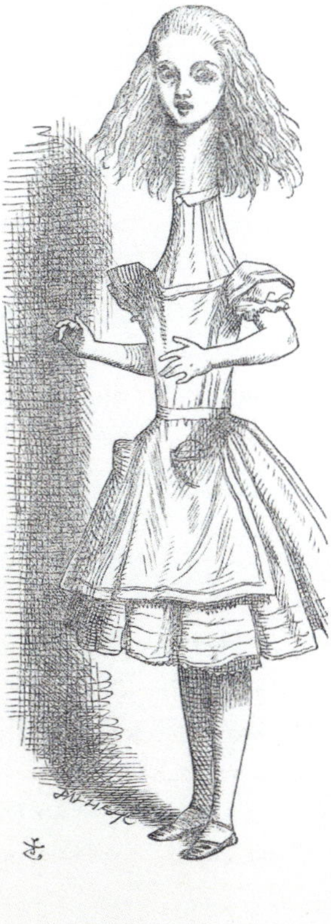

CHAPTER II.

THE POOL OF TEARS.

"Curiouser and curiouser!" cried Alice (she was so much surprised, that for the moment she quite forgot how to speak good English); "now I'm opening out like the largest telescope that ever was! Good-bye, feet!" (for when she looked down at her feet, they seemed to be almost out of sight, they were getting so far off) "Oh, my poor little feet, I wonder

NEWTON, Isaac (1643–1727).
Philosophia naturalis principia mathematica.
London, 1687.

The most important book on mathematical physics before Einstein.

Newton's *Philosophiæ Naturalis Principia Mathematica* (1687) laid out the laws of motion and universal gravitation, establishing the foundation for classical mechanics. Special Collections' copy of the first edition, part of the Posner Memorial Collection, survives in its original, seventeenth-century binding.

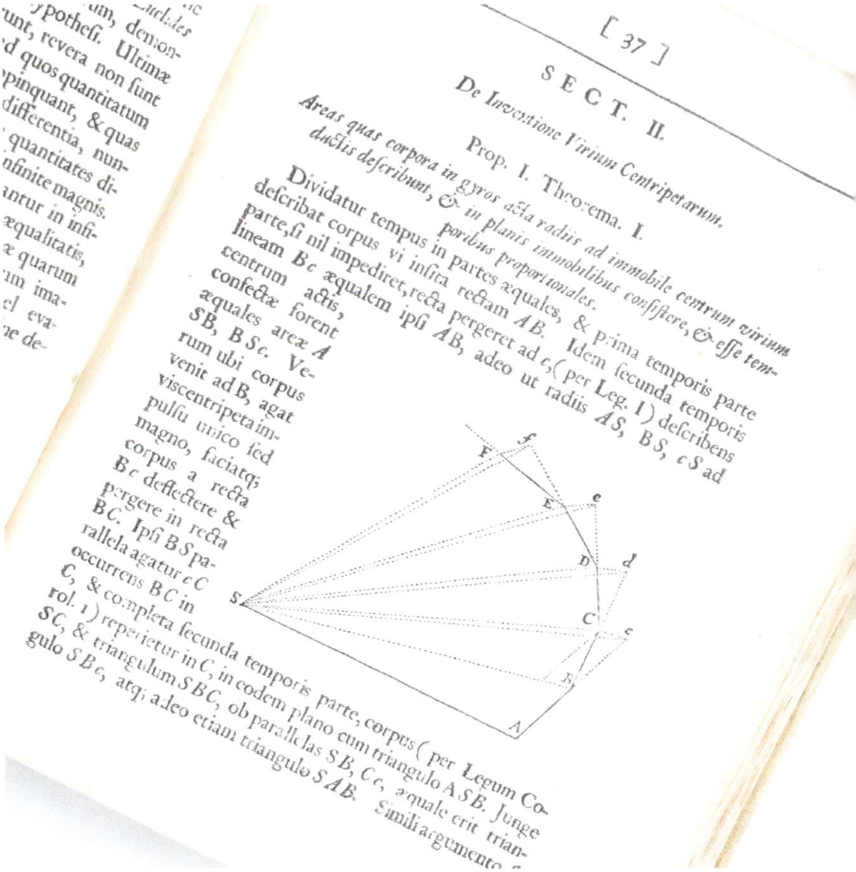

[37]

SECT. II.

De Inventione Virium Centripetarum.

Prop. I. Theorema. I.

Areas quas corpora in gyros acta radiis ad immobile centrum virium ductis describunt, & in planis immobilibus consistere, & esse temporibus proportionales.

Dividatur tempus in partes æquales, & prima temporis parte describat corpus vi insita rectam *AB*. Idem secunda temporis parte, si nil impediret, recta pergeret ad *c*, (per Leg. I) describens lineam *Bc* æqualem ipsi *AB*, adeo ut radiis *AS*, *BS*, *cS* ad centrum actis, confectæ forent æquales areæ *ASB*, *BSc*. Verum ubi corpus venit ad *B*, agat vis centripeta impulsu unico sed magno, faciatq; corpus a recta *Bc* deflectere & pergere in recta *BC*. Ipsi *BS* parallela agatur *cC* occurrens *BC* in *C*, & completa secunda temporis parte, corpus (per Legum Corol. 1) reperietur in *C*, in eodem plano cum triangulo *ASB*. Junge *SC*, & triangulum *SBC*, ob parallelas *SB*, *Cc*, æquale erit triangulo *SBc*, atq; adeo etiam triangulo *SAB*. Simili argumento

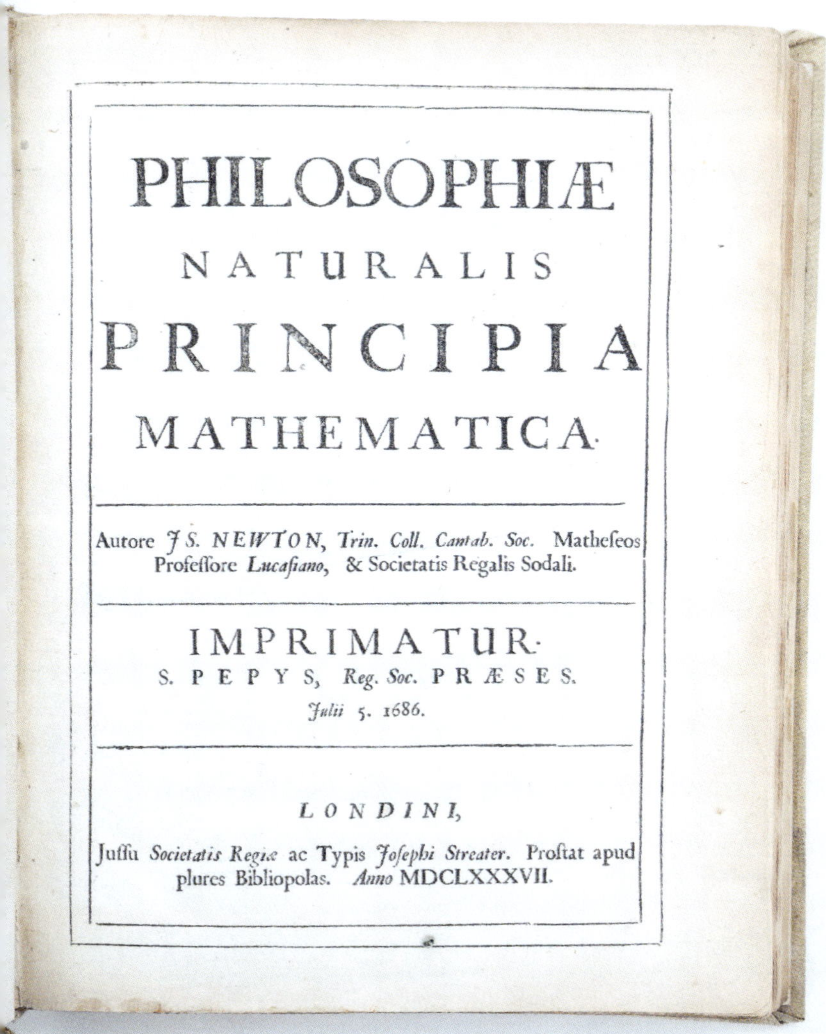

PHILOSOPHIÆ
NATURALIS
PRINCIPIA
MATHEMATICA.

Autore *JS. NEWTON*, *Trin. Coll. Cantab. Soc.* Matheseos Professore *Lucasiano*, & Societatis Regalis Sodali.

IMPRIMATUR.
S. PEPYS, *Reg. Soc.* PRÆSES.
Julii 5. 1686.

LONDINI,

Jussu *Societatis Regiæ* ac Typis *Josephi Streater*. Prostat apud plures Bibliopolas. *Anno* MDCLXXXVII.

SCHEDEL, Hartmann (1440–1514).
Liber chronicarum. Nuremberg, 1495.

A history of the known world (before 1492).

Forming an illustrated history of the known world up to 1492, the *Nuremberg Chronicle* was one of the most ambitious publishing projects of its time. Compiled by the Nuremberg physician and humanist Hartmann Schedel, the *Chronicle* combines biblical events, classical history, and contemporary European affairs in one richly illustrated volume (the book contains some 1,800 woodcuts).

The *Chronicle* was also designed to grow beyond its printed scope: many copies contain blank sheets of paper that early readers used to record personal histories, world events, or genealogical data that postdated the work's 1493 publication.

Special Collections holds two copies of the 1495 edition, a reissue of unsold copies of the original 1493 Latin printing.

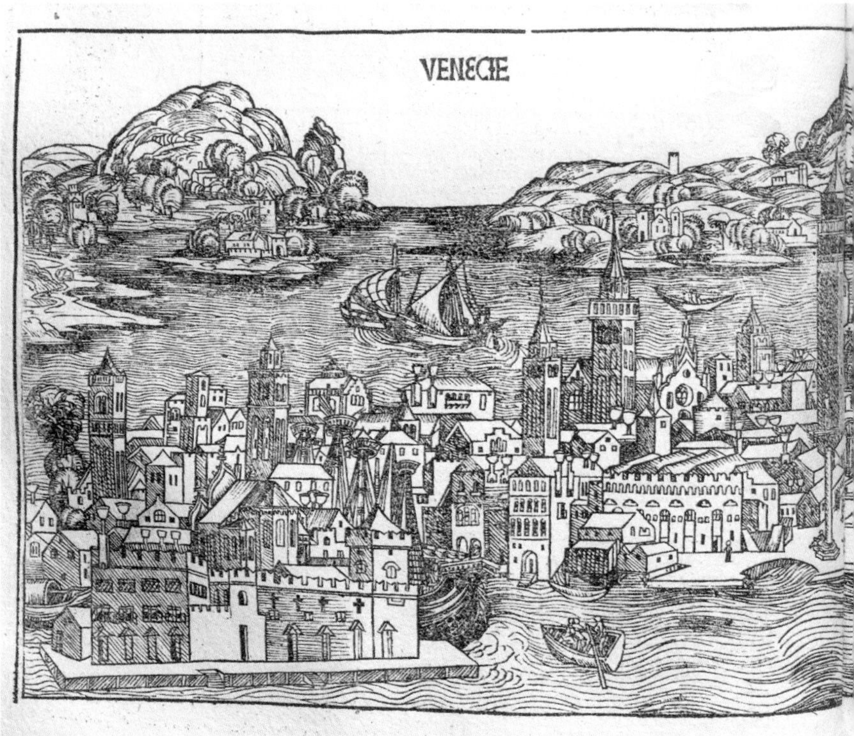
VENECIE

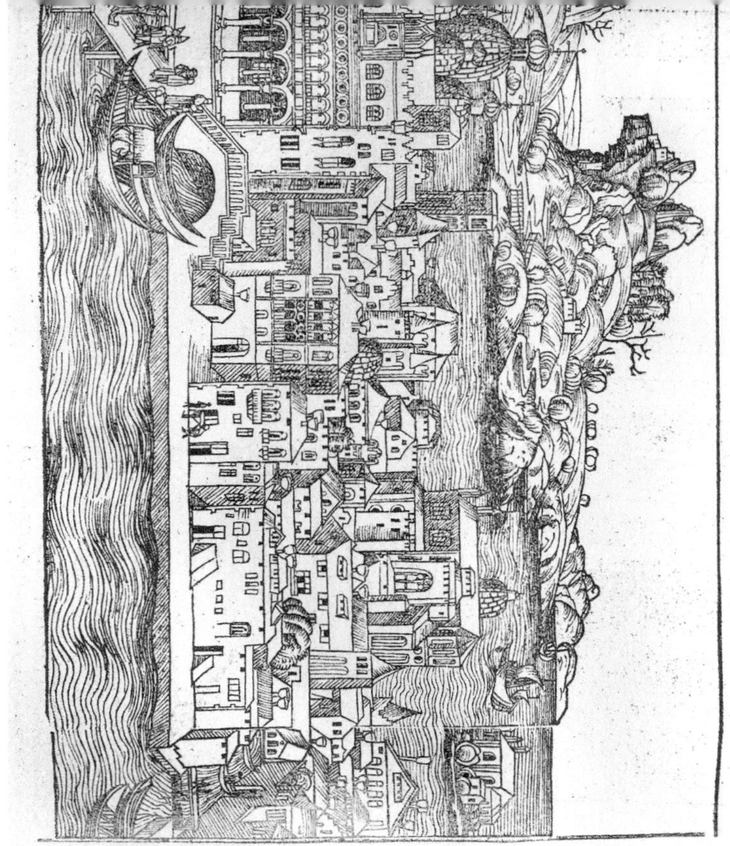

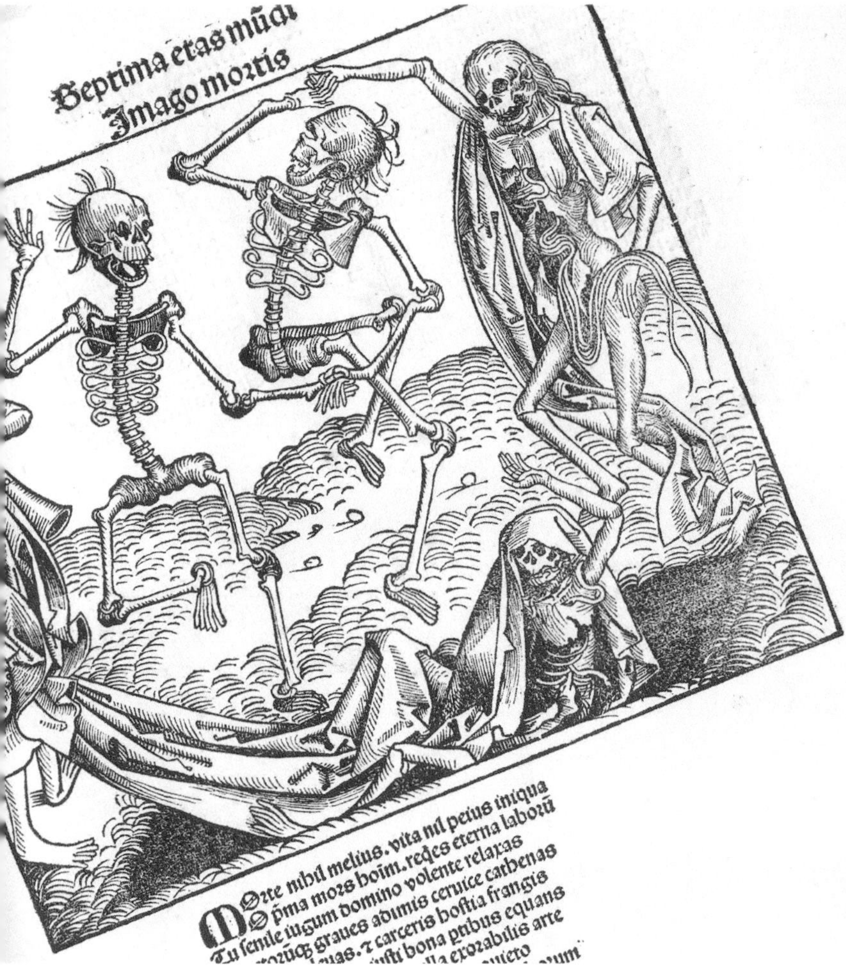
Septima etas mūdi
Imago mortis
Morte nihil melius. vita nil peius iniqua
O pma mors hoim. reqes eterna laborū
Tu senile iugum domino volente relaxas
Vinctorūq graues adimis ceruice cathenas

COLONNA, Francesco (1433–1527).
Hypnerotomachia Poliphili. Venice, 1499.

"The most beautiful book of the 15th century" — and the weirdest.

Attributed to the obscure Dominican friar Francesco Colonna, the *Hypnerotomachia Poliphili* recounts the journey of its hero, Poliphilo, to recover Polia, his beloved. Falling into a recursive dream, Poliphilo moves through a surreal landscape of antiquities, dark forests, and ruins, where he encounters allegorical scenes, classical deities, and mythical beasts. Called "the most beautiful book of the 15th century," the work is illustrated with more than 150 woodcuts that are sometimes attributed (largely without evidence) to Botticelli.

Special Collections holds two copies. One bears evidence of censorship: several woodcuts depicting erotic subjects and scenes were at one point obscured beneath pasted slips of paper or blobs of colored wax.

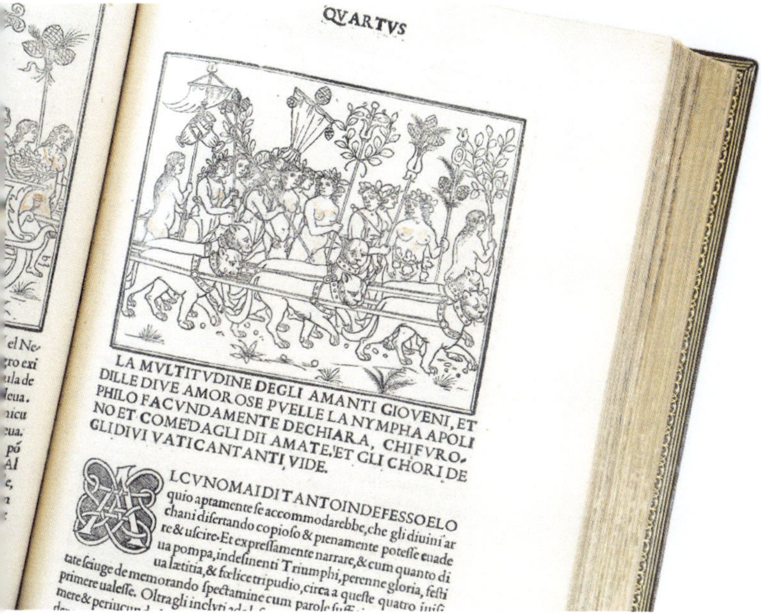
QVARTVS
LA MVLTITVDINE DEGLI AMANTI GIOVENI, ET
DILLE DIVE AMOROSE PVELLE LA NYMPHA APOLI
PHILO FACVNDAMENTE DECHIARA, CHI FVRO
NO ET COME DAGLI DII AMATE, ET GLI CHORI DE
GLIDIVI VATICANTANTI, VIDE.
LCVNO MAI DI TANTO INDEFESSO ELO
quio aptamente se accommodarebbe, che gli diuini ar
chani disertando copioso & pienamente potesse enade
re & uscire. Et expressamente narrare, & cum quanto di
ua pompa, indefinenti Triumphi, perenne gloria, festi
ua lætitia, & fœlice tripudio, circa a queste quatro inuic
tate seiuge de memorando spectamine cum parole suffi

HEVELIUS, Johannes (1611–1687).
Selenographia, sive, Lunae descriptio.
Gdańsk, 1647.

The origin point of exogeography with breathtakingly beautiful engravings of the lunar surface.

Hevelius's *Selenographia* is a pioneering work on exogeography — the study and description of the surface of celestial bodies beyond Earth. Published in 1647, the *Selenographia* offered an unprecedented level of detail in its depictions of the Moon. The work also features interactive volvelles — rotating discs of paper — that allowed early readers to simulate lunar libration, the apparent rocking or oscillatory motion of the Moon that Hevelius was the first to observe and describe.

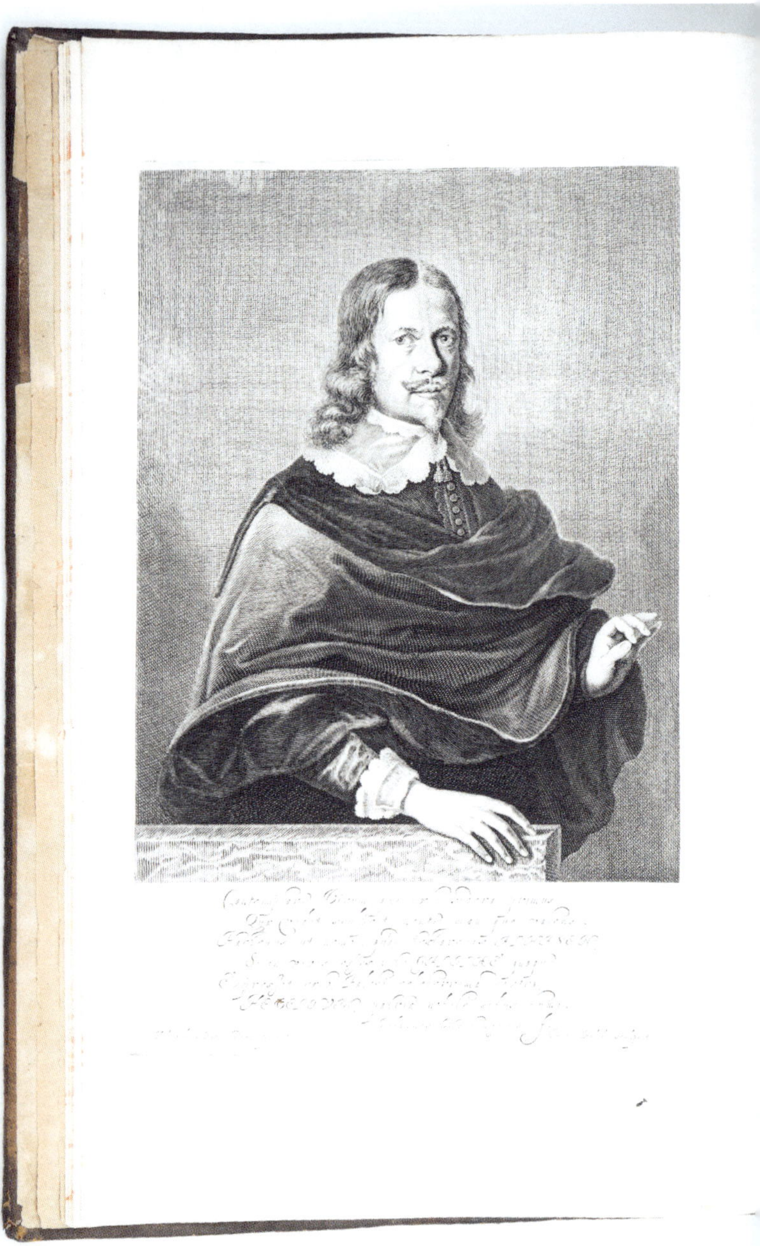

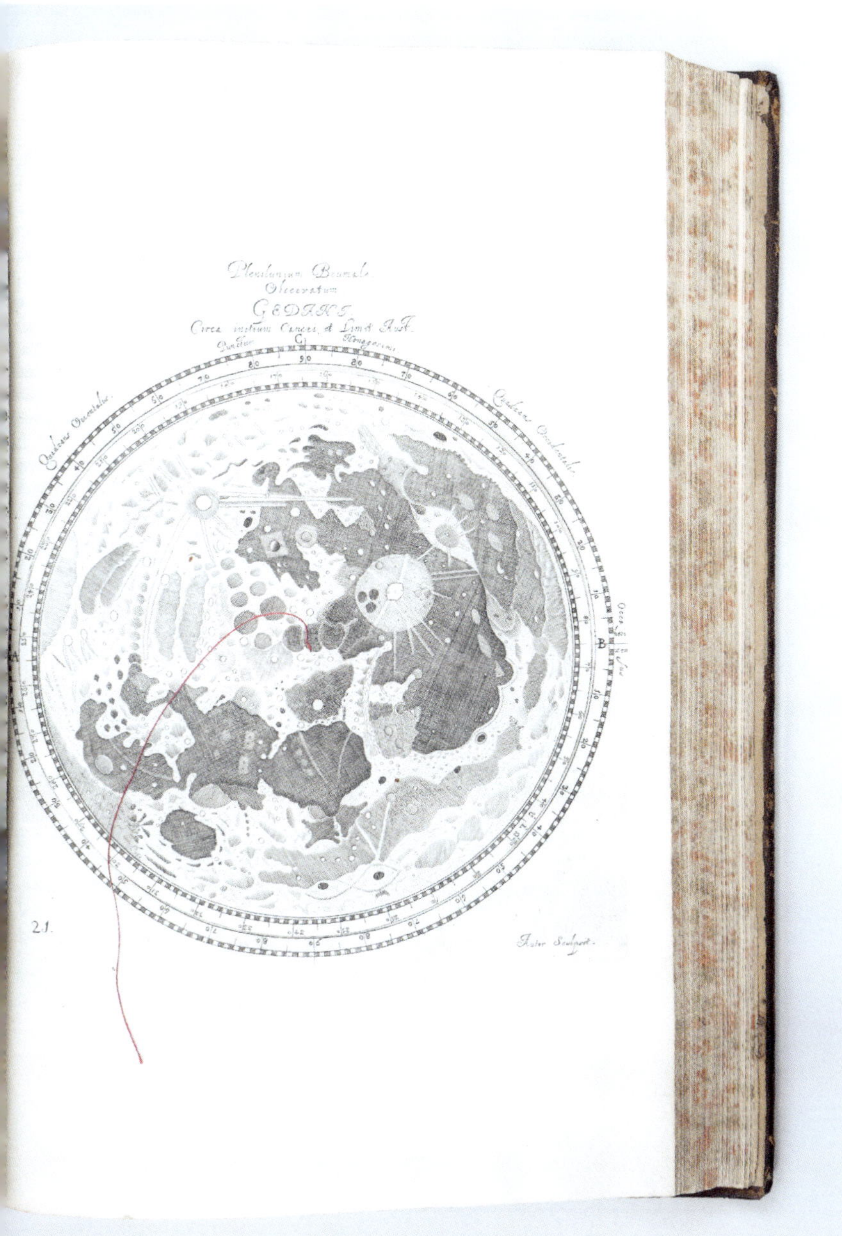
Plenilunium Boreale
Observatum
GEDANI
Circa initium Cancri, et Limit. Aust.
C
A
B
Quadrans Orientalis
Quadrans Occidentalis
21.
Autor Sculpsit.

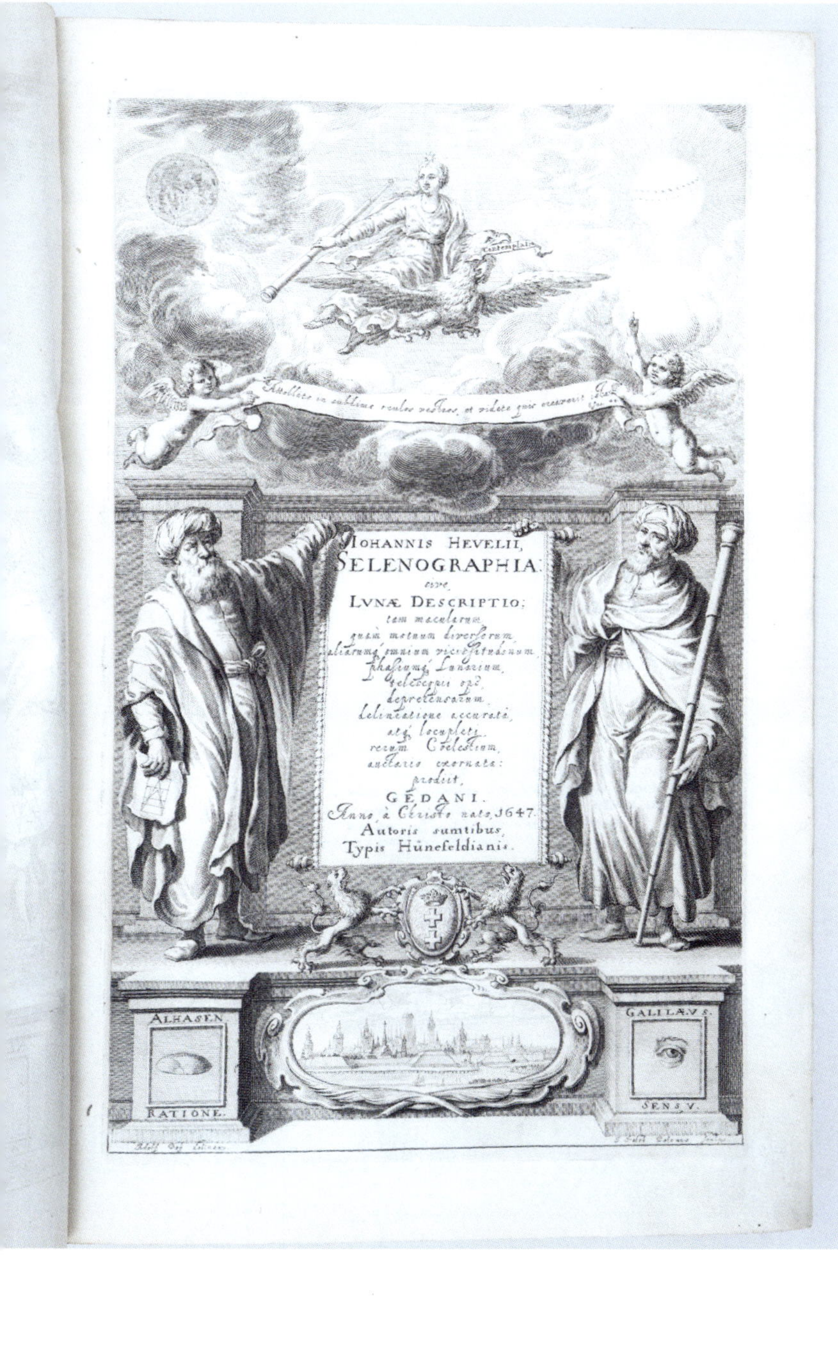
IOHANNIS HEVELII,
SELENOGRAPHIA:
sive,
LVNÆ DESCRIPTIO;
tam macularum
quàm motuum diversorum
aliarumq; omnium vicissitudinum
phasiumq; Lunarium,
telescopii ope
deprehensarum,
delineatione accuratè,
atq; locupleti
rerum Coelestium,
auctario exornata:
prodiit,
GEDANI.
Anno, à Christo nato, 1647.
Autoris sumtibus,
Typis Hünefeldianis.
ALHASEN
RATIONE.
GALILÆVS.
SENSV.

GREENAWAY, Kate (1846–1901). *Portrait of Frederick Locker-Lampson* (1821–1895). Watercolor on paper (mounted, 177 × 152 mm), 1880s.

This painting is part of the Frances Hooper Collection of Kate Greenaway — a trove of primary source materials, including sketches, paintings, woodblocks, correspondence, and notebooks by the Victorian artist known for her idealized depictions of childhood. The painting's subject, Frederick Locker-Lampson, was an important bibliophile and literary figure in nineteenth-century England. Formed by advertising executive, journalist, and bibliophile Frances Hooper, the collection documents Greenaway's creative output and her correspondence with artists and writers of the period, including John Ruskin (1819–1900).

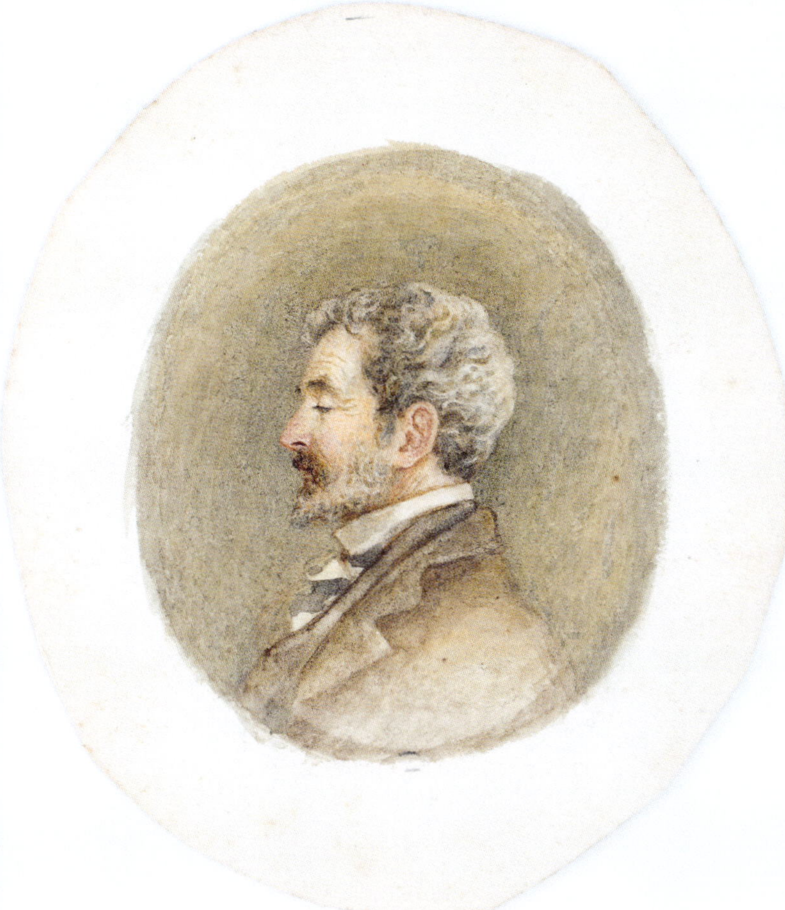

JOHNSON, Samuel (1709–1784).
A Dictionary of the English Language.
London, 1755.

"Johnson's writings had, in philology, the effect which Newton's discoveries had in mathematics."

— Noah Webster

Samuel Johnson's *Dictionary of the English Language* was the most comprehensive and influential wordlist of its time. Compiled over a decade, the *Dictionary* comprises some 42,000 definitions and etymologies and 114,000 quotations drawn from Johnson's encyclopedic knowledge of English literature. Far from dry and unimaginative, Johnson's definitions are often colored by his wit — his self-deprecating entry for *lexicographer* defines the word as "a harmless drudge."

Johnson's *Dictionary* remained a standard work of reference until the completion of the Oxford English Dictionary in 1928. Special Collections holds two copies of the first edition, both bound in two volumes. The copy shown here was part of the 1974 bequest of Charles J. Rosenbloom. It survives in its original paperboard binding.

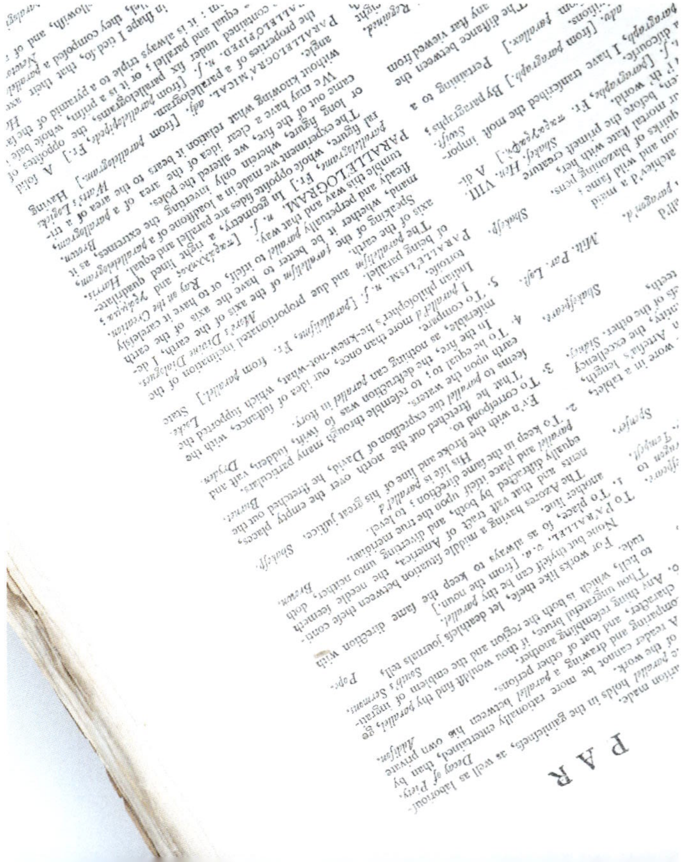

LOCKE, John (1632–1704).
An Essay Concerning Humane Understanding.
London, 1690.

Enlightenment theory of mind.

In *An Essay Concerning Humane Understanding*, John Locke argued that the human brain is an impressionable void — a blank slate, or *tabula rasa*, shaped entirely by experience. This theory not only challenged existing notions of innate knowledge, but it also foreshadowed debates in early artificial intelligence and computational heuristics. Meanwhile, Locke's emphasis on observation and psychological individuality resonated far beyond philosophy, influencing education, political theory, and science.

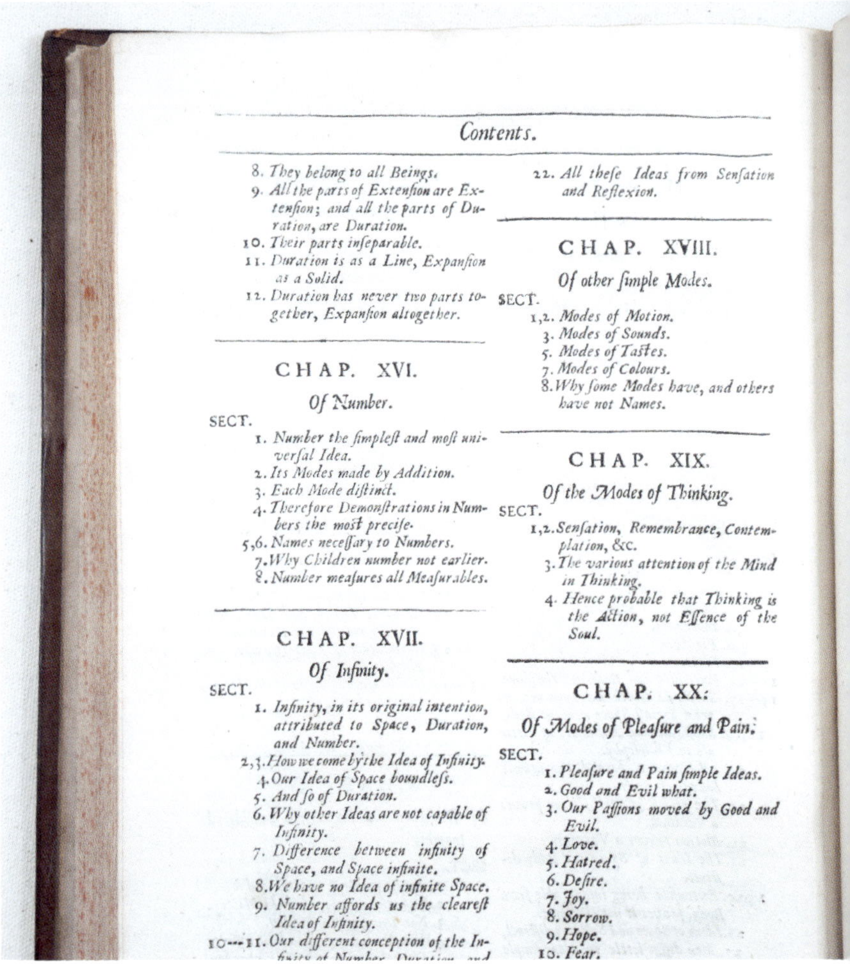

Contents.

8. They belong to all Beings.
9. All the parts of Extension are Extension; and all the parts of Duration, are Duration.
10. Their parts inseparable.
11. Duration is as a Line, Expansion as a Solid.
12. Duration has never two parts together, Expansion altogether.

CHAP. XVI.

Of Number.

SECT.
1. Number the simplest and most universal Idea.
2. Its Modes made by Addition.
3. Each Mode distinct.
4. Therefore Demonstrations in Numbers the most precise.
5,6. Names necessary to Numbers.
7. Why Children number not earlier.
8. Number measures all Measurables.

CHAP. XVII.

Of Infinity.

SECT.
1. Infinity, in its original intention, attributed to Space, Duration, and Number.
2,3. How we come by the Idea of Infinity.
4. Our Idea of Space boundless.
5. And so of Duration.
6. Why other Ideas are not capable of Infinity.
7. Difference between infinity of Space, and Space infinite.
8. We have no Idea of infinite Space.
9. Number affords us the clearest Idea of Infinity.
10---11. Our different conception of the Infinity of Number, Duration, and

22. All these Ideas from Sensation and Reflexion.

CHAP. XVIII.

Of other simple Modes.

SECT.
1,2. Modes of Motion.
3. Modes of Sounds.
5. Modes of Tastes.
7. Modes of Colours.
8. Why some Modes have, and others have not Names.

CHAP. XIX.

Of the Modes of Thinking.

SECT.
1,2. Sensation, Remembrance, Contemplation, &c.
3. The various attention of the Mind in Thinking.
4. Hence probable that Thinking is the Action, not Essence of the Soul.

CHAP. XX.

Of Modes of Pleasure and Pain.

SECT.
1. Pleasure and Pain simple Ideas.
2. Good and Evil what.
3. Our Passions moved by Good and Evil.
4. Love.
5. Hatred.
6. Desire.
7. Joy.
8. Sorrow.
9. Hope.
10. Fear.

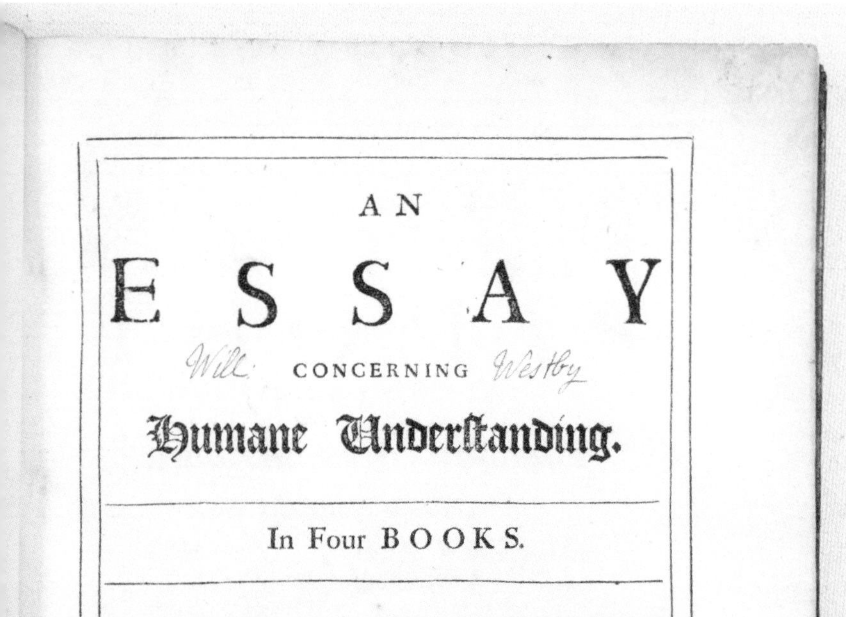

AN

ESSAY

CONCERNING

Humane Understanding.

In Four BOOKS.

Quam bellum eſt velle confiteri potius neſcire quod neſcias, quam iſta effutientem nauſeare, atque ipſum ſibi diſplicere! Cic. de Natur. Deor. *l.* 1.

FUCHS, Leonhart (1501–1566).
De historia stirpium. Basel, 1542.

Botanical specimens on the page.

Leonhart Fuchs' *On the History of Plants* marked a turning point in the botanical sciences. Adapting ancient herbal traditions, it brought empirical observation to the study of plants. Published in 1542, the book contains over five hundred meticulous woodcut illustrations, many drawn from life. These images inaugurated a new accuracy in botanical art.

A composite portrait at the end of the volume depicts the book's artist (Albrecht Meyer), draughtsman (Heinrich Füllmaurer), and woodcutter (Veit Rudolph Speckle), who are shown practicing their respective crafts: Meyer paints a botanical specimen, Füllmaurer transfers Meyer's images onto wood, and Speckle — apparently waiting for his collaborators to finish their work — will cut the woodblock in preparation for printing. This is one of the earliest known instances of a scientific book crediting its illustrators rather than its author alone.

LEONHARTVS FVCHSIVS
ÆTATIS SVÆ ANNO XLI.

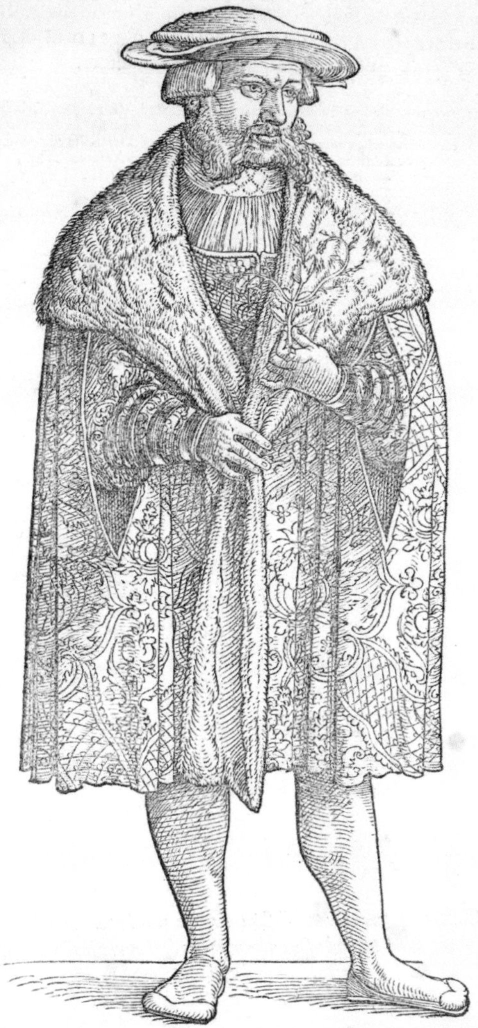

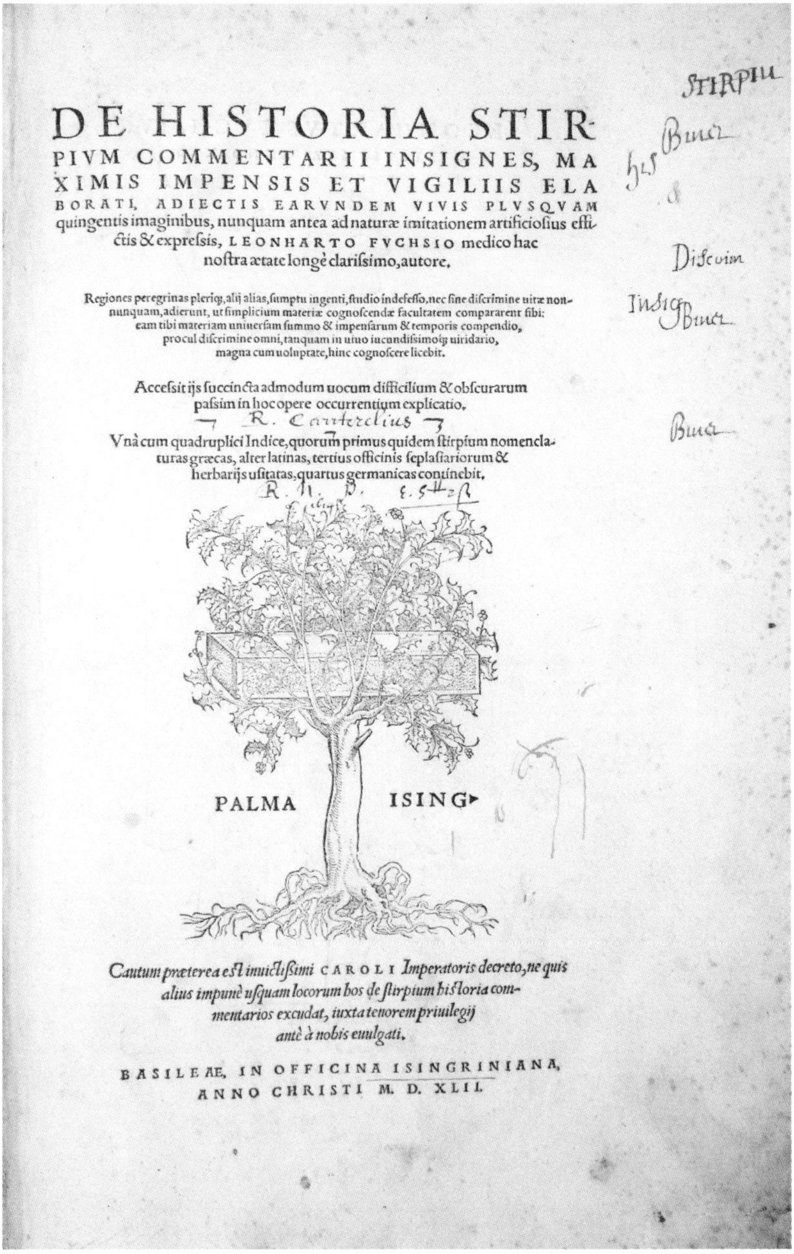

DE HISTORIA STIRPIVM COMMENTARII INSIGNES, MAXIMIS IMPENSIS ET VIGILIIS ELABORATI, ADIECTIS EARVNDEM VIVIS PLVSQVAM quingentis imaginibus, nunquam antea ad naturæ imitationem artificiosius effictis & expreſsis, LEONHARTO FVCHSIO medico hac noſtra ætate longè clariſsimo, autore.

Regiones peregrinas plericp, alij alias, ſumptu ingenti, ſtudio indefeſſo, nec ſine diſcrimine uitæ nonnunquam, adierunt, ut ſimplicium materiæ cognoſcendæ facultatem compararent ſibi: eam tibi materiam uniuerſam ſummo & impenſarum & temporis compendio, procul diſcrimine omni, tanquam in uno iucundiſsimoqꝫ uiridario, magna cum uoluptate, hinc cognoſcere licebit.

Acceſsit ijs ſuccincta admodum uocum difficilium & obſcurarum paſsim in hoc opere occurrentium explicatio.

Vnà cum quadruplici Indice, quorum primus quidem ſtirpium nomenclaturas græcas, alter latinas, tertius officinis ſeplaſiariorum & herbarijs uſitatas, quartus germanicas continebit.

Cautum præterea eſt inuictiſsimi CAROLI *Imperatoris decreto, ne quis alius impunè uſquam locorum hos de ſtirpium hiſtoria commentarios excudat, iuxta tenorem priuilegij antè à nobis euulgati.*

BASILEÆ, IN OFFICINA ISINGRINIANA, ANNO CHRISTI M. D. XLII.

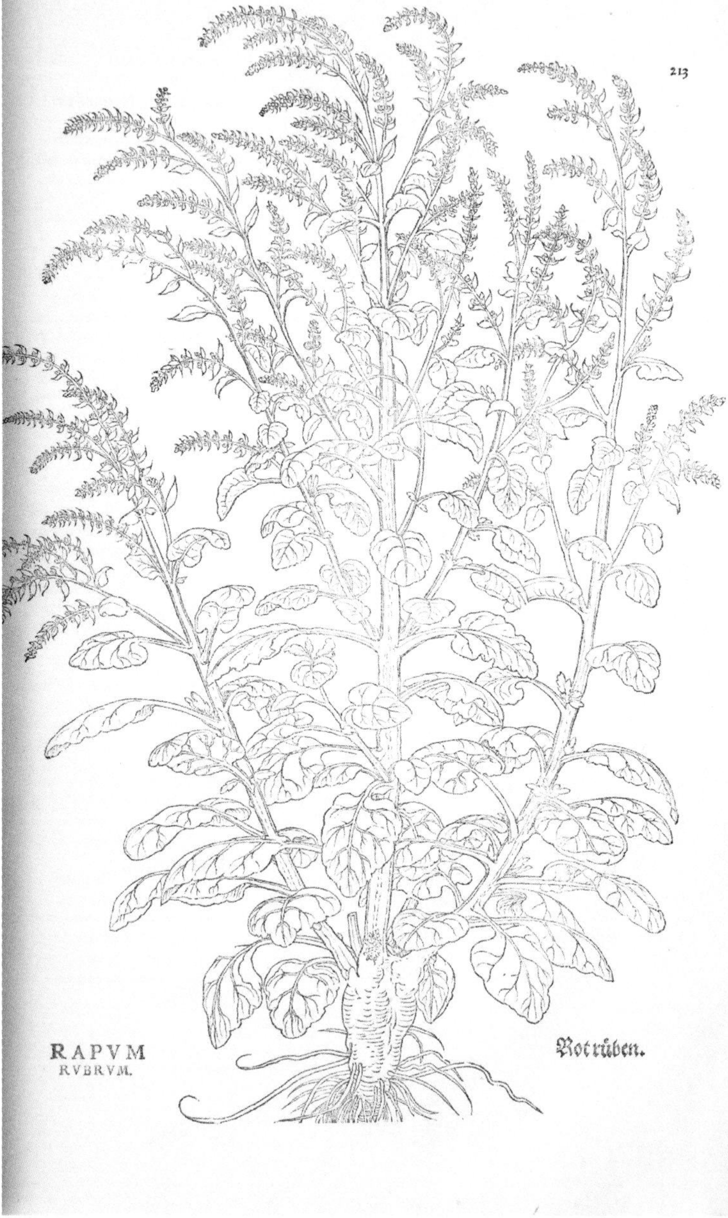

RAPVM RVBRVM.

Rotrüben.

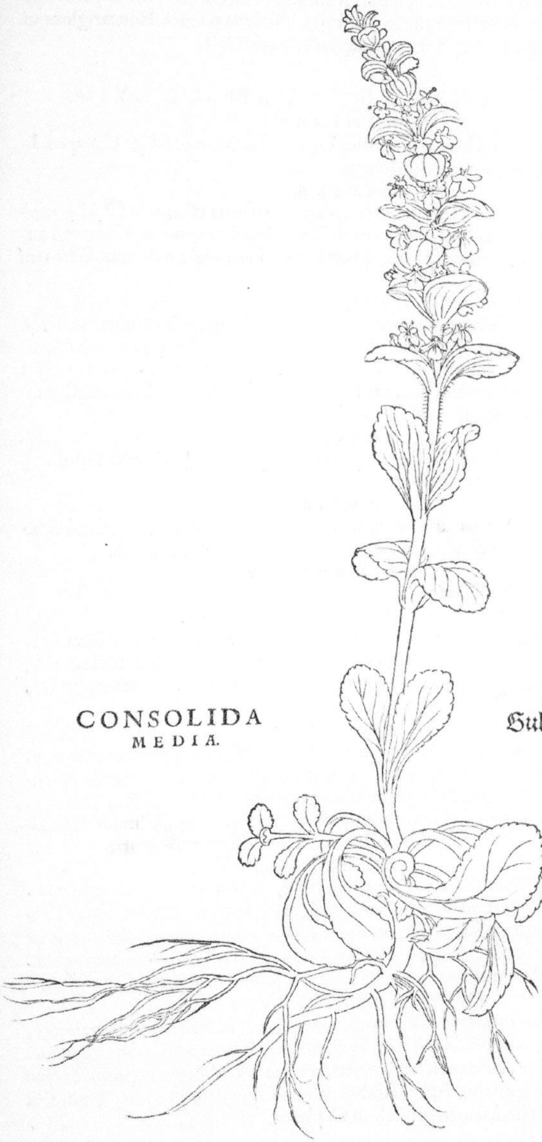

CONSOLIDA
MEDIA.

Guldeguntzel.

k 4

PICTORES OPERIS,

Heinricus Füllmaurer.

Albertus Meyer.

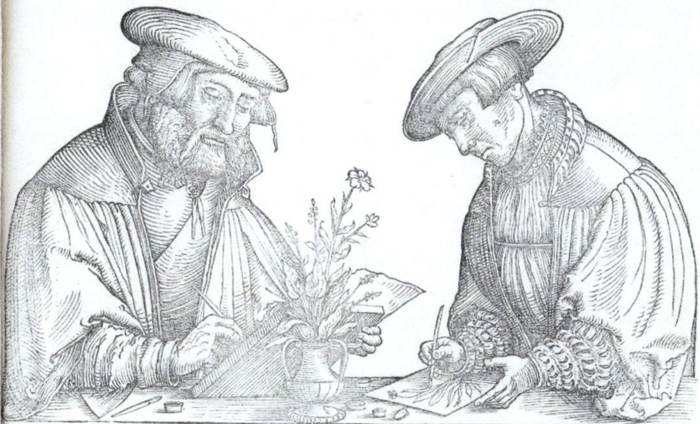

SCVLPTOR

Vitus Rodolph. Speckle.

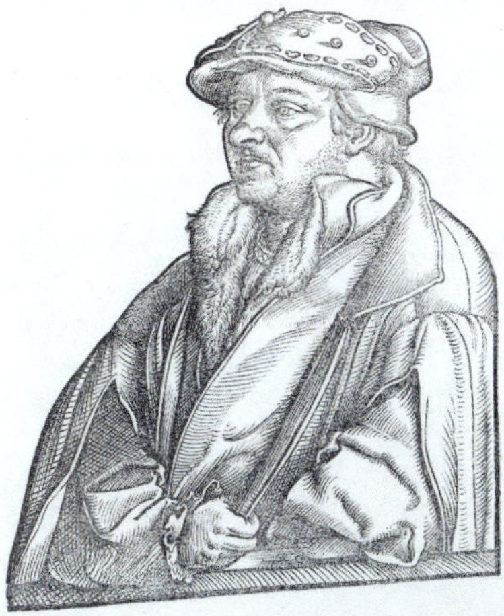

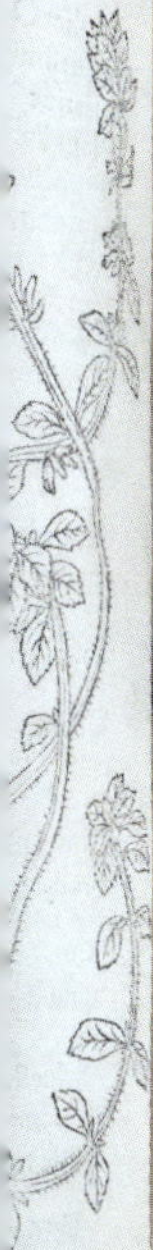

Wild Basilg

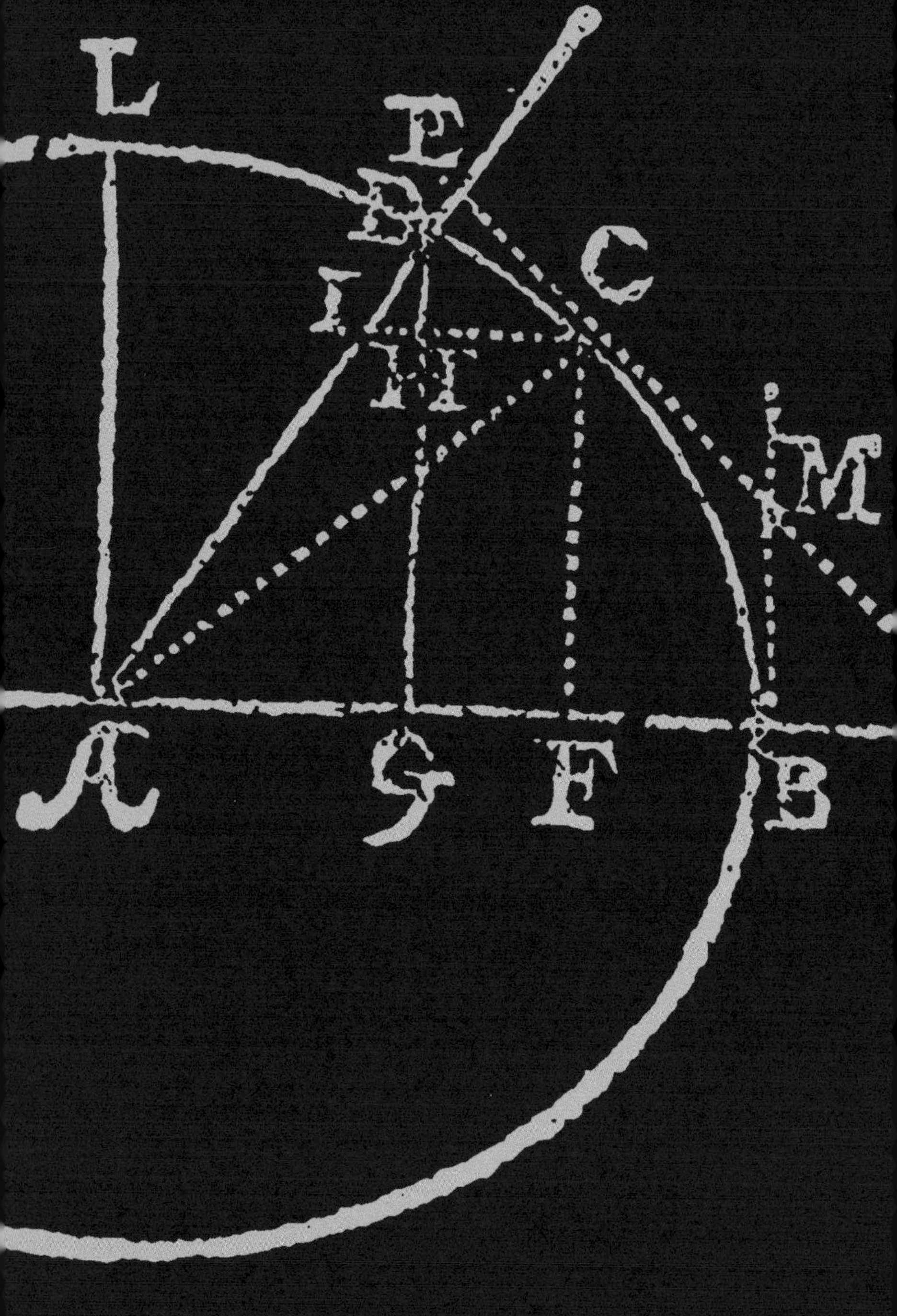

L
E
D
C
I
H
M
A
G
F
B

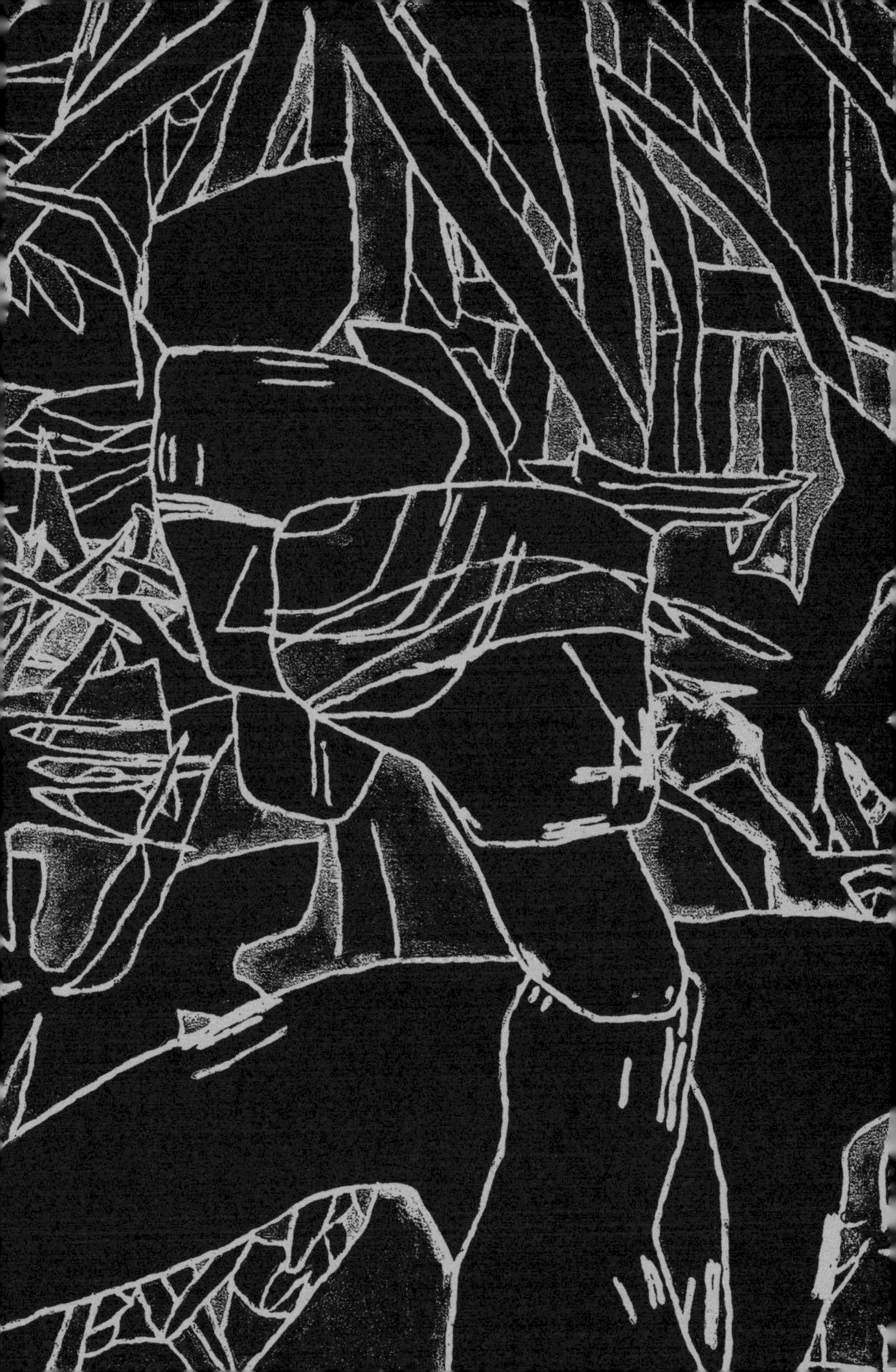

The Hunt Botanical Library & Hunt Institute for Botanical Documentation

1 Hunt, Rachel McMasters Miller. 1958–1961. *Catalogue of Botanical Books in the Collection of Rachel McMasters Miller Hunt*. Pittsburgh: Hunt Botanical Library.

Special Collections' copy of Fuchs' *De historia stirpium* (see previous entry) is held in the Posner Memorial Collection. Another hand-colored copy is held in the Library of CMU's Hunt Institute for Botanical Documentation. Founded as the Hunt Botanical Library in 1961, the Hunt Institute forms one of the largest and most important collections of rare botanical books in the world. A two-volume catalog of 767 of the Library's rare botanical books was published in 1958-1961.[1]

The Institute brings together Rachel McMasters Miller Hunt's collections — botanical books, art, manuscripts, portraits, and archives — into a unified program of research and preservation. Much more was collected following Hunt's death, and in 1971 the center was renamed Hunt Institute for Botanical Documentation with four programmatic departments: Archives, Art, Bibliography, and Library.

As a research center at Carnegie Mellon University, the Institute continues to collect and preserve rare and modern botanical books, manuscripts, artworks, archival materials, and bibliographic files related to the plant sciences.

AUDUBON, John James (1785–1851).
The Birds of America: From Drawings Made in the United States and Their Territories.
New York, 1840-1844.

First octavo edition of Audubon's birds, with five hundred hand-colored plates.

Audubon's *Birds of America* was the first comprehensive study of bird species native to the North American continent. Better known in its earlier and larger "double-elephant folio" edition (printed 1827–1838), the work was later issued in a smaller octavo format, as here. Issued in one hundred parts and seven volumes, the work comprises five hundred full-page color plates (sixty-five more than appeared in the earlier edition). This edition was designed to be more affordable and accessible than its larger precursor: while the illustrations in the folio edition were colored in by hand, the octavo edition used more economical color lithography. Special Collections' copy is preserved in its original printed paper wrappers.

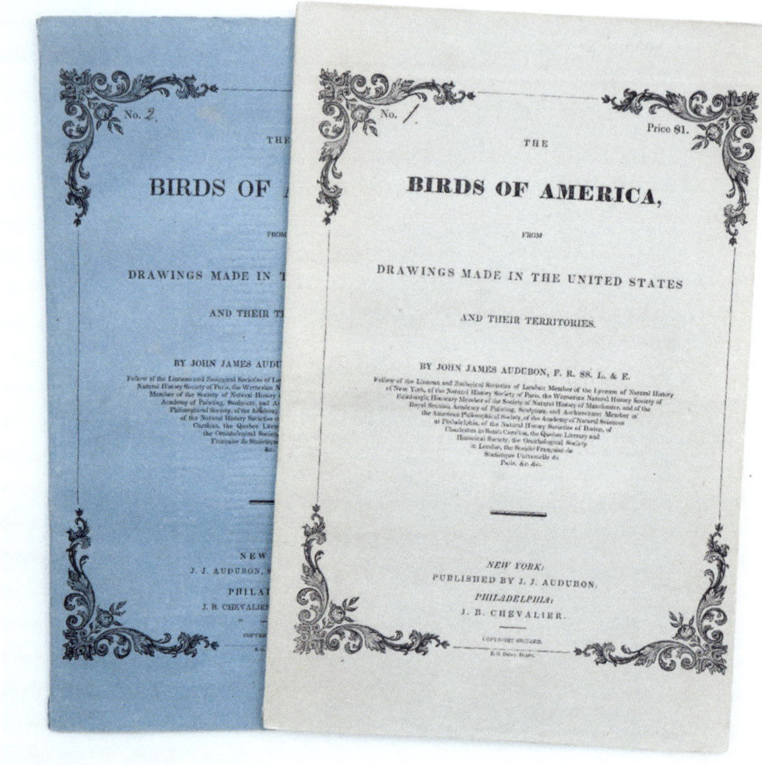
No. 2.
No. 1.
Price $1.
THE
BIRDS OF AMERICA,
FROM
DRAWINGS MADE IN THE UNITED STATES
AND THEIR TERRITORIES.
BY JOHN JAMES AUDUBON, F. R. SS. L. & E.
NEW YORK:
PUBLISHED BY J. J. AUDUBON.
PHILADELPHIA:
J. B. CHEVALIER.

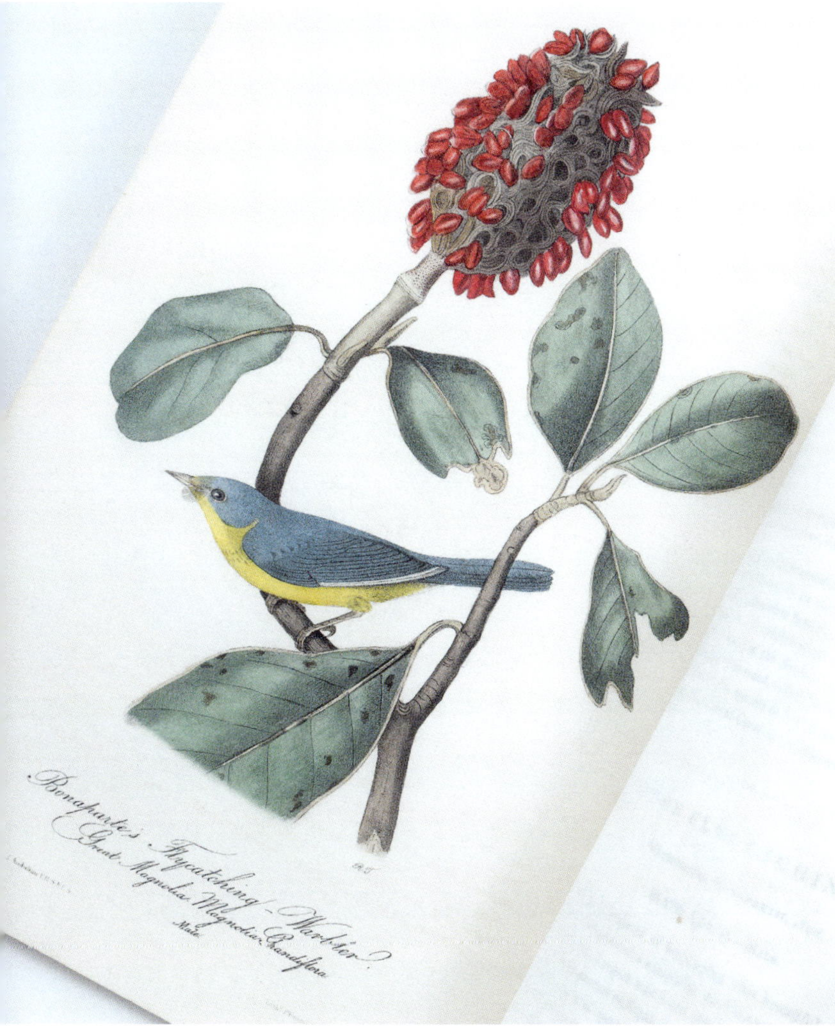
Bonaparte's Flycatching-Warbler.
Great Magnolia Magnolia Grandiflora
Male

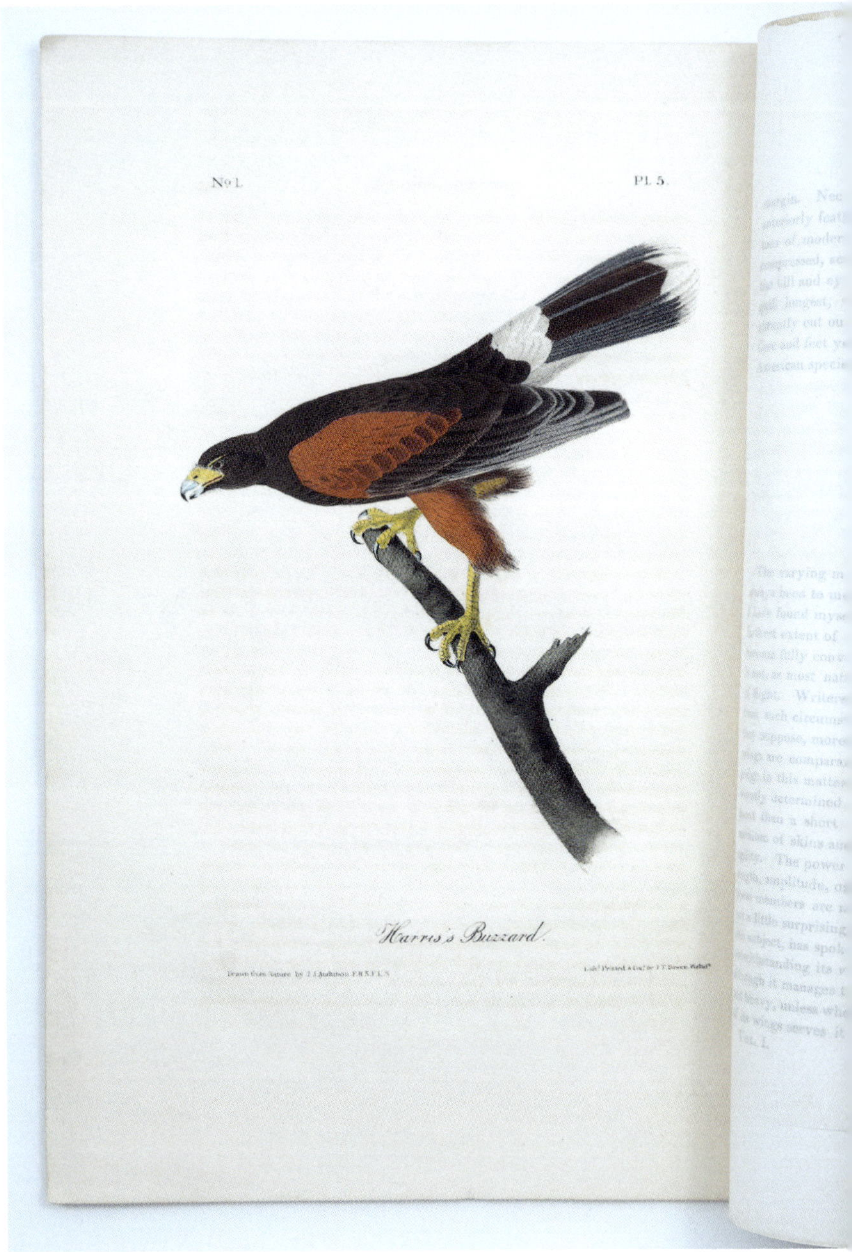
No 1.
Pl. 5.
Harris's Buzzard.

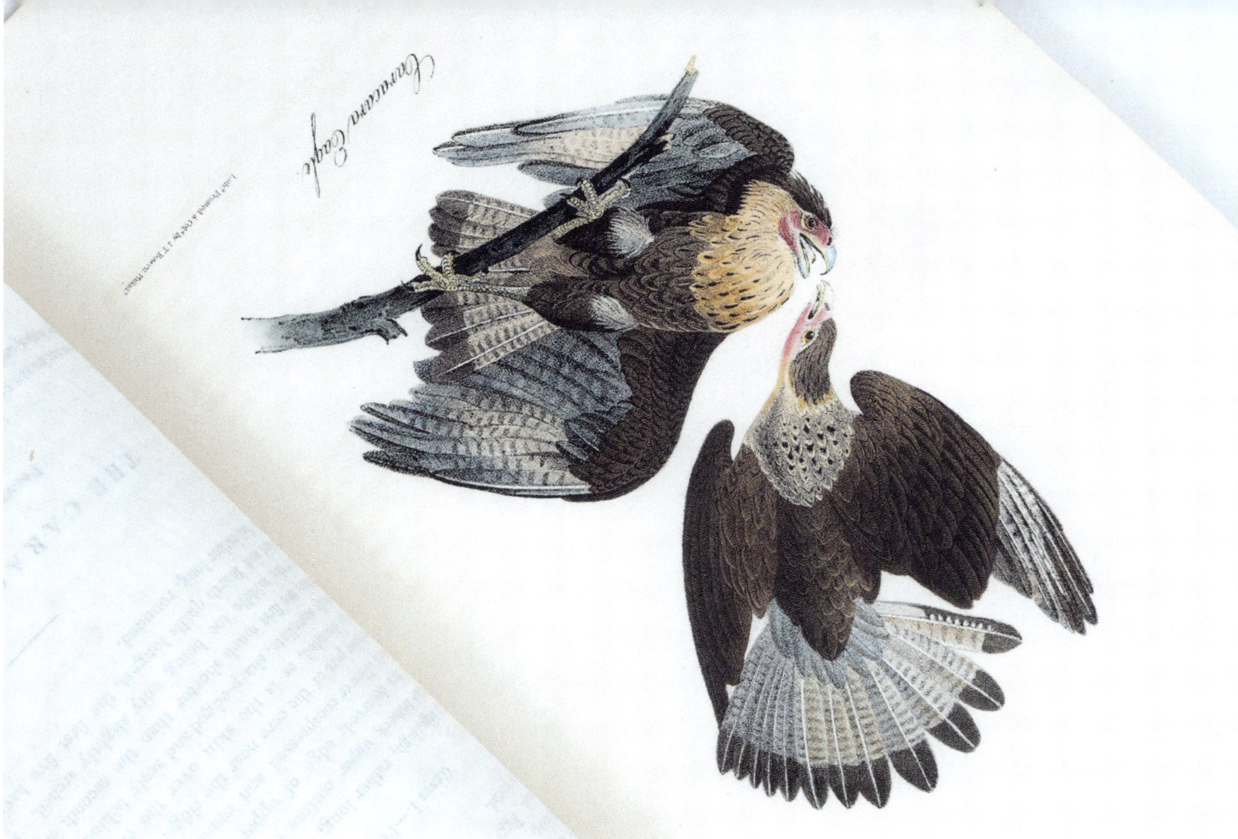
Caracara Eagle

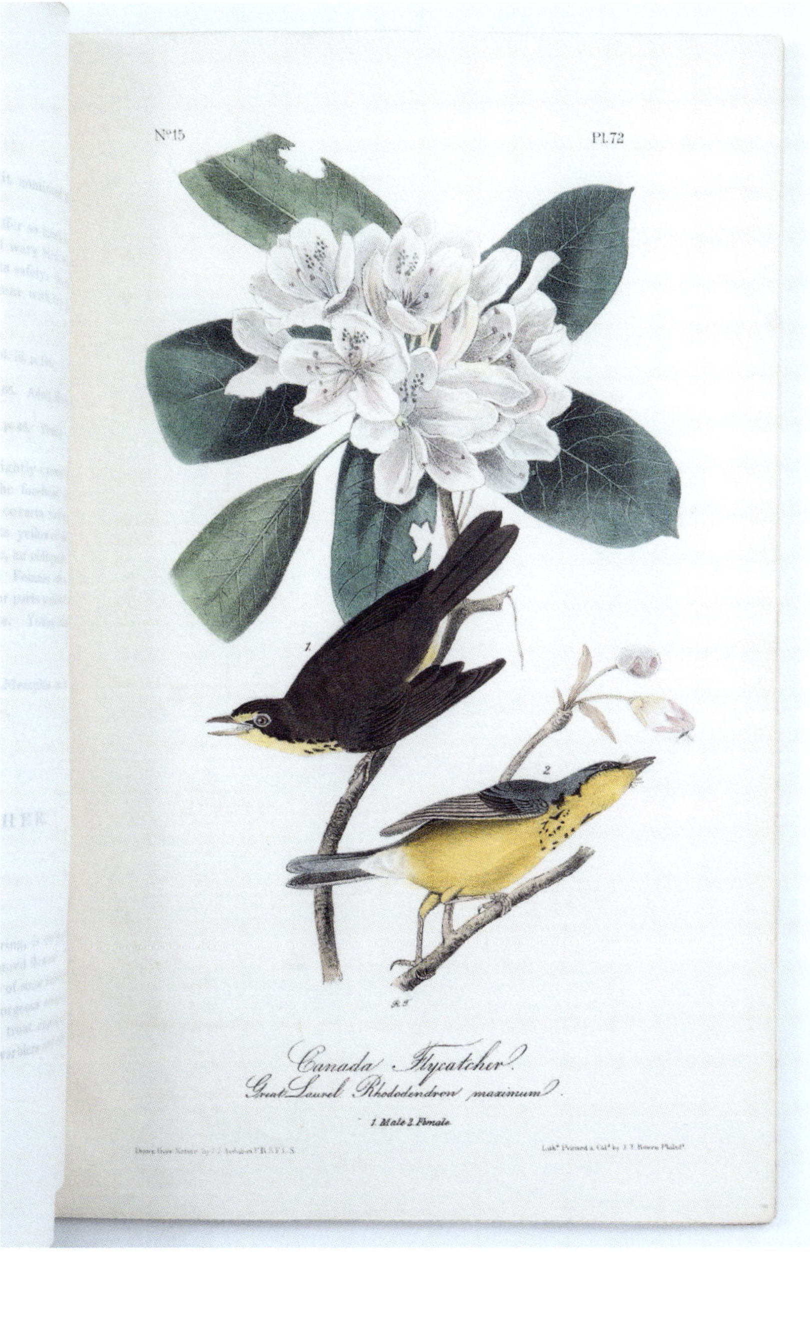
No15
Pl.72
1.
2
Canada Flycatcher.
Great Laurel Rhododendron maximum.
1 Male 2 Female

BERNOULLI, Jacob (1655–1705).
Ars conjectandi. Basel, 1713.

The inspiration for the first computer algorithm.

In *Ars Conjectandi* ("The Art of Conjecturing"), Jacob Bernoulli introduced several key concepts in probability theory and statistics. The book also introduced Bernoulli numbers, a series of rational numbers important in mathematical analysis and combinatorics. Ada Lovelace (1815–1852) wrote an algorithm — the first of its kind — to compute Bernoulli numbers on the Analytical Engine, an unrealized mechanical computer designed by Charles Babbage (1791–1871).

JACOBI BERNOULLI,
Profeff. Bafil. & utriufque Societ. Reg. Scientiar. Gall. & Pruff. Sodal.
MATHEMATICI CELEBERRIMI,

ARS CONJECTANDI,

OPUS POSTHUMUM.

Accedit

TRACTATUS
DE SERIEBUS INFINITIS,

Et EPISTOLA Gallicè fcripta
DE LUDO PILÆ RETICULARIS.

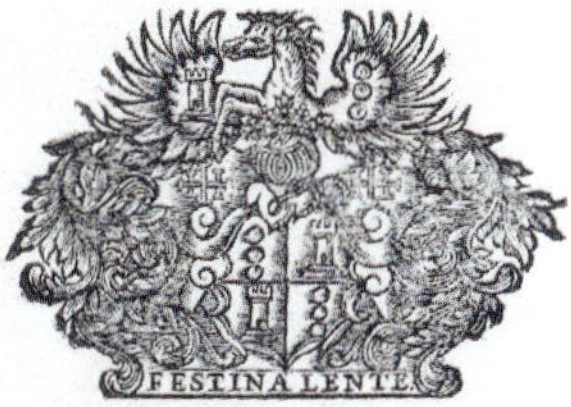

BASILEÆ,
Impenfis THURNISIORUM, Fratrum.

cIↄ Iↄcc XIII.

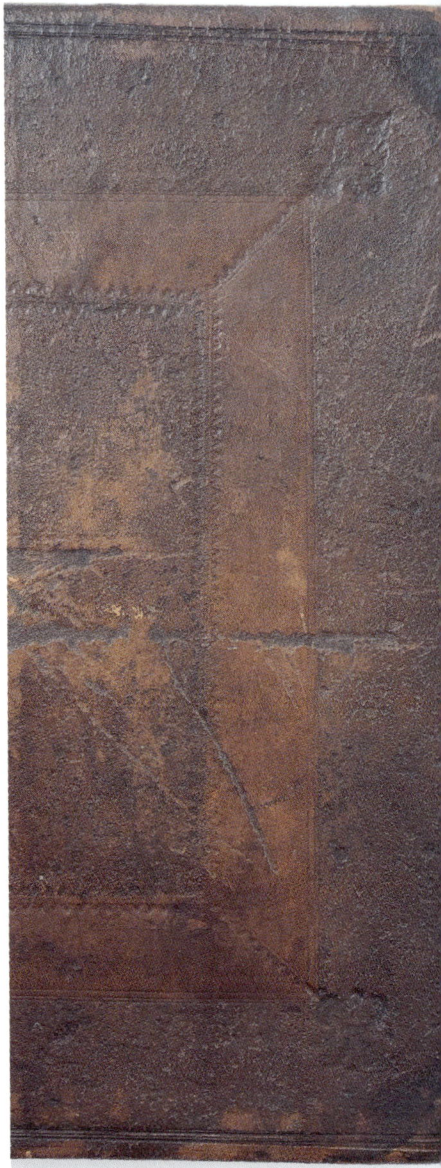

Ad pag. 306.

fig. 2

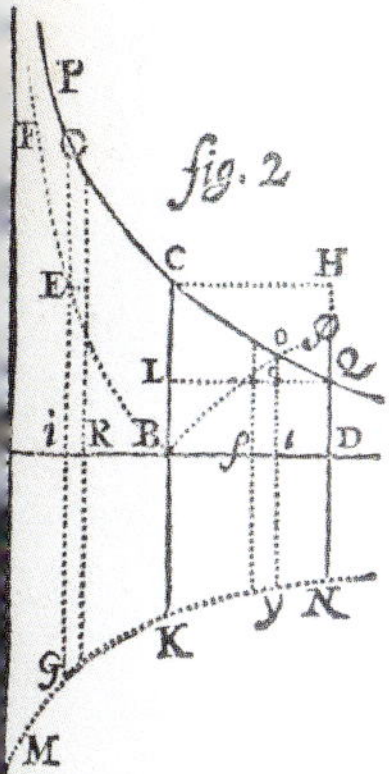

Fig. 6.

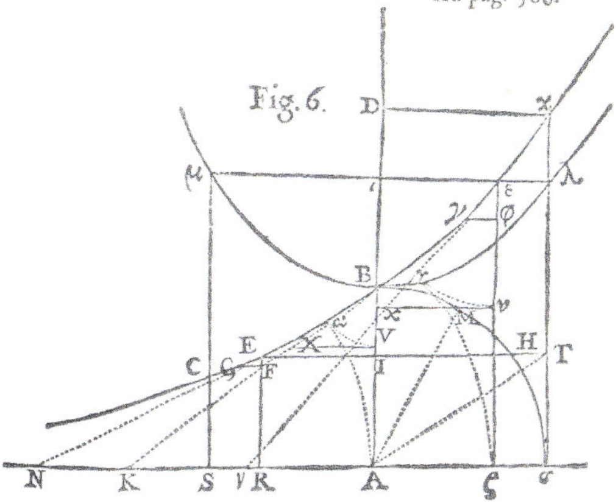

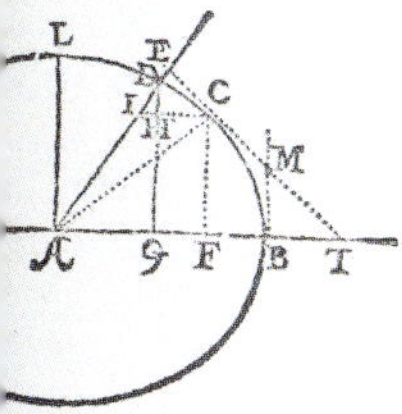

Fig. 7.

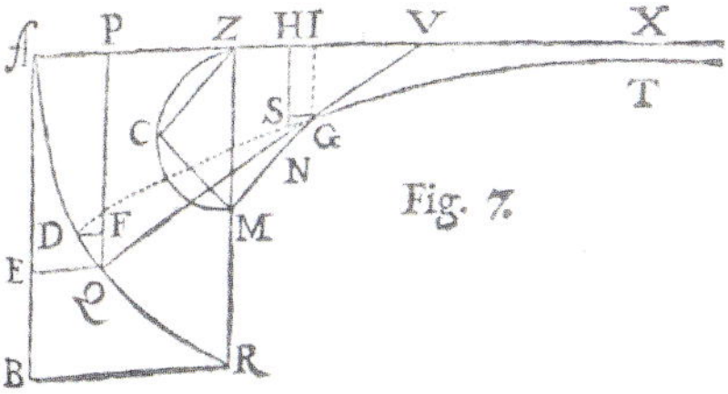

COHEN, Harold (1928–2016).
In Search of the Sources.
Computer-generated underdrawing
with hand-painted oil on canvas
(framed, 77" × 54"), 1988.

After early success as an abstract expressionist in the 1960s, Harold Cohen developed AARON — the first artificial intelligence program used to make art. Special Collections holds ten works that Cohen made in partnership with AARON. The paintings include coral-like abstractions characteristic of AARON's early period, as well as later, more figural works that show AARON's gradual evolution as Cohen made refinements to its code. AARON's programming relied on rule-based coding languages to guide its delineation of form and line, figure and ground. Cohen then overpainted AARON's compositions in color, making each work a human-computer collaboration.

This painting is part of a group of Cohen's works gifted to CMU Libraries by Pamela McCorduck in 2018. McCorduck (1940–2021) was a historian of artificial intelligence and a friend of CMU AI pioneers Allen Newell and Herb Simon. In 1990 McCorduck published Aaron's Code: Meta-Art, Artificial Intelligence and the Work of Harold Cohen, the first monograph to examine Cohen's place and influence in the early history of AI.

> I know exactly what AARON knows, but I can still be surprised. When you work on a program as I've worked on AARON, you make the program the heir to some subset of your own knowledge. When it plays that knowledge back to you, you can find yourself saying, 'Hey, where did that come from? I didn't realize that hat is what I believe.' In that sense the whole endeavor is quite a shocking and remarkable experience...
>
> — Harold Cohen reflecting on AARON in 1995

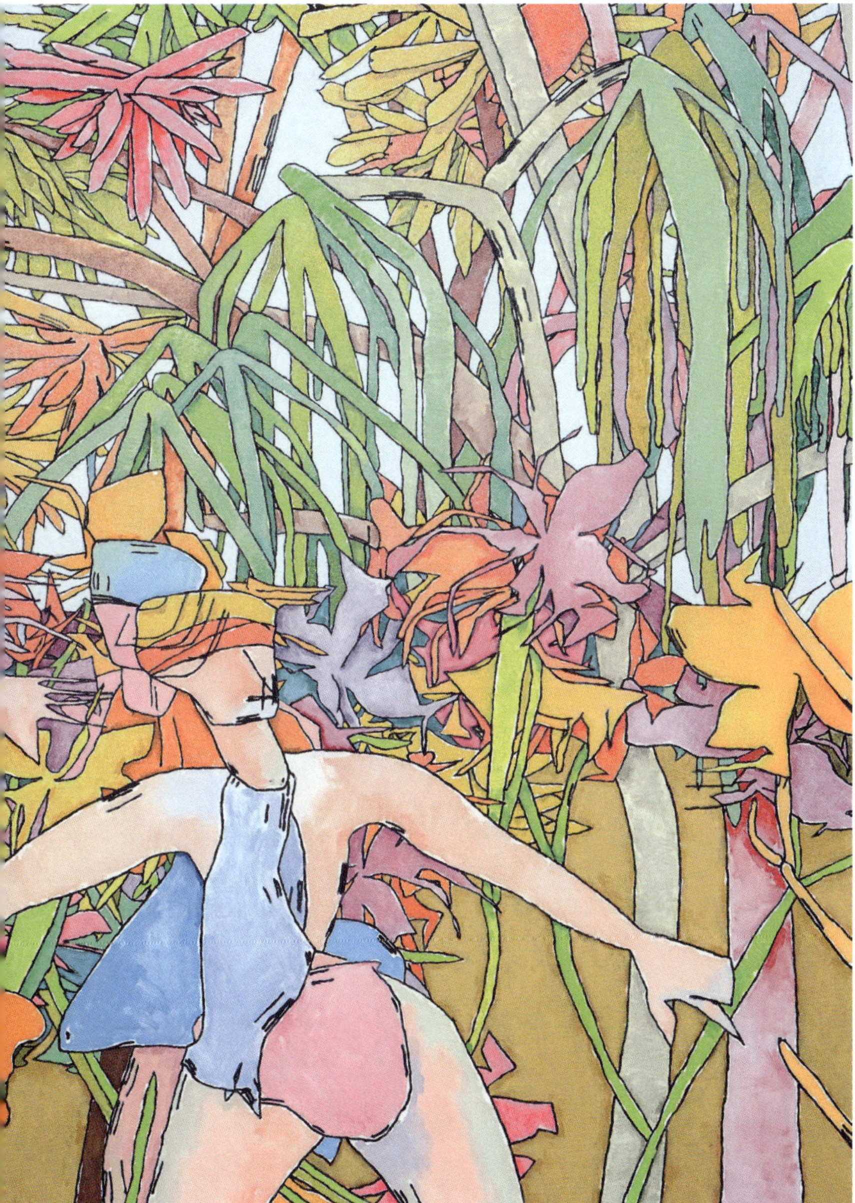

Essay

On the Four Shakespeare Folios, 1623–2023[1]

The word *folio* appears only once in Shakespeare's plays. In the final lines of the first act of *Love's Labour's Lost*, a comedy, Don Adriano de Armado invokes the instruments of his literary art: "devise wit, write pen, for I am for whole volumes in folio." The reference is not a flattering one. Armado is a braggart, enamored of his own stilted way of speaking, and his oath to fill folios with the strained devising of his wit is not one we hope he keeps.

The irony here is that *Love's Labour's Lost*, a play obsessed with literary fame — that elusive object that can "make us heirs of all eternity," according to the King (1.1.7) — puts *folio*, the word perhaps most closely tied to Shakespeare's own literary legacy, in the mouth of one of Shakespeare's most mercilessly rendered fools.

Strictly speaking, "folio" describes a bibliographic format, or shape of book: early printed books were made up of sheets of paper that, once printed,

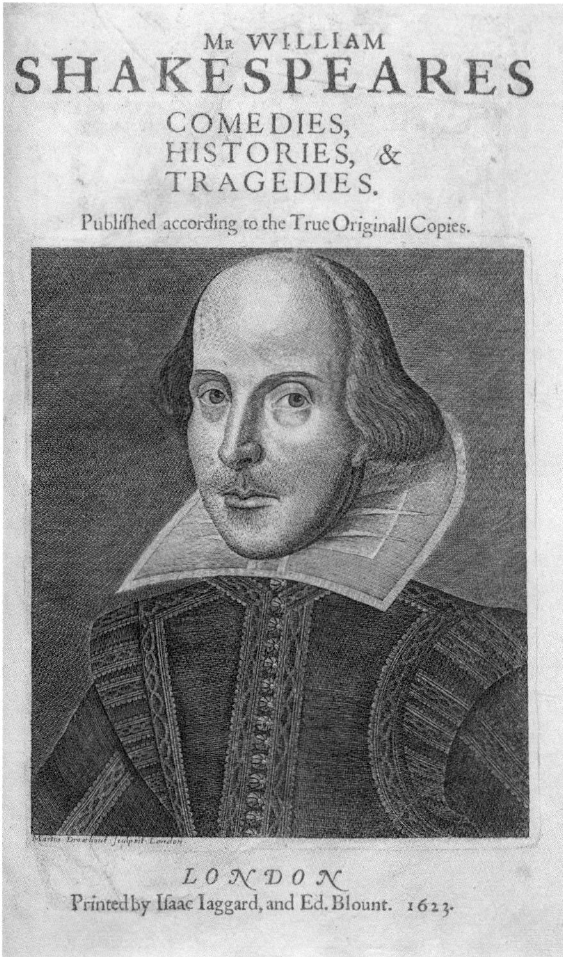

Figure 1 ↓ Title page, *Mr William Shakespeares Comedies, Histories, & Tragedies* (1623). By permission of the Folger Shakespeare Library (Call number STC 22273 Fo.1 no.05).

gathered, and folded, were stitched together in sequence and bound into volumes. A folio is simply a book made up of sheets of paper folded once, making two leaves and four pages, usually of a large size. But *folio* is more than a bibliographical term of art. When Shakespeare's acting company, the Lord Chamberlain's Men, first performed *Love's*

Labour's Lost in 1594, *folio* was already a word with considerable cultural freighting. And by the end of the seventeenth century, as Francis X. Connor observes, folios could represent "completeness, cultural prominence, and [...] literary immortality" — a fitting target, in other words, for Shakespeare's satire and an apt vessel for Armado's literary pretensions (Connor, 2014, 177).

But for Shakespeare, the word has gained other, more specific meanings. *Mr William Shakespeares Comedies, Histories, and Tragedies*, more widely known as the First Folio, contains thirty-six of the thirty-eight plays attributed in whole or part to Shakespeare — that is, those written either by himself or in collaboration with other playwrights. When it was issued from the London print shop of Isaac Jaggard in November 1623, the Folio represented the first successful attempt to gather Shakespeare's dramatic corpus in one volume. And despite the fact that it was published after Shakespeare's death, the First Folio has attained a kind of mythic status: if folios stand for completeness, cultural prominence, and literary immortality, the First Folio exemplifies the format. Standing first for completeness (its editors, the actors John Heminge and Henry Condell, claimed in their preface to have gathered the plays and printed them "absolute in their numbers, as [Shakespeare] conceived them") it has since become perhaps the most culturally prominent book in the English speaking world and now functions as a byword for Shakespeare's literary immortality.

In a sign of the First Folio's longstanding cultural status, there exists in Shakespeare studies — or at least has existed until recently — an enduring fixation on origins: the authorship and origin of individual plays (so-called "attribution studies"), the origins of Shakespeare's plots, many of which

Shakespeare borrowed and only rarely invented, and the textual origins of Shakespeare in print. In this last category, the Folio has loomed large: of the thirty-six plays it contains, eighteen were printed in no other edition. This means that without the Folio, these eighteen plays — including *The Tempest, Julius Caesar,* and *Macbeth* — would have been lost, and the Shakespeare canon would be reduced by half. Because of this, three subsequent Shakespeare Folios — a Second, printed in 1632, a Third printed in 1663, and a Fourth printed in 1685 — have, as Jeffrey Todd Knight has put it, long inhabited "a critic's no-man's land" (2017, 4). The later Folios were long viewed as textually negligible, mere derivatives of the First. And worse than that, each subsequent Folio muddled the First with textual impurities introduced by their printers' ignorance or carelessness. The later Folios therefore carried, according to Fredson Bowers — bibliographer and Shakespeare scholar — no authority in the editing of Shakespeare's plays (1951, 241). Stanley Wells' *An A-Z Guide to Shakespeare* (Oxford, 2013) follows suit and describes the later Folios as "reprints [with] no independent authority." More recently scholars have troubled this account with new evidence, but the premise holds: the First Folio's proximity to Shakespeare and to those who knew him lends it an authority that the later Folios lack. One exhibition held to mark the 400th anniversary of Shakespeare's death in 2016 enshrined this idea, describing the First Folio as "the book that gave us Shakespeare" ("First Folio!").

The idea that the First Folio is somehow uniquely Shakespearean has a long history. Samuel Johnson (1709–1784) — eighteenth-century lexicographer and famously voluble subject of James Boswell's *Life* — suggested that "the first [Folio] is equivalent to all others, and the rest only deviate from it by the printer's negligence. Whoever has any of the folios has all, excepting those diversities which mere reiteration of editions will produce" (quoted in Murphy 2021, 112). George Steevens (1736–1800), who edited Shakespeare's plays with Johnson, was even more dismissive, likening the later Folios to "mere waste paper" (quoted in Hooks 2016, 190). Implied by Johnson and Steeven's language is the idea of sameness, even equivalence between the Folios, and a generally untroubled sequence that led from the First to the Fourth.

But the survival of Shakespeare's plays does not depend solely, or even primarily, on the survival of the First Folio, 228 complete copies of which survive. Rather, the survival of the plays — their continued reception by new and increasingly varied communities of readers — is also a result of the fact that the Folio was reprinted in three subsequent editions. The four Folios, all published in the seventeenth century, stand for Shakespeare's appeal in the period, and each contributed to the gradual elevation of Shakespeare in the English literary canon. It is fitting, then, that a short poem by John Milton that describes the Folio and its contents as "a lasting Monument" was printed in the opening pages of the Second Folio (1632) and not the First. In other words, literary legacy is a fragile thing made durable by reprinting.

Yet Chris Laoutaris's 2023 book, *Shakespeare's Book: The Story Behind the First Folio and the Making of Shakespeare*, reminds us of the obstinacy of what Adam Hooks has called "Folio as fetish" (2016, 185). Laoutaris's title alone contains two tropes that seem to resurface unbidden during Shakespearean centenaries — namely, the idea that the First Folio somehow *made* (or gave us) Shakespeare and that the First Folio is "Shakespeare's book" (Laouartis finds "enticing" evidence for the involvement in the Folio's publication of Shakespeare's London lodger, a man of murky identity who perhaps acted on the deceased Shakespeare's behalf — Shakespeare has long provided fertile ground for speculative histories) (341). But if the First Folio can be said to be Shakespeare's, it has become so only posthumously and by way of wishful thinking and mere assertion. As Adam Hooks writes, "the Folio does not give us Shakespeare. Rather, it gives us a particular version of Shakespeare" — one among many, Hooks implies (2016, 185).

Comparatively under-examined, the later Folios offer up fresh and ample quarry for bibliographical work. The number of extant copies of the later Folios gives some sense of the variety and scale of the evidence that awaits analysis: some 183 copies of the Third Folio, 379 copies of the Second, and 342 copies of the Fourth are known to survive (the online *Shakespeare Census* [http://www.shakespearecensus.org] adds to these figures several dozen fragments). In this trove there are no doubt discoveries to be made.

This should not suggest that the later Folios have gone unexamined; far from it. Scholars have considered the progress of Shakespeare in print *after* the First Folio with far reaching implications for our understanding of the reception and evolution of the Shakespeare canon (see, for instance, Sonia Massai's *Shakespeare and the Rise of the Editor* (2007), which found evidence of editorial work being done in the Fourth Folio, Eric Rasmussen's ongoing investigations into the Fourth Folio's printing history (1998, 2017), and Emma Depledge and Peter Kirwan's volume, *Canonising Shakespeare: Stationers and the Book Trade*, 1640–1740 (2017), which conspicuously excludes the First Folio from its ambit). But against the exhaustive and often repetitious work done on the First Folio, the later Folios are relative enigmas. And as bibliographical scholarship turns away from textual authority and the priority of editions and toward more discursive, contingent, and embodied topics and kinds of evidence (bindings, marginal annotations, histories of book use and reception), we are well positioned to consider the later Folios and the stories they tell.

⋆ ⋆ ⋆

This essay and the book from which it is drawn reflect the themes and preoccupations of two exhibitions held to mark the quatercentenary (1623–2023) of the First Folio at Carnegie Mellon (Lemley 2024). Organized in collaboration with Pittsburgh's Frick Art Museum, these exhibits ran concurrently in two venues. One exhibit, *Inventing Shakespeare: Text, Technology, and the Four Folios*, was installed in the Hunt Library at Carnegie Mellon University; and the other, *From Stage to Page: 400 Years of Shakespeare in Print*, was installed in the Frick's Jacobean Gallery — an octagonal room paneled in seventeenth-century oak that evoked the Folios' cultural and historical origins. Spare by design, *From Stage to Page* spotlighted all seven Shakespeare Folios held in Special Collections: one copy of the First, and two copies each of the Second, Third, and Fourth Folios. *From Stage to Page* offered an unusual and, in Pittsburgh, unprecedented opportunity to view the First Folio alongside copies of each of the later Folios — it was also the first time that all seven Shakespeare Folios held at Carnegie Mellon were put on simultaneous display.

Beyond exhibiting the richness of Special Collections' holdings in early Shakespeare editions, *From Stage to Page* offered a synoptic view of how the four Folios brought about Shakespeare's literary canonization in the seventeenth century. The second exhibit, *Inventing Shakespeare*, illustrated the ways in which the Folios have prompted new and inventive methods for research. Rightly called "the most thoroughly studied early modern book," the First Folio has inspired an array of scholarly tools, techniques, and technologies (Galbraith 2010, 46). From Charlton Hinman's optical-mechanical collating machine (a functioning example of which is held in Special Collections [see "HINMAN COLLATOR," page 407]), to the *Shakespeare Virtual Reality (VR)* project (https://shakespeare-vr.library.cmu.edu/), to digital and photographic facsimiles, to ongoing experiments in bibliographical analysis using computer vision and machine learning, *Inventing Shakespeare* ranged widely.

What united both exhibitions was their interest in how the First Folio evolved through the Second, Third, and Fourth. Far from being mere imperfect imitations of the First Folio, the three later Folios offer up fascinating idiosyncrasies, some of which have only recently been studied or fully noted. The Third Folio, for instance, exists in two forms. One, dated 1663 on its title page, is an otherwise untroubled reprint of the Second Folio (1632), which preceded it by three decades. The second form (properly an "issue") of the Third Folio, however, bears a 1664 date on its reprinted title page and contains seven plays (six apocryphal) that do not appear in either the First or Second Folios. Surviving copies of the Third Folio therefore represent two distinct issues, each marked by the presence or absence of the seven plays that the 1664 title page trumpets as "never before Printed in Folio." The reason for the Third Folio's reissue is believed to have been commercial rather than literary: its booksellers sought a way to differentiate the Third Folio from both the Second and First, making it more salable.

The Fourth Folio, printed in 1685, embraces oddities of its own. Seventeen of its sheets (a full sixty-eight pages) were reprinted as late as 1700 — fifteen years after the edition's first appearance. The reasons for the reprinting of these seventeen sheets remain obscure. Scholars (Dawson, 1952; Hansen and Rasmussen, 2017) have suggested that the printer of the deficient 1685 sheets failed to deliver the required quantity, leaving the bookseller who inherited the unsold copies short-sheeted. The missing sheets may also have been damaged — torn, sodden, soiled — or simply lost while in the bookseller's warehouse and thus required replacement. Giles Dawson (1952) proposed that copies of the Fourth Folio that contain these reprinted sheets "constitute a fifth folio," different enough in kind to earn their own classification. What's more, the identities of the Fourth Folio's printers were long unknown, making the circumstances of its publication and printing more enigmatic than those of the First (Lemley et al. 2024; Warren et al. 2023). The Fourth Folio also contains one of the most glaring and comical typesetting errors in the history of English literature: the printed subtitle for Hamlet reads, "RPINCE OF DENMARK."

The texts of Shakespeare's plays in all four Folios offer up other, "literal" variants besides: these are errors introduced in the process of printing and range from inverted or damaged pieces of type to the omission or alteration of entire lines of dialogue. Such misprintings were corrected occasionally during presswork — provided, they were noticed during proofreading. Offending pieces of type were removed and replaced while the page bearing the error was at press. These tardy interventions mean that copies containing sheets with corrections make up new, recognizable "states" of the text. These many hundreds of small aberrations are masked behind the nominal tidiness of the word "copy." In truth, no two copies of any of the four Folios are known to be exactly alike — each copy differs in some way from the next. The definite article is thus misplaced: there is

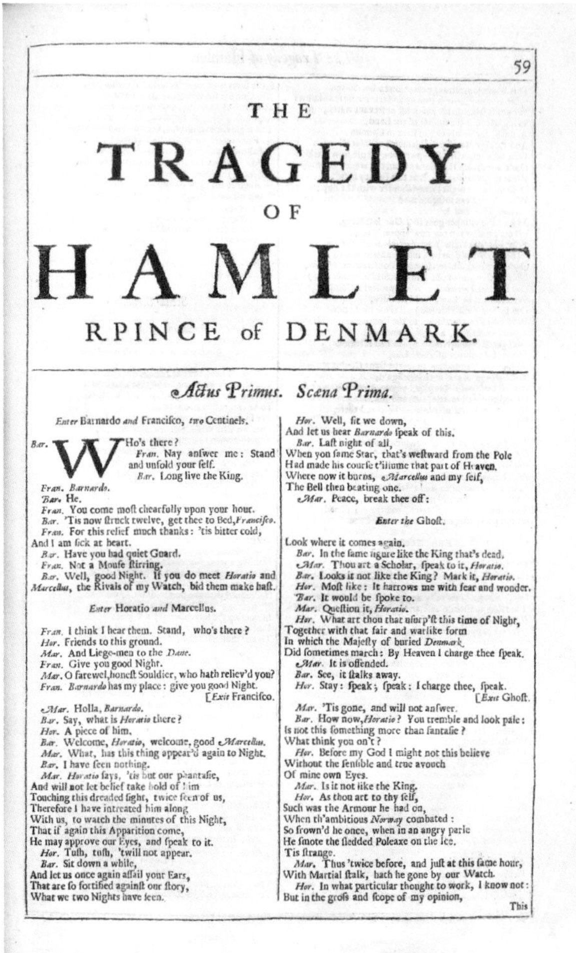

59

THE
TRAGEDY
OF
HAMLET
RPINCE of DENMARK.

Actus Primus. Scæna Prima.

Enter Barnardo *and* Francisco, *two* Centinels.

Bar. WHo's there?
Fran. Nay anſwer me: Stand and unfold your ſelf.
Bar. Long live the King.
Fran. *Barnardo.*
Bar. He.
Fran. You come moſt chearfully upon your hour.
Bar. 'Tis now ſtruck twelve, get thee to Bed, *Franciſco.*
Fran. For this relief much thanks: 'tis bitter cold,
And I am ſick at heart.
Bar. Have you had quiet Guard.
Fran. Not a Mouſe ſtirring.
Bar. Well, good Night. If you do meet *Horatio* and *Marcellus*, the Rivals of my Watch, bid them make haſt.

Enter Horatio *and* Marcellus.

Fran. I think I hear them. Stand, who's there?
Hor. Friends to this ground.
Mar. And Liege-men to the *Dane.*
Fran. Give you good Night.
Mar. O farewel, honeſt Souldier, who hath reliev'd you?
Fran. *Barnardo* has my place: give you good Night.
[*Exit* Franciſco.
Mar. Holla, *Barnardo.*
Bar. Say, what is *Horatio* there?
Hor. A piece of him.
Bar. Welcome, *Horatio*, welcome, good *Marcellus.*
Mar. What, has this thing appear'd again to Night.
Bar. I have ſeen nothing.
Mar. *Horatio* ſays, 'tis but our phantaſie,
And will not let belief take hold of him
Touching this dreaded ſight, twice ſeen of us,
Therefore I have intreated him along
With us, to watch the minutes of this Night,
That if again this Apparition come,
He may approve our Eyes, and ſpeak to it.
Hor. Tuſh, tuſh, 'twill not appear.
Bar. Sit down a while,
And let us once again aſſail your Ears,
That are ſo fortified againſt our ſtory,
What we two Nights have ſeen.
Hor. Well, ſit we down,
And let us hear *Barnardo* ſpeak of this.
Bar. Laſt night of all,
When yon ſame Star, that's weſtward from the Pole
Had made his courſe t'illume that part of Heaven,
Where now it burns, *Marcellus* and my ſelf,
The Bell then beating one.
Mar. Peace, break thee off:

Enter the Ghoſt.

Look where it comes again.
Bar. In the ſame figure like the King that's dead.
Mar. Thou art a Scholar, ſpeak to it, *Horatio.*
Bar. Looks it not like the King? Mark it, *Horatio.*
Hor. Moſt like: It harrows me with fear and wonder.
Bar. It would be ſpoke to.
Mar. Queſtion it, *Horatio.*
Hor. What art thou that uſurp'ſt this time of Night,
Together with that fair and warlike form
In which the Majeſty of buried *Denmark*
Did ſometimes march: By Heaven I charge thee ſpeak.
Mar. It is offended.
Bar. See, it ſtalks away.
Hor. Stay: ſpeak; ſpeak: I charge thee, ſpeak.
[*Exit* Ghoſt.
Mar. 'Tis gone, and will not anſwer.
Bar. How now, *Horatio*? You tremble and look pale:
Is not this ſomething more than fantaſie?
What think you on't?
Hor. Before my God I might not this believe
Without the ſenſible and true avouch
Of mine own Eyes.
Mar. Is it not like the King.
Hor. As thou art to thy ſelf,
Such was the Armour he had on,
When th'ambitious *Norway* combated:
So frown'd he once, when in an angry parle
He ſmote the ſledded Poleaxe on the Ice.
Tis ſtrange.
Mar. Thus 'twice before, and juſt at this ſame hour,
With Martial ſtalk, hath he gone by our Watch.
Hor. In what particular thought to work, I know not:
But in the groſs and ſcope of my opinion,

This

Figure 2 → "HAMLET RPINCE OF DENMARK." By permission of Carnegie Mellon University Libraries, Special Collections (Call number PR2751 .A4 1685).

no "the" First Folio. Rather, each of the four Folios is plural, every copy conveying a subtly unique form of Shakespeare's plays.

Beyond variants introduced by way of the proofreader's correcting pen, there are also variants at the level of individual copies introduced by the Folios'

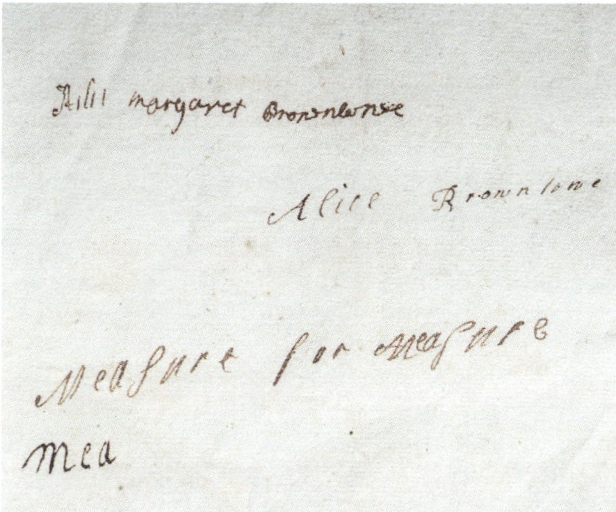

Figure 3 ↗ The inscribed names of Alice and Margaret Brownlowe, the earliest documented users of this copy of the Fourth Folio. By permission of Carnegie Mellon University Libraries, Special Collections (Call number PR2751 .A4 1685).

earliest readers and owners: inscriptions, marginal annotations, sketches made by children, or bits of bookkeeping or arithmetic made in the margins. These traces of a book's provenance (what William Sherman has likened to a fossil record of a bookish "ecology of use and reuse" [Sherman 2010, 6]) offer up poignant clues of the many lives that books live. One of the copies of the Fourth Folio in Special Collections, for instance, hints at its origins at Belton House, the Lincolnshire estate of the Brownlowe family. On one of the book's front flyleaves appears a pair of inscribed names: Alice and Margaret Brownlowe. We can be reasonably confident that Alice is Alice Brownlowe (1659–1721), wife of John Brownlowe (1659–1697), who undertook a campaign of renovation at Belton House in the 1680s and funded the expansion of its library (including, presumably, the acquisition of the Fourth Folio) in the same period. The second name is more obscure: it appears that the writer began inscribing "Alice" in a style that imitates the other, complete inscription. Changing tack, however, the writer instead spelled out her own name in full: Margaret Brownlowe. What stories can this pair of names tell? Is this evidence of the elder Alice teaching her daughter Margaret (1687–1710) to spell her name? The Shakespearean play title written out below the two names, *Measure for Measure*, suggests as much: a less confident hand has attempted to copy out the title, giving up three letters in. Or perhaps it's another daughter, also Alice (or Alicia) (1684–1727), named after her mother, conspiring with her sister to mark the Folio's blank leaves with their names. Either way, these inscriptions show that Shakespeare Folios functioned as objects of familial and matrilineal record-keeping; they also suggest that John Brownlowe was this copy's first owner, though his daughters were, it seems, its earliest users. A portrait that survives at Belton House shows Margaret at eight years old, allowing us to put a face to at least one of these haltingly inscribed names (even while the identity of the Black child attending her in the portrait remains unknown).

The bindings of the Folios also vary, telling stories of their own. Before the introduction of edition binding — a method by which every copy of a particular print run is bound up in the same or similar style — books were kept in sheets or only loosely stitched together until they were sold. The buyer then had the sheets bound in a style of their choosing. Both of CMU's copies of the Third Folio are in their original bindings, exhibiting a striking similarity in style: plain calf with a double line of embossed rules outlining the edge of each board and a mottled pattern of red decorating the edges of the text block. Separated for nearly four centuries, both of these books — perhaps bound in the same workshop in London with the same tools and from the same stock of materials (animal hide, paper, twine, pigment, and glue) — have been reunited by chance in Pittsburgh.

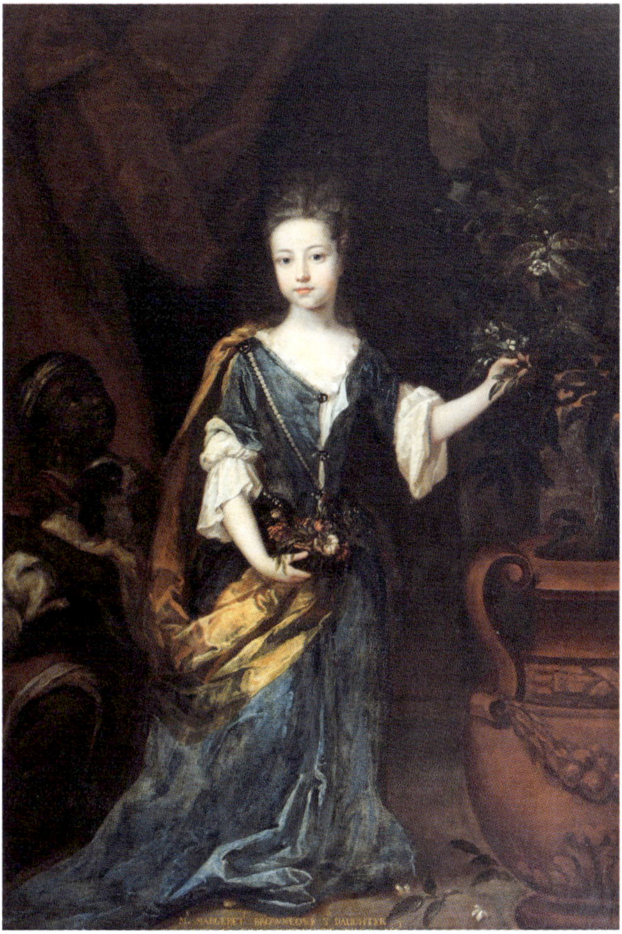

Figure 4 ↑ Henry Tilson, portrait of Margaret Brownlowe (1687–1710), aged eight, 1695. By permission of The National Trust (Image number 986143, Belton House).

In their piecewise scrutiny of the Folios, CMU Libraries' 2023 exhibitions adopted a metaphor from the First Folio's prefatory "Address to the Readers," written by the Folio's editors. In the Address, Heminge and Condell note that the Folio presents Shakespeare's plays, "cured, and perfect of their limbs," an image that suggests the posthumous care of Shakespeare's literary body. In attempting to reconcile and join the

Four Folios into a shared narrative — to cure and perfect points of disjunction between them — the exhibitions lingered over several aspects of Shakespeare's plays in folio, both material and cultural. The exhibits aimed, in other words, to reintegrate all four Folios into an unbroken history of publication, reception, and rarity across the seventeenth century and into the present.

1 This essay is an edited excerpt from the introduction to Samuel V. Lemley, ed. *The Four Shakespeare Folios, 1623–2023: Copy, Print, Paper, Type* (Penn State University Press, 2024).

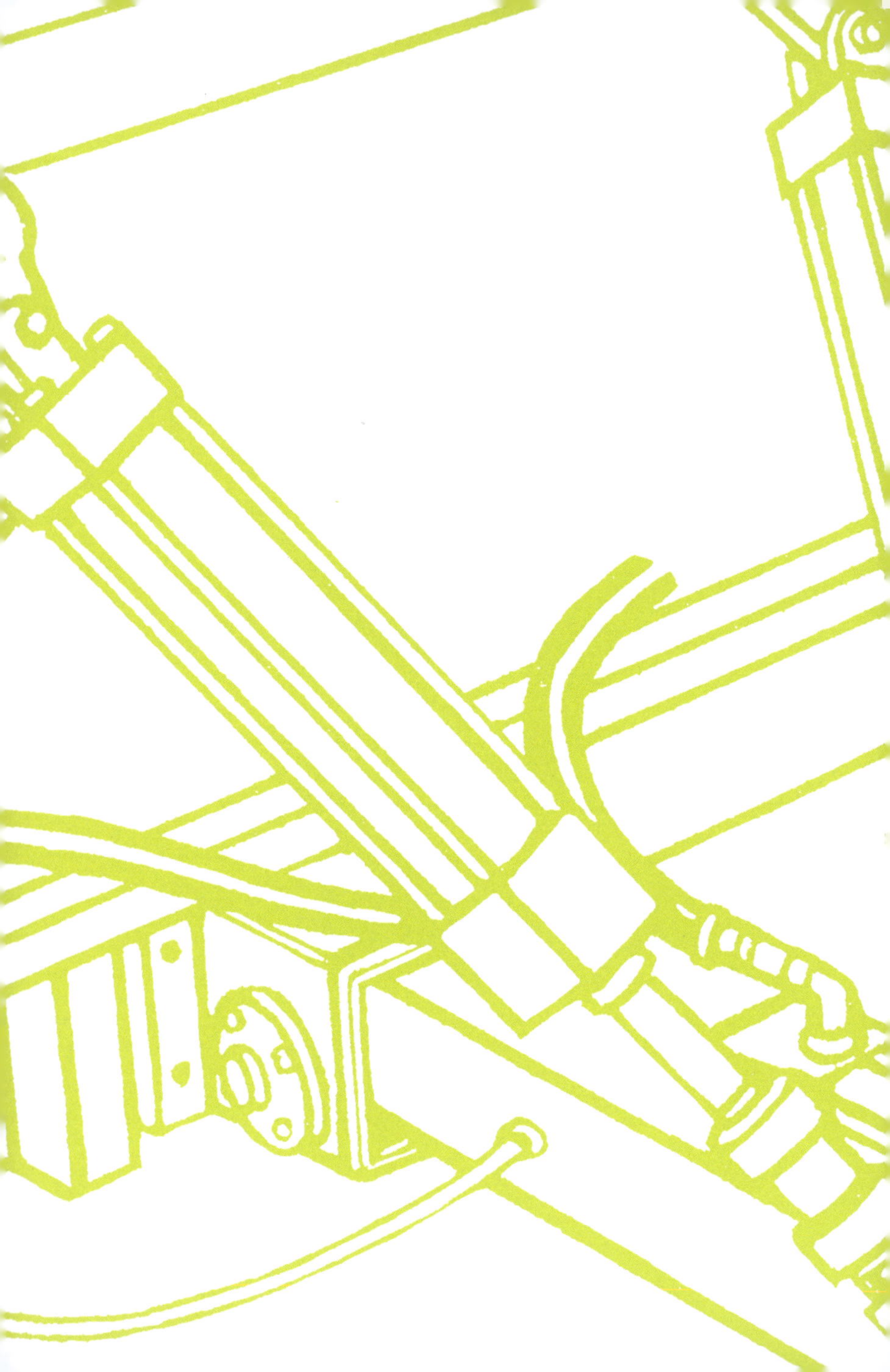

ὑπὸ αδζ γω
λειδομένη ἐ-
πεὶ δὲ ἡ αγ
ὁ τῆς αε πρὸς
πλῶν τῆς αβ,
αε ἔσται δο-
κῶ τῆς δζ λε-
ζ γωνία, τοῦ
ὥστε καὶ ἡ τε
ἡ βγ, ὅπερ ἔ
περὶ κέντρου
ειλήφθω τρία
μ αβ αγ περι
ναι ἀπὸ τῶν ἑ-
ὸ ὅμοιον ὑπὸ
ὶ ἡ γβ, ἐκβε-
ὸ ε σημεῖον,
ἡ γα περιφε
ς, οὕτως ἡ γε
ροτέρω λημ-
ν καθέτους,

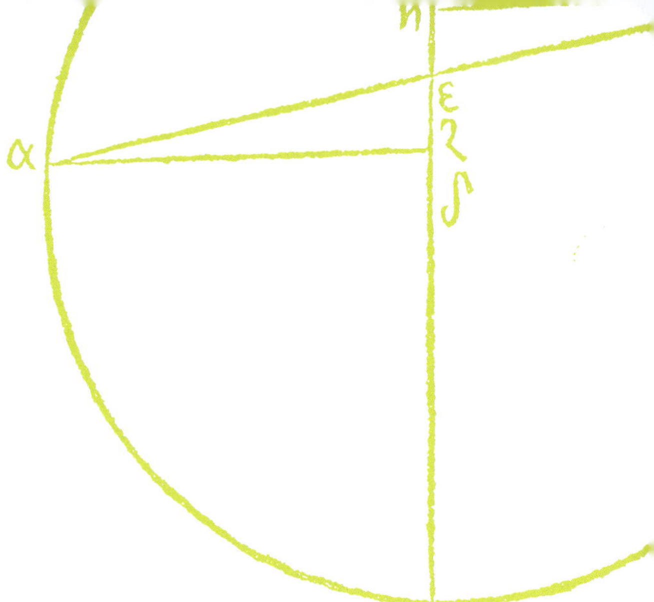

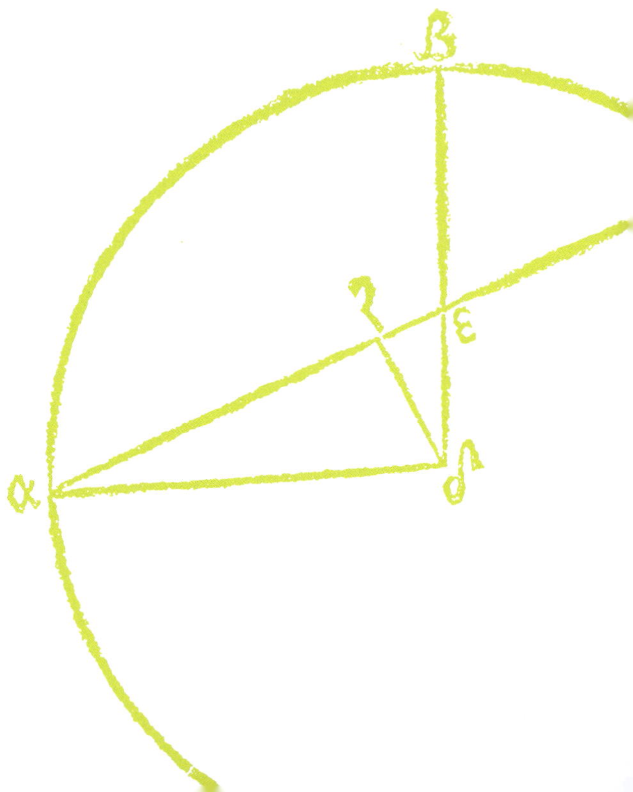

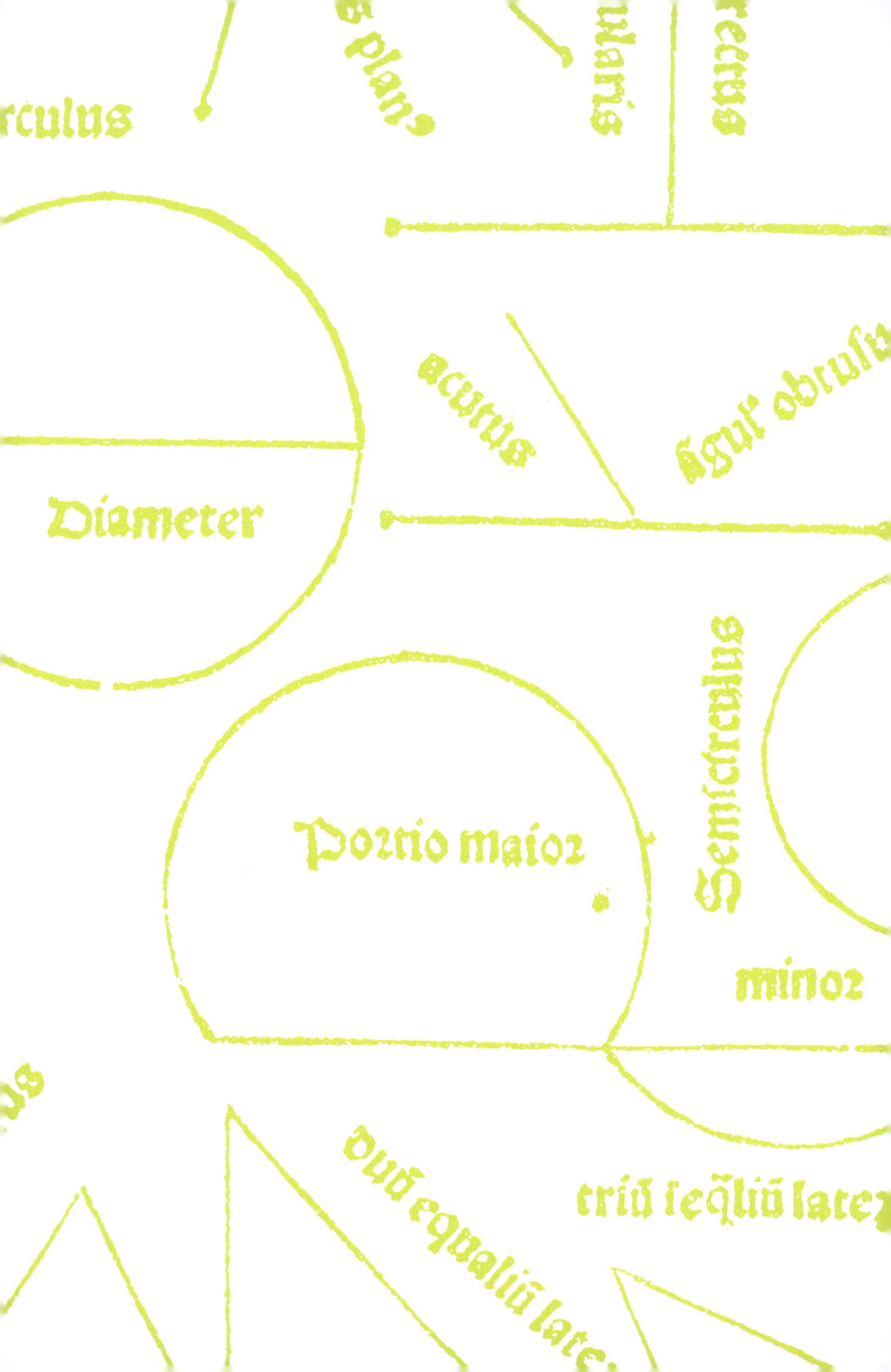
acutus
Diameter
Portio maior
Semicirculus
minor

BIBLIOGRAPHY

Agüera y Arcas, Blaise. 2002. "Computational Analytical Bibliography." In *Proceedings of the Bibliopolis Conference*. The Hague: Koninklijke Bibliotheek.

———. 2003. "Temporary Matrices and Elemental Punches in Gutenberg's DK Type." In *Incunabula and Their Readers: Printing, Selling, and Using Books in the Fifteenth Century*, edited by Kristian Jensen. London: British Library.

———. 2025. "What Is Intelligence? | Antikythera." Interlude: The Prehistory of Computing. https://whatisintelligence.antikythera.org/chapter-02.5/.

Akkerman, Nadine, and Pete Langman. 2024. *Spycraft: Tricks and Tools of the Dangerous Trade from Elizabeth I to the Restoration.* New Haven: Yale University Press.

Baxandall, David, and Jane Pugh. 1975. *Calculating Machines and Instruments: Catalogue of the Collections in the Science Museum.* London: Science Museum.

Biss, Eula. 2014. *On Immunity: An Inoculation.* Minneapolis, Minnesota: Graywolf Press.

Blayney, Peter W. M. 1982. *The Texts of King Lear and Their Origins.* Cambridge: Cambridge University Press.

Borelli, Giovanni Alfonso. 1680. *De Motu Animalium.* Romae: Ex typographia Angeli Bernabo. http://archive.org/details/McGillLibrary-osl_motu-animalium_B731m1680-81_v1-20166.

Bourne, Claire M. L., ed. 2021. *Shakespeare / Text: Contemporary Readings in Textual Studies, Editing and Performance Readings in Textual Studies, Editing and Performance.* London: Bloomsbury.

Bowers, Fredson. 1951. "Robert Roberts: A Printer of Shakespeare's Fourth Folio." *Shakespeare Quarterly* 2 (3): 241. https://doi.org/10.2307/2866657.

———. 1977. "News & Comment: Charlton Joseph Kadio Hinman." *The Book Collector* 26 (3): 388–91.

Carter, Harry. 1969. *A View of Early Typography up to about 1600.* Oxford: Clarendon Press.

Connor, Francis X. 2014. "Epilogue: Henry Herringman's Restoration Folios." In *Literary Folios and Ideas of the Book in Early Modern England. History of Text Technologies,* by Francis X. Connor, 167–78. New York: Palgrave Macmillan US. https://doi.org/10.1057/9781137438362_6.

Cordell, Ryan. 2020. "Speculative Bibliography." *Anglia* 138 (3): 519–31. https://doi.org/10.1515/ang-2020-0041.

Daston, Lorraine. 2024. *Rules: A Short History of What We Live By.* Princeton University Press.

Dawson, Giles E. 1951. "Some Bibliographical Irregularities in the Shakespeare Fourth Folio." *Studies in Bibliography* 4:93–103.

Depledge, Emma, and Peter Kirwan, eds. 2017. *Canonising Shakespeare: Stationers and the Book Trade, 1640–1740.* Cambridge: Cambridge University Press. https://doi.org/10.1017/9781316650752.

Descartes, René. 1999. *Discourse on Method and Meditations on First Philosophy.* Translated by Donald A. Cress. Indianapolis: Hackett Publishing Company.

Desmond, Allen, and James Moore. 1991. *Darwin: The Life of a Tormented Evolutionist.* London: Viking Penguin.

"First Folio! Shakespeare's American Tour Exhibition Material - Folgerpedia." n.d. Accessed January 24, 2025. https://folgerpedia.folger.edu/First_Folio!_Shakespeare%27s_American_Tour_Exhibition_Material.

Galbraith, Steven K. 2010. "English Literary Folios, 1593–1623: Studying Shifts in Format." In *Tudor Books and Readers: Materiality and the Construction of Meaning*, edited by John N. King, 46–67. Cambridge: Cambridge University Press.

Grafton, Anthony. 2002. *Leon Battista Alberti: Master Builder of the Italian Renaissance.* Cambridge, Massachusetts: Harvard University Press.

Gravell, Thomas L. 2002. *American Watermarks, 1690–1835.* New Castle: Oak Knoll.

Grillo, Francesco. 2019. "Hero of Alexandria's Automata: A Critical Edition and Translation, Including a Commentary on Book One." PhD Dissertation, Glasgow: University of Glasgow.

Hansen, Lara, and Eric Rasmussen. 2017. "Shakespeare Without Rules: The Fifth Shakespeare Folio and Market Demand in the Early 1700s." In *Canonising Shakespeare: Stationers and the Book Trade, 1640–1740*, edited by Emma Depledge and Peter Kirwan, 38–54. Cambridge: Cambridge University Press.

Hinman, Charlton. 1947. "Mechanized Collation: A Preliminary Report." *The Papers of the Bibliographical Society of America* 41 (2): 99–106. https://doi.org/10.1086/pbsa.41.2.24298708.

———. 1953. "Variant Readings in the First Folio of Shakespeare." *Shakespeare Quarterly* 4 (3): 279–88.

———. 1955. "Mechanized Collation at the Houghton Library." *Harvard Library Bulletin* 9:132–34.

———. 1963. *The Printing and Proof-Reading of the First Folio of Shakespeare.* Oxford: Clarendon Press.

Hodges, Andrew. 1983. *Alan Turing: The Enigma.* New York: Simon & Schuster.

Hooks, Adam G. 2016. *Selling Shakespeare: Biography, Bibliography, and the Book Trade.* Cambridge: Cambridge University Press. https://doi.org/10.1017/CBO9781316488836.

Hopper, Grace Murray. 2021. “The Education of a Computer (1952).” In *Ideas That Created the Future: Classic Papers of Computer Science*, edited by Harry R. Lewis. Cambridge: MIT Press.

Hrdlicka, James. 2020. *Colonists, Citizens, Constitutions: Creating the American Republic.* New York: Rizzoli.

Hunt, Rachel McMasters Miller. 1958. *Catalogue of Botanical Books in the Collection of Rachel McMasters Miller Hunt.* Pittsburgh: Hunt Botanical Library.

James, Kathryn, and Aaron T. Pratt. 2019. *Collated & Perfect.* New Haven: Beinecke Rare Book & Manuscript Library, Yale University.

Jazarī, Ismāʿīl ibn al-Razzāz. n.d. “Kitāb fī maʿrifat al-ḥiyal al-handasīyah كتاب في معرفة الحيل الهندسية Jazarī, Ismāʿīl ibn al-Razzāz جزري، إسماعيل بن الرزّاز.” Manuscript. Or 116. British Library: Oriental Manuscripts. Accessed March 6, 2025. https://www.qdl.qa/en/archive/81055/vdc_100023421307.0×000001.

Kahn, David. 1996. *The Codebreakers: The Comprehensive History of Secret Communication from Ancient Times to the Internet.* New York: Scribner.

Kaminski, John P., Thomas H. Linley, Elizabeth M. Schoenleber, Timothy D. Moore, Oindrila Chattopadhyay, Dustin M. Cohan, Anna M. Biermeier, Christopher S. Berry, Sarah K. Danforth, and Daniel J. Hoefs, eds. 2020. *The Documentary History of the Ratification of the Constitution and the Bill of Rights, Volume 37: The Bill of Rights, No. 1.* Madison: Wisconsin Historical Society Press.

Kazuo, Murakami. 2012. *Japanese Automata: Karakuri Zui, An Eighteenth-Century Japanese Manual of Automatic Mechanical Devices*. Sakamoto (Japan): Published by the author.

Knight, Jeffrey Todd. 2017. “Economies of Scale: Shakespeare and Book History.” *Literature Compass* 14 (6): e12393. https://doi.org/10.1111/lic3.12393.

Laoutaris, Chris. 2023. *Shakespeare’s Book: The Story Behind the First Folio and the Making of Shakespeare.* New York: Simon & Schuster.

Lemley, Samuel. 2021. "Fascinating Illustrations of 18th-Century Inoculations by the Inventor of the Smallpox Vaccine." *Hyperallergic*. 2021. http://hyperallergic.com/650913/illustrations-of-18th-century-inoculations/.

Lemley, Samuel V., ed. 2024. *The Four Shakespeare Folios, 1623–2023: Copy, Print, Paper, Type*. University Park, Pennsylvania: Penn State University Press.

———. 2025. "Stab-Stitch Holes, Deckles, and Broken Types: Some Notes on the First Official Printing of the US Bill of Rights (1792)." *The Library* [forthcoming].

Lemley, Samuel V., Nikolai Vogler, Christopher N. Warren, D.J. Schuldt, Laura S. DeLuca, Kari Thomas, Taylor Berg-Kirkpatrick, Elizabeth Dieterich, and Max G'Sell. 2024. "Everything There Is to Be Learned About Seventeenth-Century Types: Computational Bibliography and the Fourth Folio's Printers." In *The Four Shakespeare Folios, 1623–2023: Copy, Print, Paper, Type*, 117–44. University Park, Pennsylvania: Penn State University Press.

Lesser, Zachary, and Whitney Trettien. 2021. "Material / Digital." In *Shakespeare / Text: Contemporary Readings in Textual Studies, Editing and Performance Readings in Textual Studies, Editing and Performance*, edited by Claire M. L. Bourne, 402–23. London: Bloomsbury.

Lewis, Sarah Elizabeth. 2020. "Opinion | Where Are the Photos of People Dying of Covid?" *The New York Times*, May 1, 2020, sec. Opinion. https://www.nytimes.com/2020/05/01/opinion/coronavirus-photography.html.

Locke, John. 1690. *An Essay Concerning Humane Understanding*. London.

Marlowe, Christopher. 2014. *Dr. Faustus: The A- and B-Texts (1604, 1616)*. Manchester: Manchester University Press.

Massai, Sonia. 2007. *Shakespeare and the Rise of the Editor*. Cambridge: Cambridge University Press.

McCorduck, Pamela. 2004. *Machines Who Think: A Personal Inquiry into the History and Prospects of Artificial Intelligence*. New York: A.K. Peters/CRC Press.

Menabrea, Luigi Federico. 1843. *Sketch of the Analytical Engine Invented by Charles Babbage [...] with Notes by the Translator*. Translated by Ada Augusta Lovelace. London: R. & J. E. Taylor.

Montagu, Mary Wortley. 1763. *Letters of the Right Honourable Lady M--y W---y M----e; Written, during Her Travels in Europe, Asia and Africa, to Persons of Distinction, Men of Letters, &c. in Different Parts of Europe.* 2nd edition. Vol. 3. London: Printed for T. Becket and P.A. De Hondt.

Murphy, Andrew. 2021. *Shakespeare in Print: A History and Chronology of Shakespeare Publishing.* Cambridge: Cambridge University Press.

Otis, Jessica Marie. 2024. *By the Numbers: Numeracy, Religion, and the Quantitative Transformation of Early Modern England.* Oxford: Oxford University Press.

Pasquinelli, Matteo. 2023. *The Eye of the Master: A Social History of Artificial Intelligence.* London: Verso.

Peterson, Mark A. 2011. *Galileo's Muse: Renaissance Mathematics and the Arts.* Cambridge, Massachusetts: Harvard University Press. https://doi.org/10.4159/harvard.9780674062979.

Pettegree, Andrew, and Arthur der Weduwen. 2021. *The Library: A Fragile History.* New York: Basic Books.

Pratt, Aaron T. 2015. "Stab-Stitching and the Status of Early English Playbooks as Literature." *The Library* 16 (3): 304–28. https://doi.org/10.1093/library/16.3.304.

Rasmussen, Eric. 1998. "Anonymity and the Erasure of Shakespeare's First Eighteenth- Century Editor." In *Reading Readings: Essays on Shakespeare Editing in the Eighteenth Century*, edited by Joanna Gondris, 318–23. Madison: Farleigh Dickinson University Press.

Rhodes, Neil, and Jonathan Sawday, editors. 2000. *The Renaissance Computer: Knowledge, Technology in the First Age of Print.* New York: Routledge.

Riskin, Jessica, ed. 2007. *Genesis Redux: Essays in the History and Philosophy of Artificial Life.* University of Chicago Press. https://doi.org/10.7208/chicago/9780226720838.001.0001.

———. 2016. *The Restless Clock: A History of the Centuries-Long Argument over What Makes Living Things Tick.* Chicago: The University of Chicago Press.

———. 2024. "'You Are My Friend': Early Androids and Artificial Speech." *The Public Domain Review*. 2024. https://publicdomainreview.org/essay/early-androids-and-artificial-speech/.

———. n.d. "Frolicsome Engines: The Long Prehistory of Artificial Intelligence." *The Public Domain Review*. Accessed March 5, 2025. https://publicdomainreview.org/essay/frolicsome-engines-the-long-prehistory-of-artificial-intelligence/.

Roose, Kevin. 2024. “Why Anthropic’s Claude Is a Hit with Tech Insiders.” *The New York Times*, December 16, 2024. https://www.nytimes.com/2024/12/13/technology/claude-ai-anthropic.html.

Saiber, Arielle. 2017. *Measured Words: Computation and Writing in Renaissance Italy.* Toronto: University of Toronto Press.

Sawday, Jonathan. 2007. *Engines of the Imagination: Renaissance Culture and the Rise of the Machine.* New York: Routledge.

Sherman, William H. 2009. *Used Books: Marking Readers in Renaissance England.* Philadelphia: University of Pennsylvania Press. http://ebookcentral.proquest.com/lib/cm/detail.action?docID=3441433.

———. 2010. “How to Make Anything Signify Anything: William F. Friedman and Birth of Modern Cryptanalysis.” *Cabinet* 40 (Winter 2010–2011). https://cabinetmagazine.org/issues/40/sherman.php.

Smith, Steven Escar. 2000. “‘The Eternal Verities Verified’: Charlton Hinman and the Roots of Mechanical Collation.” *Studies in Bibliography* 53:129–61.

———. 2002. “‘Armadillos of Invention’: A Census of Mechanical Collators.” *Studies in Bibliography* 55:133–70.

Swift, Jonathan. 1726. *Travels into Several Remote Nations of the World. In Four Parts.* London: For Benjamin Motte.

Tanselle, G. Thomas. 1989. “Reproductions and Scholarship.” *Studies in Bibliography* 42:25–54.

Thorndike, Lynn. 1958. *A History of Magic and Experimental Science: Volume VII, The Seventeenth Century.* New York: Columbia University Press.

Tolonen, Mikko, Mark J. Hill, Ali Zeeshan Ijaz, Ville Vaara, and Leo Lahti. 2021. “Examining the Early Modern Canon: The English Short Title Catalogue and Large-Scale Patterns of Cultural Production.” In *Data Visualization in Enlightenment Literature and Culture*, edited by Ileana Baird, 63–119. Cham: Springer International Publishing. https://doi.org/10.1007/978-3-030-54913-8_3.

Turing, A. M. 1950. “Computing Machinery and Intelligence.” *Mind* 59 (236): 433–60.

Turing, Alan M. 1951. “‘Intelligent Machinery, a Heretical Theory’, a Lecture given to ‘51 Society’ at Manchester.” The Turing Digital Archive. https://turingarchive.kings.cam.ac.uk/publications-lectures-and-talks-amtb/amt-b-4.

Warren, Christopher N., Samuel V. Lemley, DJ Schuldt, Elizabeth Dieterich, Laura S. DeLuca, Max G'Sell, Taylor Berg-Kirkpatrick, Kari Thomas, Kartik Goyal, and Nikolai Vogler. 2023. "Who Rpinted Shakespeare's Fourth Folio?" *Shakespeare Quarterly* 74 (2): 139–46. https://doi.org/10.1093/sq/quad021.

Warren, Christopher N., Pierce Williams, Shruti Rijhwani, and Max G'Sell. 2020. "Damaged Type and Areopagitica's Clandestine Printers." *Milton Studies* 62 (1): 1–47. https://doi.org/10.1353/mlt.2020.0005.

Warren, Christopher N., Avery Wiscomb, Pierce Williams, Samuel V. Lemley, and Max G'Sell. 2021. "Canst Thou Draw Out Leviathan with Computational Bibliography? New Angles on Printing Thomas Hobbes' 'Ornaments' Edition." *Eighteenth-Century Studies* 54 (4): 827–59. https://doi.org/10.1353/ecs.2021.0094.

Weiss, Adrian. 1988. "Reproductions of Early Dramatic Texts as a Source of Bibliographical Evidence." *Text* 4:237–68.

———. 1990. "Font Analysis as a Bibliographical Method: The Elizabethan Play-Quarto Printers and Compositors." *Studies in Bibliography* 43:95–164.

———. 1991. "Bibliographical Methods for Identifying Unknown Printers in Elizabethan/Jacobean Books." *Studies in Bibliography* 44:183–228.

———. 1992. "Shared Printing, Printer's Copy, and the Text(s) of Gascoigne's 'A Hundreth Sundrie Flowres.'" *Studies in Bibliography* 45:71–104.

Wells, Stanley. 2013. *An A-Z Guide to Shakespeare.* Oxford: University Press.

Werner, Sarah. 2016. "Digital First Folios." In *The Cambridge Companion to Shakespeare's First Folio*, edited by Emma Smith, 170–84. Cambridge Companions to Literature. Cambridge: Cambridge University Press. https://doi.org/10.1017/CCO9781316162552.012.

Williams, Kim, Lionel March, and Stephen R. Wassell, eds. 2010. *The Mathematical Works of Leon Battista Alberti.* Basel: Springer Basel. https://doi.org/10.1007/978-3-0346-0474-1.

Wootton, David. 2015. *The Invention of Science: A New History of the Scientific Revolution.* London: Allen Lane.

INDEX OF BOOKS & OBJECTS

Adder, The
(ca. 1880)
300

Adix Adding Machine
(ca. 1905)
301

Alberti, Leon Battista.
Opuscoli morali (1568)
117

Archimedes Calculator
(ca. 1910)
299

Arithmomètre
(ca. 1910)
299

Arithmomètre
(Serial No. 502)
(ca. 1863)
297

Arithmomètre,
Model T1858 A
(Serial No. 11)
(ca. 1860)
287, 296

AT&T Picturephone,
Mod II (1970)
403

Audubon, John James.
The Birds of America
(1840–1844)
537

Babbage, Charles.
Autograph letter
(1852)
298

Babbage, Charles.
The Ninth Bridgewater Treatise (1841)
298

Baldi, Bernardino.
De gli automati
(1589)
165

Bardeen, John, & Brattain,
Walter Houser. *"Physical Principles Involved in Transistor Action"* (1949)
211

Bernoulli, Jacob.
Ars conjectandi (1713)
545

Boole, George.
An Investigation of the Laws of Thought (1854)
483

Brahe, Tycho.
Astronomiæ instauratæ mechanica (1602)
479

Brass ruler,
"Pied de Roy"
(ca. 1700)
300

Brontë, Charlotte.
Jane Eyre (1847)
465

Brunswiga 13Z
(ca. 1935)
295

Burroughs Class 1, Model 3
(ca. 1905)
302

Byrne, Oliver.
The First Six Books of the Elements of Euclid (1847)
371

Calcumeter, The
(ca. 1907)
301

Calculating Slide Rule
(ca. 1880)
297

Carroll, Lewis.
Alice's Adventures in Wonderland (1866)
487

Cheng Dawei.
Xin zeng suan fa tong zong da quan (1883)
261

Cohen, Harold.
In Search of the Sources (1988)
551

Colonna, Francesco.
Hypnerotomachia Poliphili (1499)
503

Congress, United States.
US Bill of Rights (1792)
55

Copernicus, Nicolaus.
De revolutionibus orbium coelestium (1543)
41

Cuneiform tablet
(ca. 2350 BCE)
353

Curie, Marie.
Traité de Radioactivité (1910)
339

Curie, Pierre, & Curie, Marie.
Sur une substance nouvelle radio-active (1898)
339

Curta Type I
(ca. 1960)
299

Darwin, Charles.
Insectivorous Plants (1875)
327

Darwin, Charles.
On the Origin of Species (1859)
323

Descartes, René.
Discours de la méthode (1637)
313

Della Porta, Giambattista.
De furtivis literarum notis (1563)
111

Della Porta, Giambattista.
De furtivis literarum notis (1591)
111

Einstein, Albert.
Die Grundlage der allgemeinen Relativitätstheorie (1916)
453

Einstein, Albert.
"Die Grundlage..." in Annalen der Physik (1916)
453

Enigma, Model M4
(ca. 1942)
295

Enigma, Model M3
(ca. 1942)
129

Euclid.
Elementa Geometriae (1482)
433

Facit, Model TK
(ca. 1940)
301

Felt & Tarrant Comptometer (ca. 1895) 297

Felt & Tarrant Comptometer, Model J (ca. 1915) 297

Felix M Arithmometer (ca. 1960) 296

Ferranti, Ltd. *Faster than Thought: The Ferranti Nimrod Digital Computer* (1951) 219

Fuchs, Leonhart. *De historia stirpium* (1542) 525

Galilei, Galileo. *Discorsi e dimostrazioni matematiche* (1638) 59

Genaille, Henri, & Lucas, Edouard. *Les réglettes multiplicatrices* (1885) 281

Greenaway, Kate. *Portrait of Frederick Locker-Lampson* (1821–1895) 513

Gutenberg Bible [fragment] (ca. 1455) 357

Hevelius, Johannes. *Selenographia* (1647) 507

Hinman, Charlton, & Arthur M. Johnson Co. *Hinman Collator* (ca. 1963) 407

Hobbes, Thomas. *Leviathan* (1690) 397

Hosokawa Hanzō Yorinao. *Karakuri zui* (1808) 159

Ibn al-Haytham. *Opticae thesaurus* (1572) 333

Isomura Yoshinori. *Zōho sanpō ketsugishō* (1684) 257

Ivory sector (ca. 1800) 300

Jacquard, Joseph Marie. *À la mémoire de J.M. Jacquard* (1839) 361

Jenner, Edward. *An Inquiry into the Causes and Effects of the Variolæ Vaccinæ* (1798) 71

Johnson, Samuel. *A Dictionary of the English Language* (1755) 517

Knuth, Donald E. *TEX and METAFONT* (1979) 375

Leibniz Calculator, or Stepped-Drum Calculator (replica) (ca. 1960) 267, 302

Leibniz, Gottfried Wilhelm von. *Brevis descriptio machinae arithmeticae* (1710) 267

Leibniz, Gottfried Wilhelm von. *Nouvelle arithmétique binaire* (1720)
365

Locke, John. *An Essay Concerning Humane Understanding* (1690)
521

M-209-B Cipher Machine (ca. 1944)
297

Mercedes Euklid M1 (ca. 1922)
300

Melville, Herman (1819–1891). *The Whale* (1851)
457

Mendel, Johann Gregor. *Versuche über Pflanzen-Hybride*n (1865)
461

Millionaire Calculator (Serial No. 2380) (ca. 1900)
296

Montagu, Mary Wortley. *Letters...* (1763)
123

Napier, John. *Napier's rods* (ca. 1690)
247

Napier, John. *Raddologia* (1623)
251

Napier, John. *Mirifici logarithmorum canonis descriptio* (1620)
295

Newton, Isaac. *Opticks* (1704)
65

Newton, Isaac. *Philosophia naturalis principia mathematica* (1687)
493

Original-Odhner Model 127 / 602 No. 5 (ca. 1937)
295

Pascal, Blaise. *Oeuvres* (1779)
277

Pascal, Blaise. *Pascaline* mechanical calculator (replica) (ca. 1960)
273, 301

Pascal, Blaise. *Traitez de l'équilibre des liqueurs, et de la pesanteur de la masse de l'air (*1663)
319

Peerless Calculator (ca. 1907)
295

Ptolemy, Claudius. *Magnae constructionis* (1538)
49

Rapid Calculator (ca. 1920)
299

Sang's Platometer (ca. 1855)
301

Schedel, Hartmann. *Liber chronicarum* (1495)
497

Schott, Gaspar.
Schola steganographica (1665)
207

Shakespeare, William.
First Folio (1623)
437

Shakespeare, William.
Second Folio (1632)
441

Shakespeare, William.
Third Folio (1663/64)
445

Shakespeare, William.
Fourth Folio (1685)
449

Shelley, Mary.
Frankenstein (1818)
203

Sutherland, Ivan Edward.
A Walking Robot (1983)
169

Swift, Jonathan.
Gulliver's Travels (1726)
197

Tate Arithmometer (ca. 1900)
302

Thacher's Calculating Instrument, Model 4012 (ca. 1910)
299

TIM ("Time is Money") Calculator (ca. 1910)
298

Trinks-Brunsviga Model A (ca. 1900)
296

Turing, Alan.
"Computing Machinery and Intelligence" (1950)
215

Vesalius, Andreas.
De humani corporis fabrica (1543)
471

Printed and bound in Lithuania

ISBN 978-0-88748-723-1

Library of Congress Control Number:
2025939052

Text & Curation
Samuel V. Lemley

Creative Director
Heidi Wiren Kébé

Managing Editor
Brad King

Publication Design
Kelsey Dusenka
Kelsey Elder

Photography
Samuel V. Lemley
Heidi Wiren Kébé
Kevin Lorenzi
Joseph Mesco
Ann Marie Mesco

Typefaces
ROM
Seb McLauchlan,
ABC Dinamo

Martina Plantjn
Kris Sowersby,
Klim Type Foundry

Printing
Balto Print

Publisher
Carnegie Mellon
University Press

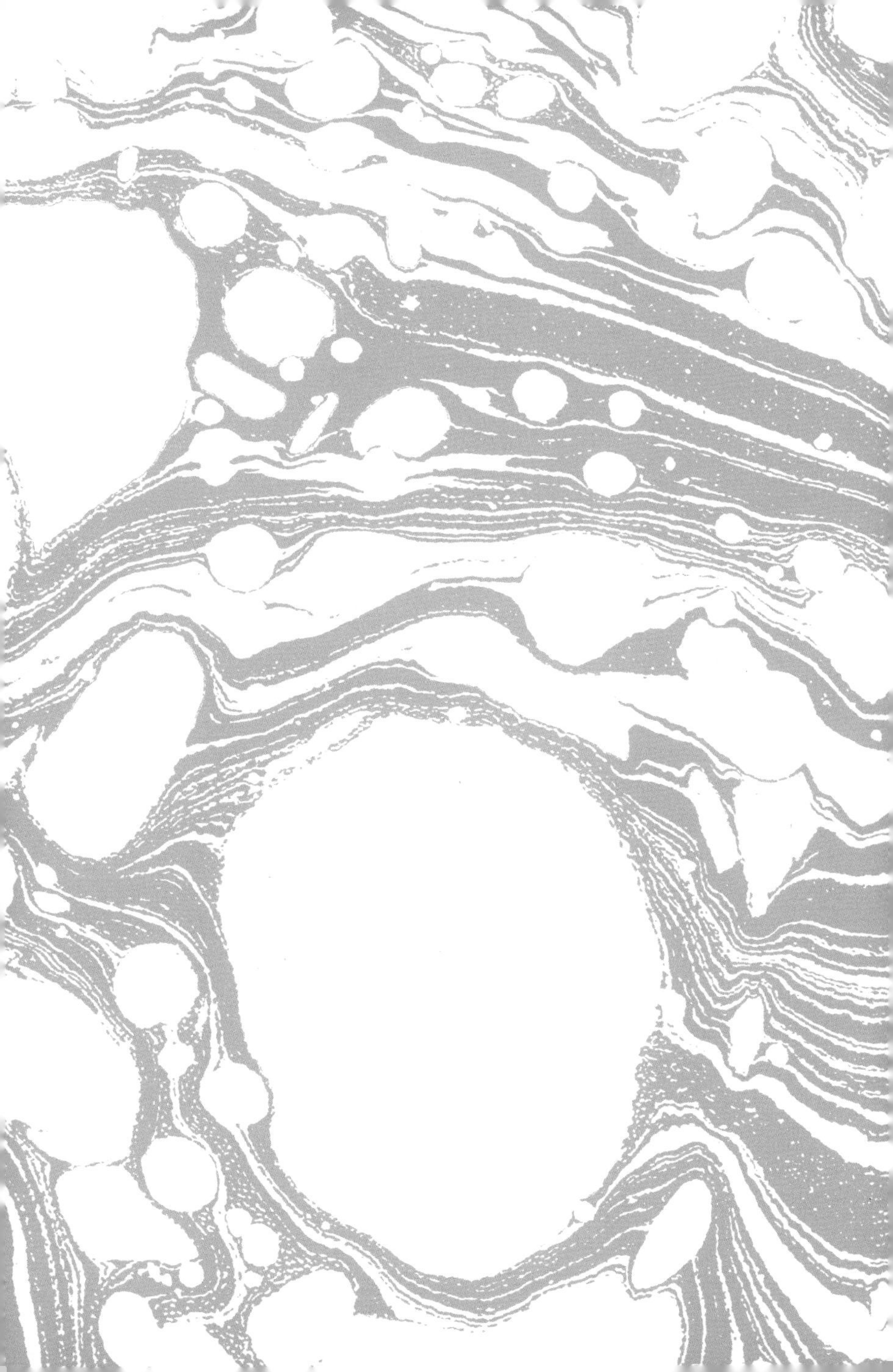